Intelligent Control and Design of Converters for Renewable Energy Sources and Electric Vehicle Applications

The book discusses aspects such as converter design, control techniques, and other functionality augmented in the controllers mandated by the grid. It further explains how the power from renewable energy is converted to electrical energy and integrated into the grid smoothly. It showcases how an electrical motor powered by a battery is controlled in electric vehicles.

Features:

- Incorporates the latest advancements in intelligent control and converter design for renewable energy and electric vehicles.
- Focuses on power electronics and control for a wide range of topics like integration of renewable energy sources, converter topologies and their design, and energy management.
- Discusses the efficient control applied to power electronic converters for improved performance.
- Presents practical implementation guidelines for intelligent control algorithms and converter design.
- Includes case studies, simulation examples, and real-world applications to facilitate the application of intelligent control techniques in renewable energy and electric vehicle systems.

It is primarily written for senior undergraduates, graduate students, and academic researchers in the fields of electrical engineering, electronics and communication engineering, computer science and engineering, and automotive engineering.

Power Electronics: Technology and Applications
Series Editors: Ramani Kannan and Dharavath Kishan

Intelligent Control and Design of Converters for Renewable Energy Sources and Electric Vehicle Applications
Edited by Ravi Raushan, Md Waseem Ahmad, Ranjan Kumar Behera, Ramani Kannan, and Anubrata Das

Intelligent Control and Design of Converters for Renewable Energy Sources and Electric Vehicle Applications

Edited by
Ravi Raushan, Md Waseem Ahmad,
Ranjan Kumar Behera, Ramani Kannan,
and Anubrata Das

CRC Press
Taylor & Francis Group
Boca Raton London New York

CRC Press is an imprint of the
Taylor & Francis Group, an **informa** business

Designed cover image: © Shutterstock Images

First edition published 2026
by CRC Press
2385 NW Executive Center Drive, Suite 320, Boca Raton FL 33431

and by CRC Press
4 Park Square, Milton Park, Abingdon, Oxon, OX14 4RN

CRC Press is an imprint of Taylor & Francis Group, LLC

ISBN: 9781032849591 (hbk)
ISBN: 9781032849584 (pbk)
ISBN: 9781003515807 (ebk)

DOI: 10.1201/9781003515807

Typeset in Sabon
by KnowledgeWorks Global Ltd.

Contents

About the editors

Ravi Raushan is currently working as an assistant professor in the Department of Electrical and Electronics Engineering at the National Institute of Technology Surathkal, Karnataka, India. His areas of research interest include multilevel inverters, power electronics, photovoltaic systems, and electrical drives. He has taught courses such as digital electronics, digital systems design, power electronics, and control systems. He has published research papers in journals and conferences of national and international repute.

Md Waseem Ahmad is working as an assistant professor in the Department of Electrical and Electronics Engineering at the National Institute of Technology Surathkal, Karnataka, India. His areas of expertise include efficient power electronic converters, fault diagnostics, and condition monitoring of power electronic converters. He has published research papers in journals and conferences of national and international repute.

Ranjan Kumar Behera is presently working as an associate professor in the Department of Electrical Engineering at the Indian Institute of Technology Patna, Bihar, India. He has published research papers in journals and conferences of national and international repute. His areas of research include power electronics converters, electrical machines and high-performance drives, microgrids and smart grids, solar and wind energy integration, and solar tree design for rural electrification. He has taught courses such as FACTS controllers in power systems, control of electric drives, distributed energy resources, and power electronics.

Ramani Kannan is working as a senior lecturer in the Department of Electrical and Electronics Engineering at the Universiti Teknologi PETRONAS (UTP), Malaysia. His areas of expertise include power electronics and drives, renewable energy, intelligent control techniques, and smart grid applications. He has published more than 200 research papers in journals and conferences of national and international repute. He is a senior member of the IEEE Power Electronics and Power Energy and a life member of the Indian Society for Technical Education.

Anubrata Das is presently working as an assistant professor in the Department of Electrical Engineering at the Indian Institute of Technology, Banaras Hindu University, India. He previously worked as a postdoctoral researcher at the University of Tennessee, Knoxville, TN, USA. His areas of research include power electronics, inverters, voltage regulation, transformers, electrical machines, and power electronic applications. He has published research papers in journals and conferences of national and international repute.

List of contributors

Md Waseem Ahmad
Department of Electrical and Electronics Engineering, National Institute of Technology Karnataka (NITK)
Surathkal, India

D. Anitha
Department of Electrical and Electronics Engineering, SRM Institute of Science and Technology
Kattankulathur, India

Mohit Bajaj
Department of Electrical Engineering, Graphic Era (Deemed to be University)
Dehradun, India
Hourani Center for Applied Scientific Research, Al-Amman University
Amman, Jordan
Graphic Era Hill University
Dehradun, India

Hemapriya B.C.
Department of CSE, R. L. Jalappa Institute of Technology
Doddaballapura, Bengaluru, India

Anandita Chowdhury
Department of Electrical Engineering, Sardar Vallabhbhai National Institute of Technology (SVNIT)
Surat, India

Chia-Chi Chu
Department of Electrical Engineering, National Tsing Hua University
Hsinchu, Taiwan (R.O.C.)

Rohini Dakulagi
Department of Electrical and Electronics Engineering, National Institute of Technology Karnataka (NITK)
Surathkal, India

Veerendra Dakulagi
Department of CSE (Data Science), Guru Nanak Dev Engineering College
Bidar, India

Anubrata Das
Department of Electrical Engineering, Indian Institute of Technology (BHU)
Varanasi, India

Raghu J.
Department of Electronics and Communication Engineering, The National Institute of Engineering
Mysuru, India

Debashisha Jena
Department of Electrical and Electronics Engineering, National Institute of Technology Karnataka (NITK)
Surathkal, India

Shashidhara K.S.
Department of Electronics and Communication Engineering, Nitte Meenakshi Institute of Technology
Bengaluru, India

Dharavath Kishan
Department of Electrical and Electronics Engineering, National Institute of Technology Karnataka (NITK)
Surathkal, India

Bhukya Hari Krishna
Department of Electrical and Electronics Engineering, National Institute of Technology Karnataka (NITK)
Surathkal, India

Prashant Kumar
Department of Electrical Engineering, Netaji Subhas University of Technology
Delhi, India

V. Vignesh Kumar
Department of Electrical and Electronics Engineering, National Institute of Technology Tiruchirappalli
Tiruchirappalli, India

Dibyendu Mandal
Bosch Global Software Technologies Pvt. Ltd.
Bangalore, India

Tukaram Moger
Department of Electrical and Electronics Engineering, National Institute of Technology Karnataka (NITK)
Surathkal, India

S. Sunil Naik
Department of Electrical and Electronics Engineering, National Institute of Technology Karnataka (NITK)
Surathkal, India

R. Rajarajeswari
Department of Electrical and Electronics Engineering, SRM Institute of Science and Technology
Kattankulathur, India

Ravi Raushan
Department of Electrical and Electronics Engineering, National Institute of Technology Karnataka (NITK)
Surathkal, India

Arvind R. Singh
Department of Electrical Engineering, School of Physics and Electronics Engineering, Hanjiang Normal University
Shiyan, China

D. Suchitra
Department of Electrical and Electronics Engineering, SRM Institute of Science and Technology
Kattankulathur, India

Anish Tiwari
Department of Electrical Engineering, Sardar Vallabhbhai National Institute of Technology (SVNIT)
Surat, India

R. Uthra
Department of Electrical and Electronics Engineering, SRM Institute of Science and Technology
Kattankulathur, India

Pradeep Vishnuram
Department of Electrical and Electronics Engineering, SRM Institute of Science and Technology
Kattankulathur, India

Udaykumar R. Yaragatti
Department of Electrical and Electronics Engineering, National Institute of Technology Karnataka (NITK)
Surathkal, India

Chapter 1

Power converter topologies for single-phase grid-tied PV microinverters

V. Vignesh Kumar

1.1 INTRODUCTION

The development of renewable energy systems has been proliferating in the last two decades across the globe due to rising environmental concerns. In India, the Ministry of New and Renewable Energy (MNRE) has set the goal of achieving 50% of the total electric power installed capacity from renewable energy sources by 2030. Hence, energy generation from the self-renewing sources like solar, geothermal, and wind is broadly promoted. Among the renewable energy systems, solar photovoltaic (PV) power generation is most popular. This is due to its attributes such as clean, omnipresence, inexhaustible, simple system structure, high-energy conversion efficiency, and distributed resources of raw materials for PV panel production.

In general, the applications of PV power generation systems are categorized into off-grid system and grid-tied system. The off-grid systems are employed in island, countryside, and other remote area power supply units. On the other hand, grid-tied systems have PV source exporting power to the utility grid via a power electronic interface. The general block diagram representing the two types of PV systems are given in Figure 1.1.

The grid-connected PV inverters are classified into central inverters, string inverters, and microinverters. In earlier days, grid-tied inverters were used in PV farms, with a capacity of usually greater than 10 kW and having a central inverter architecture, as shown in Figure 1.2a. The major drawback of central inverters is the large amount of power interruption during inverter failure. Another class of PV inverter, the string inverter, is an updated version of central inverter in which a string of series-connected PV modules are fed into the inverter, as seen in Figure 1.2b. The string inverters are widely used for power capacity less than 10 kW and they could eliminate the major problems of central inverter. Though the central inverters and string inverters are economical and are widely employed in large-scale PV plants, they are prone to performance degradation under partial shading conditions and have reduced fault tolerance capability. Further, it cannot be guaranteed that string-level maximum power point tracking (MPPT) performed in the case of string and central inverters ensures the tracking of maximum power point for each

DOI: 10.1201/9781003515807-1

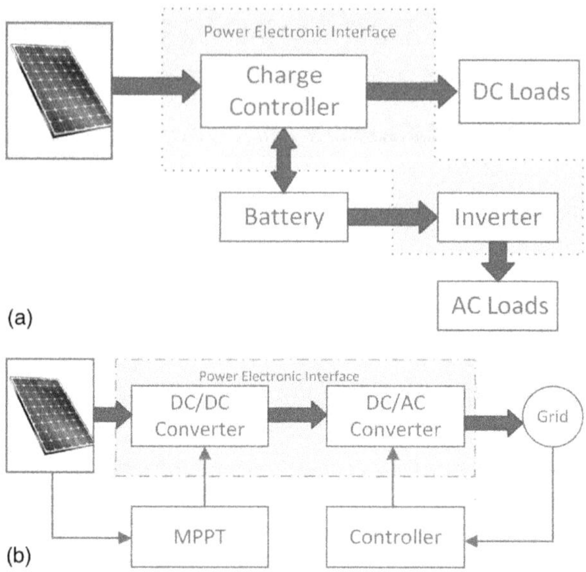

Figure 1.1 Block diagram of (a) off-grid PV system and (b) grid-tied PV system.

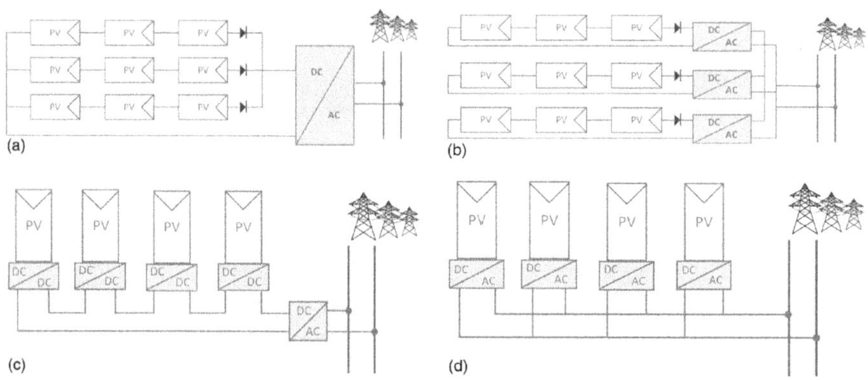

Figure 1.2 (a) Central inverter. (b) String inverter. (c) DC optimizer. (d) Microinverter topology.

individual panel's characteristics. Therefore, panel-level power electronics interface-based architecture such as DC–DC optimizers and microinverters, as shown in Figure 1.2c and d, is developed to avoid huge mismatch losses. Though the DC–DC optimizer can reduce mismatch losses to a great extent and yield efficiency of around 98%, it also requires a central inverter to integrate PV output to grid. Consequently, it leads to huge power interruption in the event of inverter failure. On the other hand, the microinverter architecture in Figure 1.2d alleviates the mismatch losses in the PV system, reduces power interruption, and achieves MPPT for each panel while integrating to grid.

This chapter provides a comprehensive review of the topologies employed in the PV microinverter system. The major focus is given to the two-stage topologies attributing to its high reliability in comparison with other variants of microinverter. The operating principles and performance comparison of selected topologies, survey on commercially available PV microinverters, a case study with simulation results for flyback-based microinverter are presented lucidly in the subsequent sections of this chapter.

1.2 PV MICROINVERTER SYSTEM

The term microinverter is referred to a low-power module-level inverter incorporated into each PV module. The concept of microinverter is developed to capture the MPPT for each panel, so that energy efficiency is enhanced during the partial shading condition, as illustrated in Figure 1.3, in comparison with the string inverter system. The benefits of employing microinverter-based architecture are manifold. It eliminates the bulky low-frequency transformer and permits a compact design for inverter topologies which can have a plug and play feature. It provides flexibility and modularity as it is easily mounted on the back side of a PV panel.

It enhances energy harvesting and improves system efficiency to a greater extent by reducing mismatch losses, partial shading losses, and suboptimal MPPT losses. Because of its modular nature, during the individual inverter unit failure scenario, it is possible to prevent the large power interruption. It improves reliability of the overall system, owing to the low power-handling capacity requirements of inverter and its reduced temperature levels leading to removal of fans. Further, it paves way for standardized system design to reduce the production cost, by driving up the manufacturing quantities. It can replace the conventional hard-switched converter with soft-switching counterparts, thereby optimizing the energy harvest and also increasing the lifetime of inverter to match with the PV panel lifetime.

The microinverters that are employed for grid integration converts the low DC voltage from the PV panel to the level suitable for AC current injected into the grid and achieve the power balance. It performs the power conversion for a wide range of PV voltage and with high efficiency. The microinverter topologies presented in the literature and available on the market

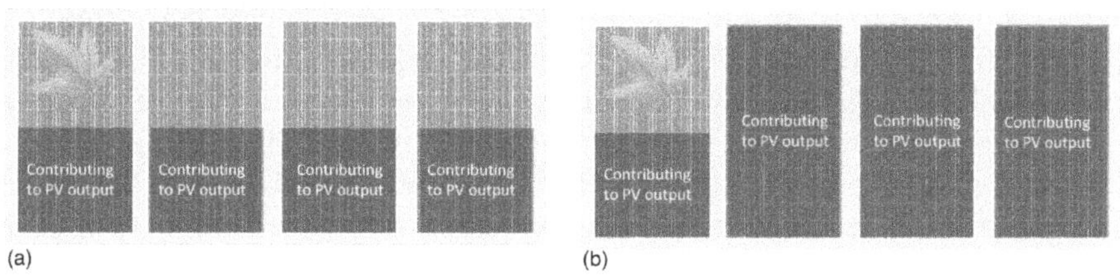

Figure 1.3 Impact of partial shading on energy harvest. (a) String inverter. (b) Microinverter.

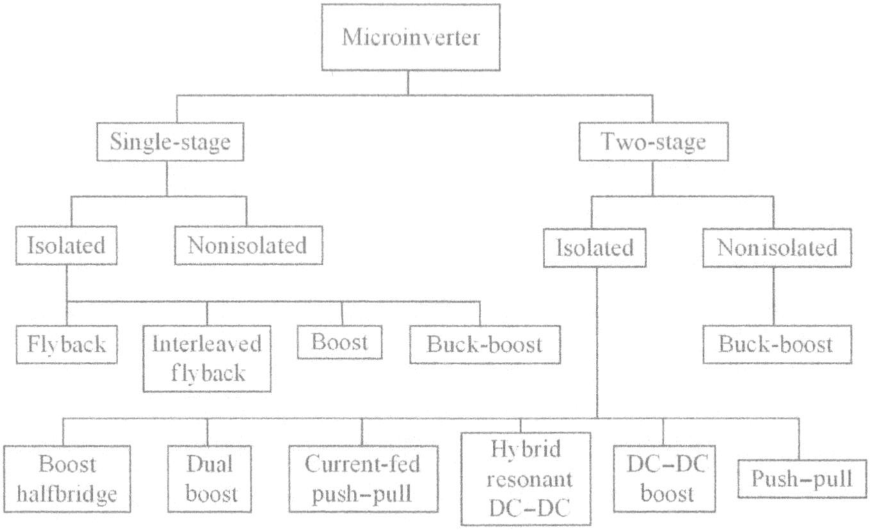

Figure 1.4 Categorization of microinverter topologies.

usually adopt either single-stage or two-stage conversion. Some of the popular microinverter topologies available in the literature are bifurcated into the above-mentioned two groups, as provided in Figure 1.4. Further, each category possesses both nonisolated and isolated variants for the microinverter architecture. The single-stage microinverter performs MPPT in unison for boosting the PV voltage and converting to AC for satisfying the grid code. On the other hand, in the two-stage microinverter, the first stage comprises DC–DC converter to execute the MPPT operation and boost the voltage to DC link value. In the second stage, AC–DC conversion is accomplished for grid integration.

This chapter mainly focuses on the two-stage isolated topologies for PV microinverters attributed to the fact that they have features such as simple control, enhanced reliability, etc. over the single-stage and nonisolated architectures. Though the single-stage nonisolated microinverters are compact with reduced components and single controller, the requirement of large DC link capacitor, leakage currents from PV, and power decoupling are the major hindrance to their widespread usage. In general, single-stage topology requires large DC link capacitor to handle the ripple magnitude of DC voltage within the limits. Further, the capacitor used as an energy storage element for power decoupling to balance the instantaneous input and output power will shorten the lifetime of the microinverter. In addition to the above-mentioned disadvantages, the reactive power compensation is the challenging task in the single-stage PV microinverters.

1.3 LITERATURE REVIEW ON ISOLATED TWO-STAGE MICROINVERTERS

There are several topologies existing in the literature for single-phase microinverters [1]. They are mainly pivoted around the number of power conversion stages, galvanic isolation methods, location of decoupling capacitor, and leakage current behavior [2]. Further, majority of the existing works in the literature recommend topologies with two-stage conversion for single-phase systems. However, they offer dissociated multiple tasks, namely stepping up of voltage, MPPT, grid current control for reactive power compensation, etc. with minimal voltage and current stress in the components.

Therefore, a detailed review of works reported in the literature on two-stage isolated microinverter topologies is presented in the subsequent texts. A module-integrated isolated microinverter based on pseudo-DC link topology is proposed in Ref. [3], as shown in Figure 1.5a. This topology possesses an isolated dual-boost converter, which is fed from an inverse buck converter. It has a voltage doubler circuit with polarity selector on the secondary side, thereby eliminating the requirement of high-voltage electrolytic capacitors in the microinverter. The experimental results were reported to show the veracity of topology in achieving high MPPT accuracy. Mamun et al. [4] had developed a flyback converter with dual switch for operating in a wide range of PV modules used in the microinverter topology. The dual-switch flyback topology presented in Figure 1.5b alleviates the voltage stress across the switch and has improved efficiency compared to the conventional topology. A boost type DC–DC converter utilizing half-bridge converter with minimum number of devices is presented in Figure 1.5c. The authors have employed full-bridge inverter with IIR filter-based current control for regulating the grid current [5]. The experimental results depicted show dynamic stiffness in the response of system for rapid change in solar irradiance. Further, the proposed controller provides high power factor, low current harmonics, and better current regulation over a wide load range. A soft-switched single-inductor-based push–pull was developed in Ref. [6] for microinverter DC–DC stage. This converter is claimed to have better efficiency than the conventional converter. The high performance is achieved by zero voltage switching (ZVS) and zero current switching (ZCS) of the primary-side power switches. The ZVS and ZCS conditions are obtained using the resonance between the high-frequency transformer leakage inductance and resonant capacitor. The theoretical analysis was authenticated using a 250-W system and the recorded results were presented by Kim et al. [6]. LaBella and Lai [7] introduced a hybrid resonant converter by adding bidirectional switch for higher efficiency under discontinuous operating mode. As shown in Figure 1.5e, this topology utilizes a phase-shifted full-bridge converter performing buck operation cascaded with boost converter for addressing wide operating

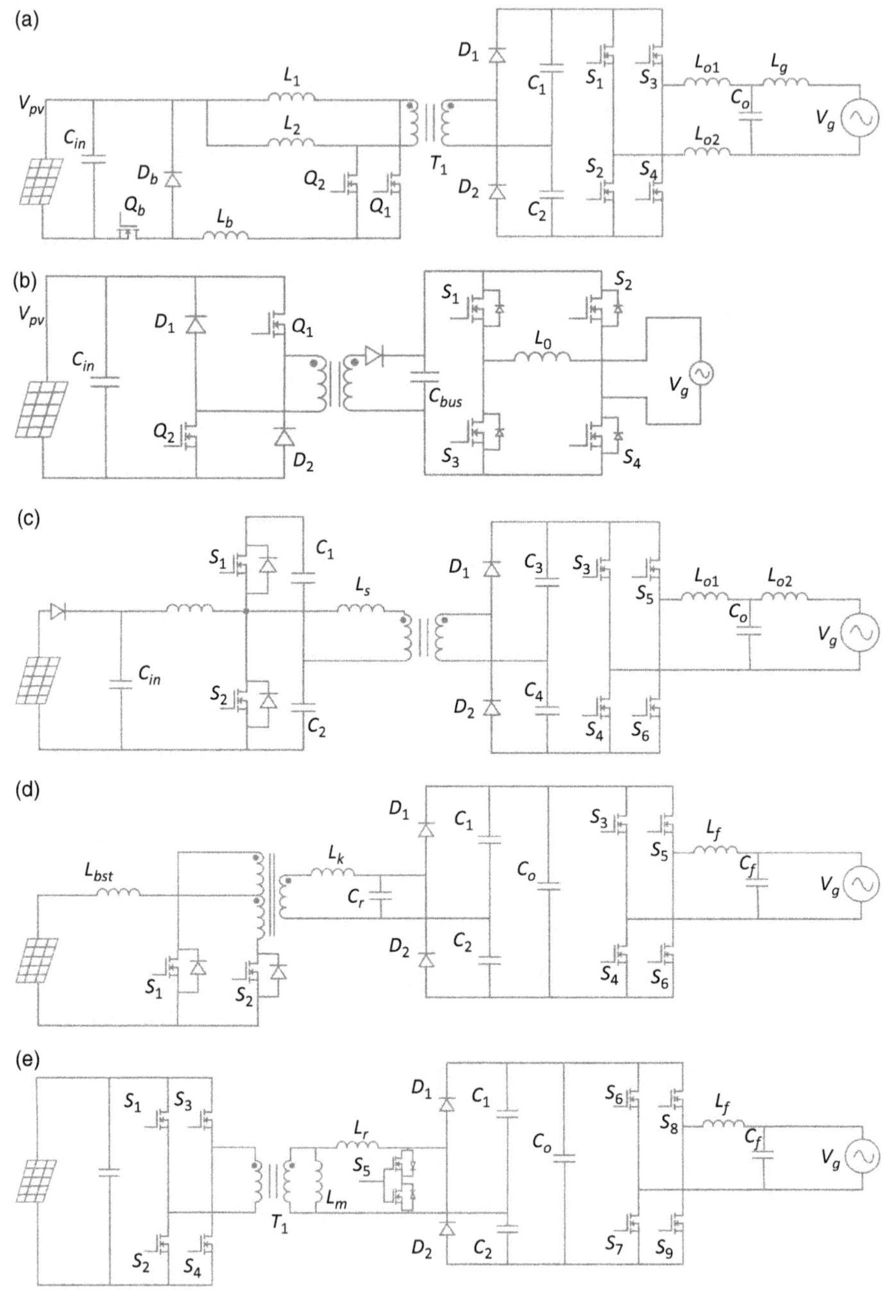

Figure 1.5 Topology proposed in (a) [3], (b) [4], (c) [5], (d) [6], (e) [7], (f) [8], (g) [9], and (h) [10]. *(Continued)*

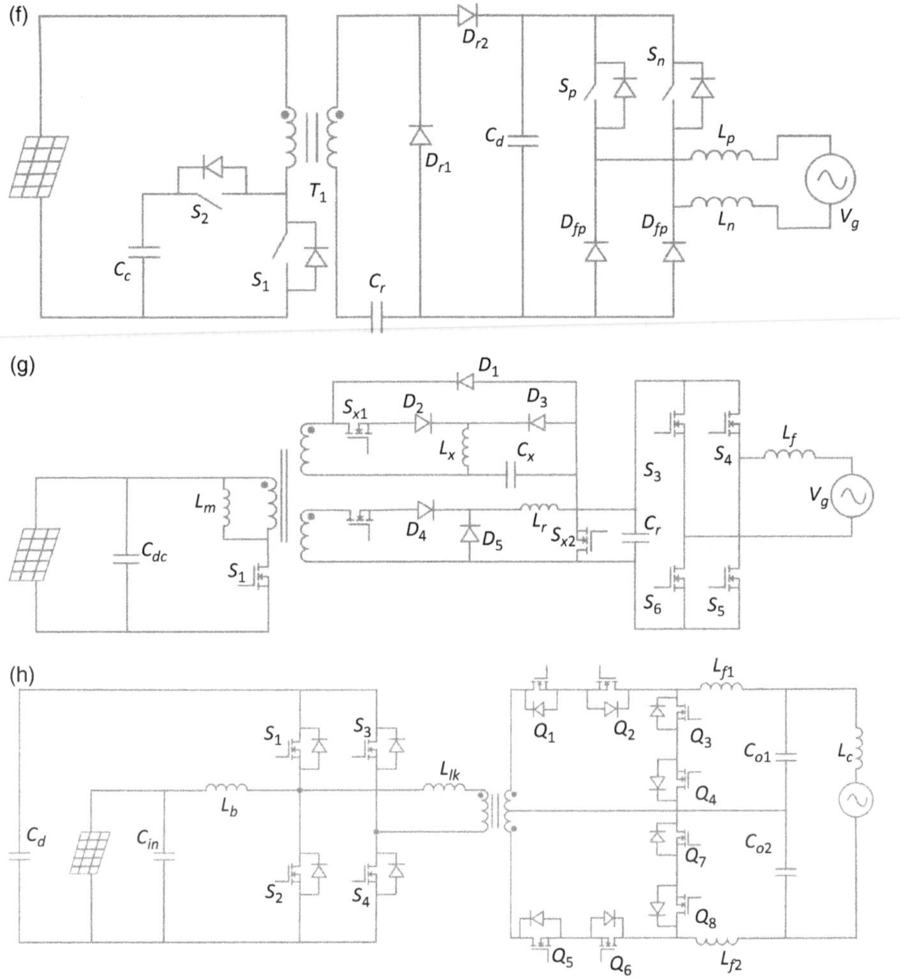

Figure 1.5 (Continued)

voltage range. Further, the control technique employed was simple and it uses two carrier waveforms for pulse-width modulation (PWM) signals to have smooth transition between modes. An active clamp step-up converter with series-resonant voltage doubler is used for microinverter in Ref. [8]. The high efficiency is achieved by recycling of leakage inductance energy in active clamp circuit and elimination of reverse recovery loss with removal of rectifier diodes in voltage doubler circuit. Liao et al. [9] proposed a micro-inverter topology with active power decoupling circuit to remove the electrolytic capacitor. The reliability of power converter is enhanced with the absence of electrolytic capacitor. Further, the converter is proven to have low voltage stress with reduced number of components. The energy loss is minimized by recycling the magnetizing energy in continuous conduction mode operation. Another microinverter topology with active power decoupling is proposed in Ref. [10]. It is reported to handle a wide input voltage range by integrating boost converter with high-frequency link. The high gain is achieved without adding extra switches. Further, the voltage-source inverter feature enables this topology to be suitable for both grid-tied and islanded operations. A step-up DC–DC stage for microinverter is developed using the switched capacitor network and boost converter in Ref. [10]. This DC–DC stage is designed particularly for systems that need to obtain high DC voltages from a low-voltage DC source such as solar microinverter. Li et al. [10] have declared that the output voltage increases by nine times even with duty of 30%. The other advantages of the topology are low turns ratio of transformer, reduced number of components, and low-voltage stress across the switches. Li et al. and Meshael et al. [11, 12] developed a next-generation microinverter belonging to 600-W class. A modulation strategy is proposed on the secondary side to obtain soft-switching characteristics for all the switches. The experimental results were depicted for dual-mode operation and the converter is estimated to have 94.6% peak efficiency.

1.4 REQUIREMENTS AND PERFORMANCE ANALYSIS OF THE TOPOLOGIES

1.4.1 Total harmonic distortion (THD) and grounding requirements

In general, the PV microinverters have to comply with the grid code standards for the benefit of both consumers and utilities. The IEEE 1547 and IEC 61727 specify the requirements for the PV microinverter topology to achieve grid compliance. The THD and grounding considerations are the predominant requisite as per the above standards [13–15]. In order to ensure the power quality, THD is expected to be less than 5% and the galvanic isolation of high-frequency transformer helps in controlling it. Further, grounding is vital in delivering safe and reliable operation. It helps in preventing the damage to system due to electromagnetic interferences and lightening. Also, the galvanic isolation plays a major role in suppressing the low-frequency leakage currents, due to stray capacitance between the primary and secondary windings. The various PWM techniques also have impact on the leakage current in the grid-connected PV microinverter system [1].

1.4.2 Efficiency

The performance of microinverter is greatly decided by the efficiency of the topology, which is a significant parameter for market acceptance of the microinverter. The nonisolated topologies with high-frequency transformers have improved efficiency compared to its low-frequency transformer counterparts.

The two-stage isolated topologies are reported to have peak efficiency of above 90%, the highest being proposed with 96.2% in Ref. [8].

1.4.3 MPPT and other control requirements

The reduction of mismatch losses due to partial shading, panel degradation, etc. can be eliminated with distributed MPPT in microinverters. The various MPPT techniques such as incremental conductance, hill climbing, ripple correlation, fuzzy logic based, perturb and observe (P&O), etc. are reported in the literature [16] for grid-tied microinverters. Among these, incremental conductance and hill-climbing methods are witnessed to have high MPPT efficiency of around 98% [17]. Further, the grid-tied microinverters connecting low PV voltage to high grid voltage are expected to offer reactive power control. In two-stage isolated topology, MPPT will be performed by DC–DC converters and inverters simply inject the active power employing d-axis current control. The reactive power control is obtained by adjusting the q-axis current.

1.4.4 Power density, lifetime, and cost

The microinverter being modular and normally mounted at the back of PV panel should possess high power density. Using high-frequency transformers in isolated topologies helps in minimizing the magnetics and filter size. The switching frequency is increased to as high as 80 kHz for 400-W prototype to achieve power density of 4.86 W/cm^3 [18]. The reliability of microinverter is generally governed by the electrolytic capacitor that has shorter expected lifetime in comparison with other components. The usual practice is to keep the decoupling capacitor at DC link to reduce the size and improve the microinverter lifespan. Finally, cost is an important deciding parameter in deployment of microinverter and it is based on the switch count, high-frequency transformer geometry, driver circuit, and auxiliary power supply requirements.

All the above-discussed isolated topologies are reviewed in detail and a comprehensive performance comparison is provided with appropriate explanations in the subsequent texts. The various features of all the eight topologies reviewed in this work are listed in Table 1.1. It can be seen from Table 1.1 that topology reported in Figure 1.5e shows higher efficiency followed by topologies in Figure 1.5d and f. The higher efficiency is due to the soft-switching attributes of these topologies that minimize the switching losses. Further, it can also be noticed from Table 1.1 that the topologies that possess soft-switching characteristics have relatively higher switching frequency of 100 kHz compared to hard-switched converters. The component count is vital to achieve enhanced performance of the isolated PV microinverter topologies. The reduced switch count in topologies of Figure 1.5d and e provide higher power density among others. Further, it eliminates

Table 1.1 Overview of reviewed isolated topologies

Topology	*Figure 1.5a*	*Figure 1.5b*	*Figure 1.5c*	*Figure 1.5d*	*Figure 1.5e*	*Figure 1.5f*	*Figure 1.5g*	*Figure 1.5h*
Power	250 W	550 W	210 W	250 W	300 W	400 W	240 W	600 W
Efficiency	93.7%	95%	97%	96.6%	97%	96.2%	95%	94%
Switching frequency	20 kHz	45 kHz	21.6 kHz	100 kHz	100 kHz	100 kHz	50 kHz	50 kHz
No. of switches	6	6	6	2	5	6	8	12
No. of diodes	3	2	2	2	2	4	5	–
Switching	Hard	Hard	Hard	Soft	Soft	Soft	Hard	Hard
THD	–	4%	2.87%	–	–	3.8%	–	5%
Lifetime	Long	Long	Long	Long	Long	Long	Long	Long
Cost	High	High	Moderate	High	High	High	High	Very high

the necessity of additional gate-drive circuits with associated power supply requirements, thereby leading to reduction in extra losses. The reduced component count also facilitates the reliability improvement and minimizes the heat sink requirements. It can be seen that the topologies in the literature [3–10] are reported to exhibit THD less than 5%. Hence, it can be concluded that isolated PV microinverter topologies with soft-switching feature offer enhanced performance in terms of efficiency, power density, reliability, and lifetime of the converter.

1.5 A CASE STUDY ON ACTIVE CLAMP FLYBACK CONVERTER DESIGN FOR PV MICROINVERTER

The flyback converter is the most widely employed topology for the PV microinverter. This can be attributed to its reduced switch count while providing galvanic isolation. But the main concern with flyback topology is handling the leakage energy. When the main primary-side switch is turned off, the leakage energy develops in the core that induces a huge voltage spike across the switch. This leads to destruction of the power switch and hence the reliability of the converter. In order to alleviate this problem, resistor-capacitance-diode (RCD) snubber can be used across the flyback transformer. However, it would result in converter efficiency degradation, as the energy is dissipated in the form of heat. Therefore, the more effective way to handle the leakage energy in flyback transformer is to employ active clamp circuit, as shown in Figure 1.6.

The active clamp circuit is basically a lossless snubber using a clamp capacitor C_{clamp} and an active switch. When the switch is turned off, voltage spike is clamped at designed value using C_{clamp}, and then the remaining leakage energy is stored in the clamping capacitor. This energy is fed to the secondary side

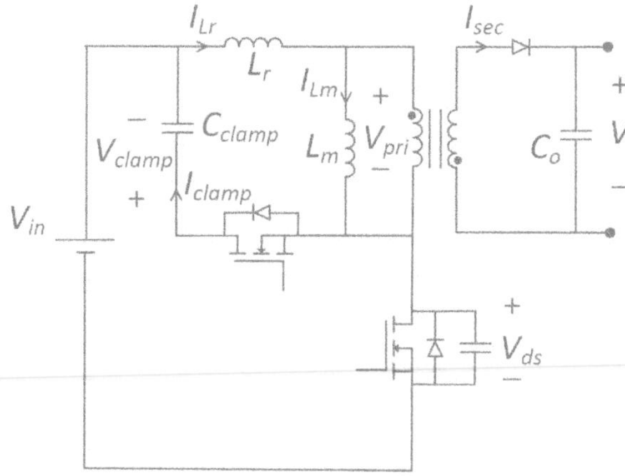

Figure 1.6 Active clamp flyback converter.

as a recycling energy. Further, the active clamp flyback circuit can offer ZVS, thereby enhancing the overall efficiency of the converter. The main critical waveforms of active clamp converter are shown in Figure 1.7.

The various equations that are employed to design the different parameters are given in the subsequent texts. By assuming the maximum duty cycle, D_{max} of the active clamp circuit, the turns ratio of the flyback transformer is provided in Equation (1.1):

$$n = \frac{N_1}{N_2} = \frac{V_{in,min}}{V_o} \frac{D_{max}}{1 - D_{max}} \tag{1.1}$$

where $V_{in,min}$ and V_o are the minimum input voltage and output voltage, respectively. The magnetizing inductance can be obtained from Equation (1.2):

$$L_m = \frac{V_{in}^2 * n * V_o^2}{K1 * P_{out} * f_s * (nV_{in} + V_o)^2} \tag{1.2}$$

where P_{out} is the output power rating of the converter and f_s is the switching frequency of the converter. The maximum magnetizing inductance current is given by Equation (1.3):

$$I_{Lmax} = \frac{V_{in} D}{K1 * L_m * f_s} \tag{1.3}$$

The value of clamp capacitor shall be obtained from Equation (1.4) for the desired clamp voltage level:

$$C_{clamp} = \frac{[(1 - D_{min})]^2}{\pi^2 f_s^2 L_r} \tag{1.4}$$

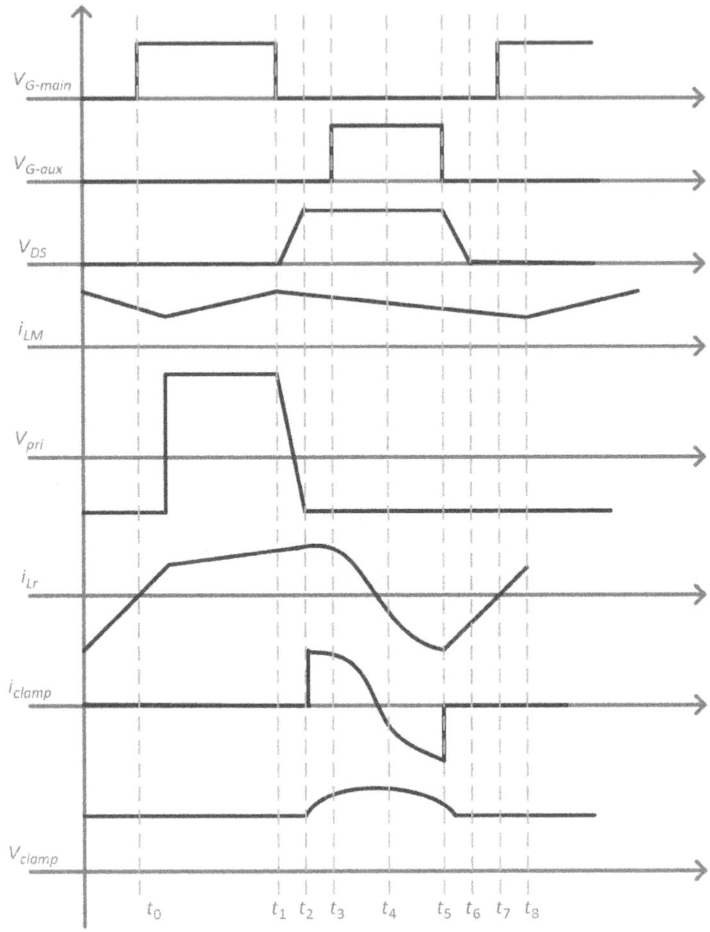

Figure 1.7 Typical waveforms of active clamp flyback converter.

where L_r is the resonant inductance. In order to ensure effective ZVS, the energy stored in this must be greater than energy in resonant capacitance, as given in Equation (1.5):

$$L_r > \frac{C_r\left(V_{in} + nV_o\right)^2}{I_{Lmax}^2} \tag{1.5}$$

The voltage across the rectifier diode V_D and peak current I_D through it at the transformer secondary side is provided in Equations (1.6) and (1.7), respectively.

$$V_D = \frac{V_{in}}{n} + V_o \tag{1.6}$$

$$I_D = \frac{2P_o}{V_o(1 - D_{max})} \tag{1.7}$$

The output filter capacitance value can be obtained using Equation (1.8):

$$C_o = \frac{D_{max} P_o}{f_s V_o \Delta V_o} \tag{1.8}$$

1.5.1 Simulation results and discussion

In order to analyze the performance of active clamp forward converter circuit connected to PV source, a 225-W PV system model shown in Figure 1.8 is developed in MATLAB Simulink environment. The parameters used in the simulation are computed from Equations (1.1)–(1.8) and the MPPT operation of DC–DC stage is verified in simulation.

The specifications of PV panel are provided in Table 1.2 and the design specifications of flyback converter are provided in Table 1.3. The duty cycle maximum value, D_{max}, is designed to be 0.4 with some margin for PV input voltage variations. Turns ratio of the transformer is computed to be 15. The voltage stress in the main switch is 153.33 V, whereas leakage inductance,

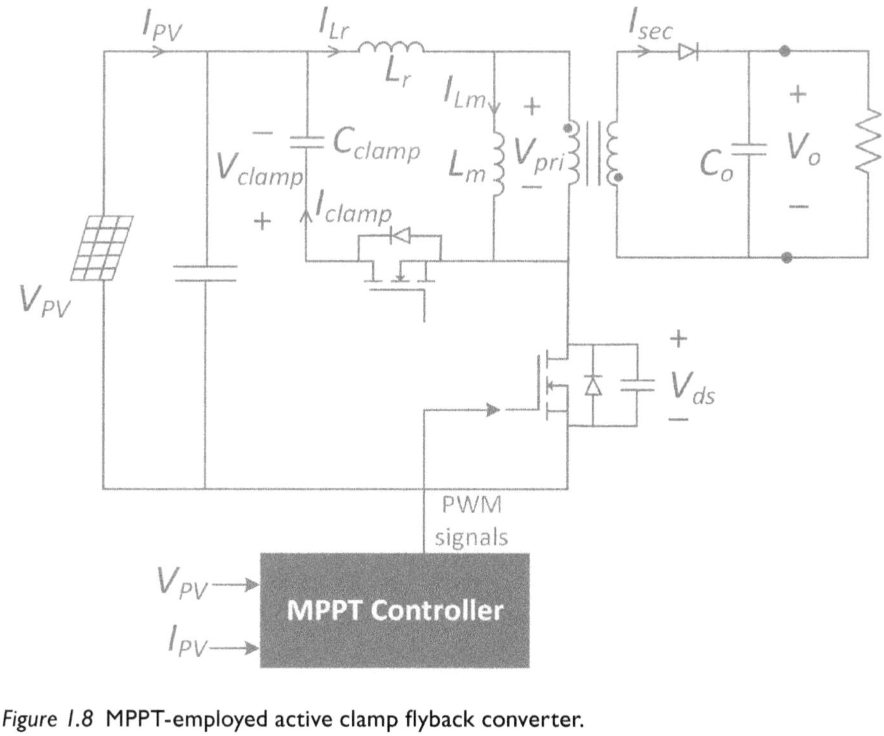

Figure 1.8 MPPT-employed active clamp flyback converter.

Table 1.2 PV module specifications

Parameter	*Value*
Open-circuit voltage V_{oc} (V)	46.104
Short-circuit current I_{sc} (A)	6
Voltage at maximum power point V_{mp} (V)	40
Current at maximum power point I_{mp} (A)	5.625
Temperature coefficient of V_{oc} (%/°C)	−0.36099
Temperature coefficient of I_{sc} (%/°C)	0.102
Cells per module (N_{cell})	36

Table 1.3 Flyback converter specifications

Parameter	*Value*
Input voltage range (V)	20–40
Output voltage (V)	400
Rated output power (W)	225
Switching frequency (kHz)	20
Circuit efficiency (%)	90
Output voltage ripple (V)	4

magnetizing inductance, and clamp capacitor values are calculated as 568.89 μH, 5.6 μH, and 41.03 μF, respectively.

The computed results for MPPT and other analysis using the MATLAB software are presented in this section. The MPPT is performed using P&O algorithm in the active clamp flyback converter. The power–voltage (P–V) curve corresponding to different insolation levels used for verifying MPPT operation is shown in Figure 1.9. A dedicated program is written for P&O algorithm, and the MPPT tracking curves obtained from simulation are presented in Figure 1.10. It can be seen from Figure 1.10 that active clamp flyback converter circuit with PV source is capable of operating at maximum power point for all the insolation levels. Initially, the system shown in Figure 1.8 is subjected to insolation level of 1,000 W/m^2 and P&O algorithm is enabled to perform MPPT operation. It successfully tracks the MPP of 225 W in 35.8 ms. At t = 200 ms, the insolation level is drastically changed from 1,000 W/m^2 to 250 W/m^2. It is evident from Figure 1.10 that the system is capable of successfully tracking the new MPP of 50 W in 23.8 ms. Further, the process is repeated at t = 400 ms and t = 517 ms for insolation levels of 800 W/m^2 and 400 W/m^2, respectively, and in this case also, that system is operating at corresponding MPP of 175.8 W and 86 W, as seen in Figure 1.10. Hence, the simulation results reveal that the active clamp flyback converter design is valid for the MPPT operation in the PV microinverter system. The system shown in Figure 1.8 can be employed for DC–DC stage in the front-end of PV microinverter followed by a H-bridge inverter in cascaded manner to grid.

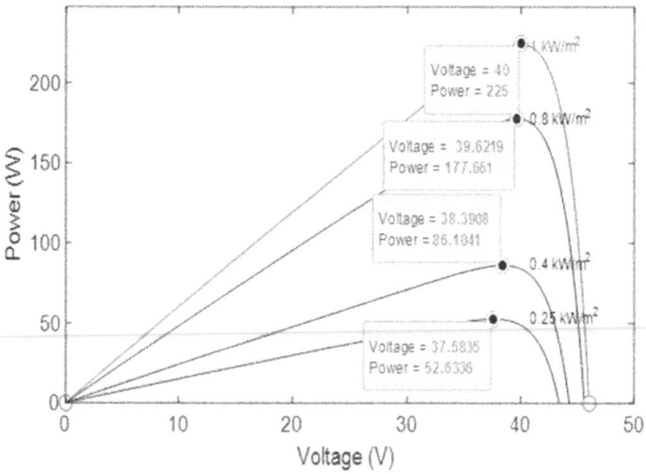

Figure 1.9 Power–voltage (P–V) curves employed for MPPT in system shown in Figure 1.8.

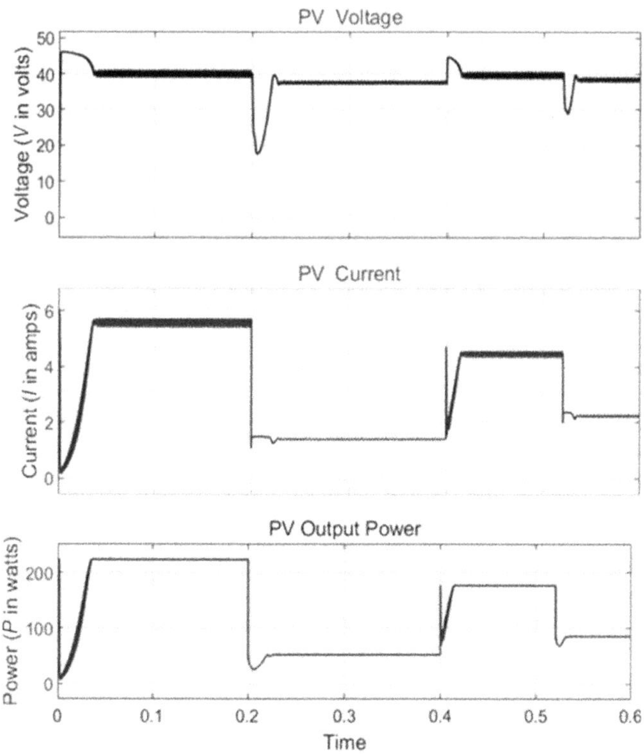

Figure 1.10 MPPT curves using P&O method for different insolation levels.

The primary side and secondary side parameter waveforms obtained from the MATLAB simulation are illustrated in Figure 1.11. It can be seen from this figure that during turn-on period of main switch, the primary side voltage is equal to supply voltage of 40 V. When the main switch is turned off, the supply voltage equals $nV_0 = 26$ V. The magnetizing current waveform in Figure 1.11b reveals that energy is transferred to the secondary side when the main switch is turned off. Further, it can also be witnessed that flux waveform is in alignment with the magnetizing current waveform.

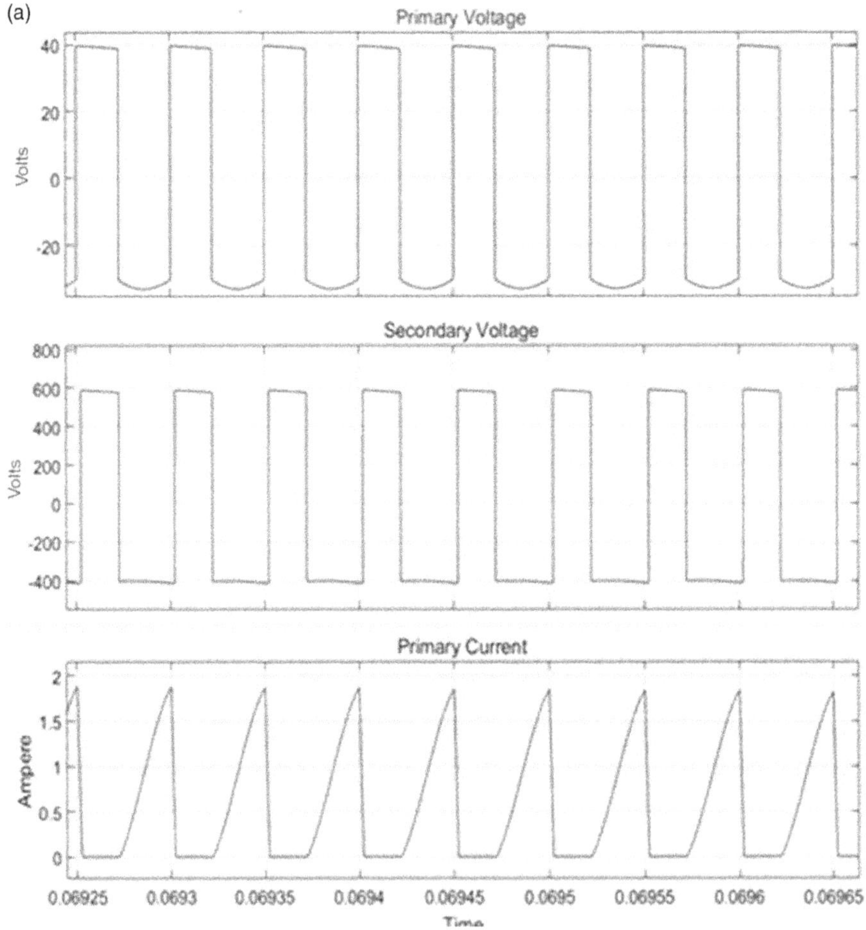

Figure 1.11 Flyback transformer parameters waveform. (a) Primary voltage, secondary voltage, and primary current. (b) Secondary current, magnetizing current, and core flux. *(Continued)*

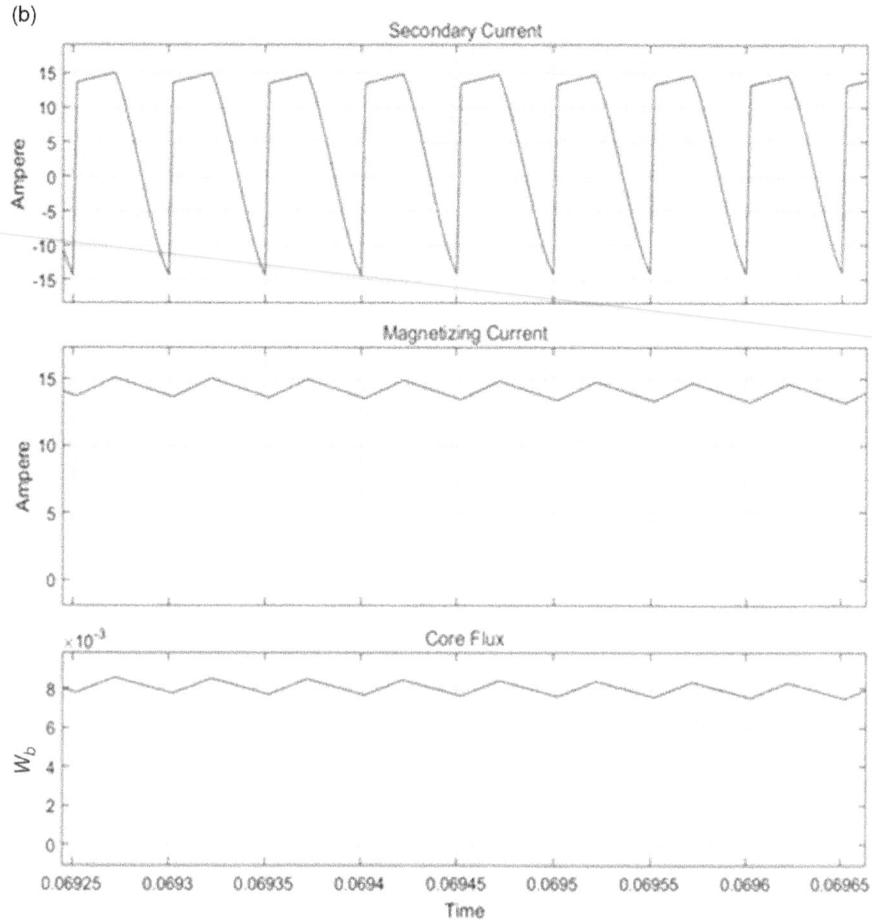

Figure 1.11 (Continued)

The simulation waveforms of various parameters in active clamp snubber circuit of system in Figure 1.8 are presented in Figure 1.12. It can be observed from Figure 1.12a that as soon as the auxiliary switch is turned on, the clamp capacitor voltage starts to build up and clamp capacitor is in resonance with the leakage inductance during the turn-on period of auxiliary switch. The resonant condition shall also be seen in Figure 1.12b from the clamp capacitor current waveform. The drain–source voltage of the main switch in Figure 1.8 is analyzed with the simulation waveforms depicted in Figure 1.13. It is observed that the drain–source voltage is clamped to the designed value of 72 V, thereby reducing the voltage stress on the switch. Therefore, from the simulation results presented in Figures 1.10–1.13, it is evident that active clamp flyback converter is the promising candidate for DC–DC stage of PV microinverter. The active clamp circuit enhances the overall energy efficiency in addition to successfully achieving the MPPT operation.

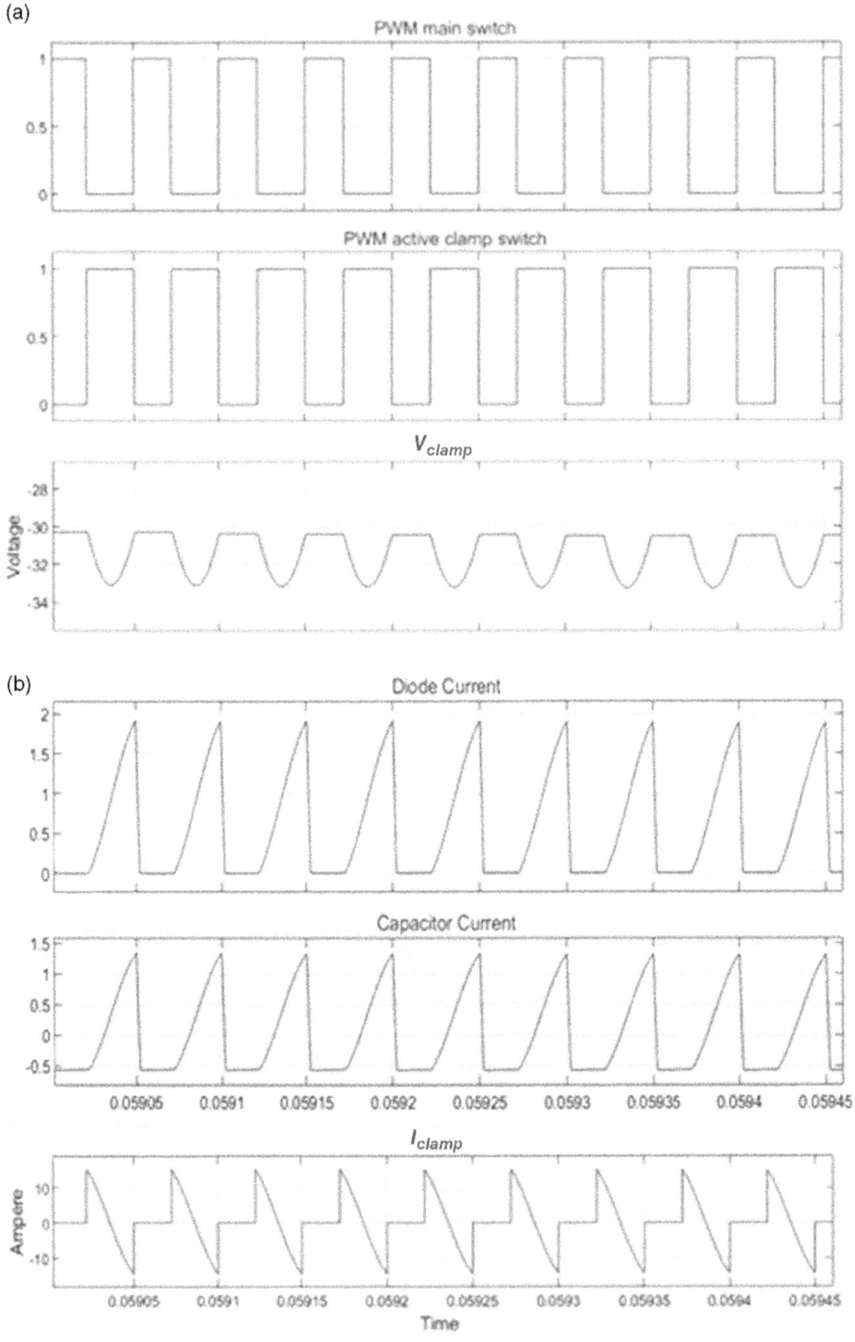

Figure 1.12 Active clamp circuit waveforms. (a) PWM signal for main switch, active clamp switch, and clamp capacitor voltage. (b) Clamp capacitor current, diode current, and output capacitor current.

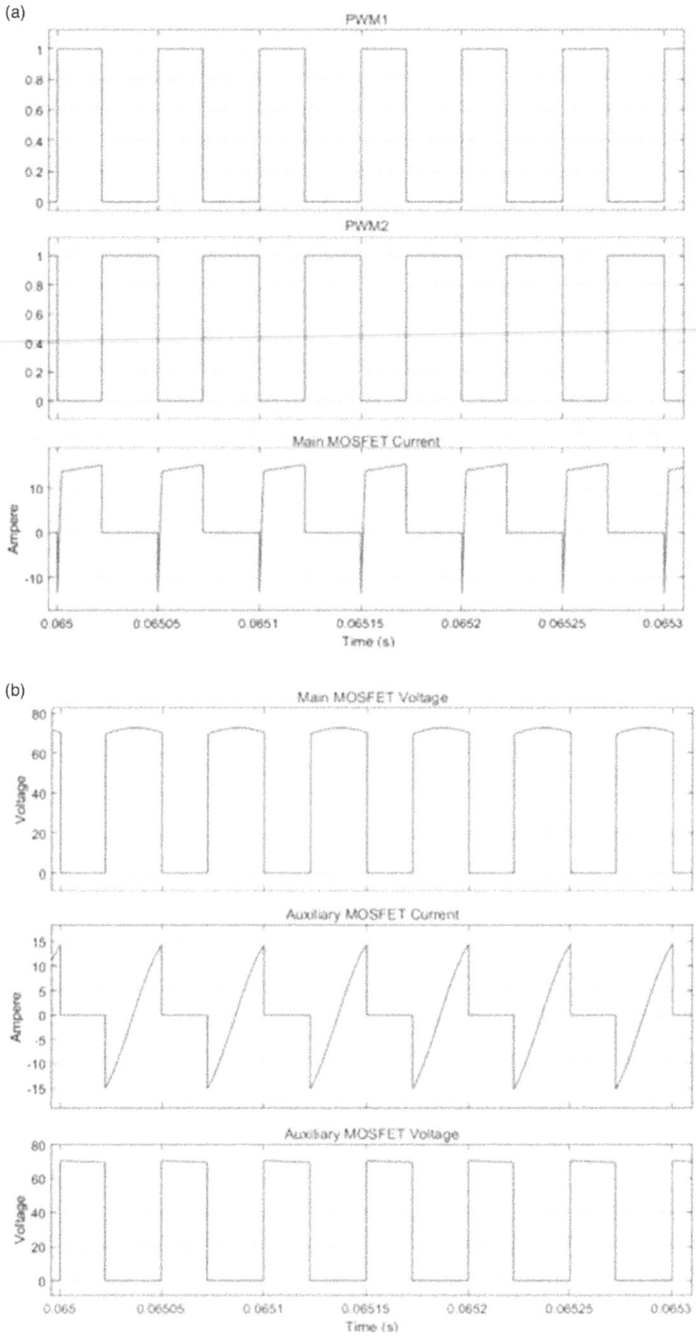

Figure 1.13 Simulation waveforms of (a) PWM signal for main switch, PWM signal for auxiliary active clamp switch, and main switch current. (b) Main switch voltage, auxiliary switch current, and auxiliary voltage.

1.6 CONCLUSION

In residential applications, PV microinverter technology is gaining rapid attention over the past decades. Hence, it is necessary to understand the challenges involved in PV microinverter and explore the various works presented in the literature to address them. In this context, a comprehensive review of isolated two-stage PV microinverter topologies is presented in this chapter. The study on experimentally verified isolate PV microinverter topologies in the literature reveals the importance of galvanic isolation and power quality. The design requirement of isolated topologies should be focused on achieving high power density, high efficiency, reduced component count, compactness, less weight, and low cost. Further, appropriate control scheme and soft-switching characteristics are to incorporated in the topology deployed for PV microinverter system to attain high efficiency. Moreover, a case study for active clamp flyback topology-based PV microinverter is reported. The design procedures for development of the active clamp flyback converter are provided in a systematic manner. The MATLAB simulation was performed with MPPT-employed closed-loop PV system possessing the designed active clamp flyback converter. The simulation results are portrayed with appropriate analysis and lucid explanations. The content presented in this chapter will serve as a quick reference for the power electronics engineers and researchers to understand the PV microinverter topologies.

REFERENCES

1. K. Alluhaybi, I. Batarseh, and H. Hu, "Comprehensive review and comparison of single-phase grid-tied photovoltaic microinverters," in *IEEE Journal of Emerging and Selected Topics in Power Electronics*, vol. 8, no. 2, pp. 1310–1329. 2020.
2. Y. Chen and D. Xu, "Review of soft-switching topologies for single-phase photovoltaic inverters," in *IEEE Transactions on Power Electronics*, vol. 37, no. 2, pp. 1926–1944. 2022.
3. H.-J. Chiu et al., "A module-integrated isolated solar microinverter," in *IEEE Transactions on Industrial Electronics*, vol. 60, no. 2, pp. 781–788. 2013.
4. A. A. Mamun, M. I. I. Sakib, K. F. I. Faruque, M. R. Uddin, and K. M. Salim, "A grid-tie microinverter design based on dual-switch flyback topology for solar PV application," 2022 International Conference on Engineering and Emerging Technologies (ICEET), pp. 1–5, 2022.
5. S. Jiang, D. Cao, Y. Li, and F. Z. Peng, "Grid-connected boost-half-bridge photovoltaic microinverter system using repetitive current control and maximum power point tracking," in *IEEE Transactions on Power Electronics*, vol. 27, no. 11, pp. 4711–4722. 2012.
6. Y.-H. Kim, S.-C. Shin, J.-H. Lee, Y.-C. Jung, and C.-Y. Won, "Soft-switching current-fed push–pull converter for 250-W AC module applications," in *IEEE Transactions on Power Electronics*, vol. 29, no. 2, pp. 863–872. 2014.
7. T. LaBella and J.-S. Lai, "A hybrid resonant converter utilizing a bidirectional GaN AC switch for high-efficiency PV applications," in *IEEE Transactions on Industry Applications*, vol. 50, no. 5, pp. 3468–3475. 2014.

8. W.-J. Cha, Y.-W. Cho, J.-M. Kwon, and B.-H. Kwon, “Highly efficient micro-inverter with soft-switching step-up converter and single switch-modulation inverter,” in *IEEE Transactions on Industrial Electronics*, vol. 62, no. 6, pp. 3516–3523. 2015.
9. C.-Y. Liao, Y.-M. Chen, and W.-H. Lin, “Forward-type microinverter with power decoupling,” in Proceedings of the Twenty-Eighth Annual IEEE Applied Power Electronics Conference and Exposition, March 17–21, 2013.
10. X. Li, J. Liu, F. Ji, X. Cao, Y. Wang, and J. Liu, “Wide-input-range high-frequency-link split-phase microinverter with active-power-decoupling ability,” 2024 IEEE 10th International Power Electronics and Motion Control Conference (IPEMC2024-ECCE Asia), pp. 2524–2529, 2024.
11. X. Li, J. Liu, F. Ji, X. Cao, Y. Wang, and J. Liu, “A single-stage high-frequency-link split-phase microinverter for both grid-tied and islanded operation,” in *IEEE Transactions on Power Electronics*, vol. 39, no. 8, pp. 10409–10423. 2024.
12. H. Meshael, A. Elkhateb, and R. Best, “High gain isolated switched-boost DC–DC converter for PV application,” 2023 IEEE International Conference on Environment and Electrical Engineering and 2023 IEEE Industrial and Commercial Power Systems Europe (EEEIC/I&CPS Europe), pp. 1–6, 2023.
13. IEEE Standard for Interconnecting Distributed Resources with Electric Power Systems, Standard 1547.2-2008. 2009.
14. Standard for Inverters, Converters, Controllers, and Interconnection System Equipment for Use with Distributed Energy Resources, Standard UL 1741. 2010.
15. IEC Standard for Interconnection of PV Systems to the Utility Distribution Systems, Standard IEC 61727. 2004.
16. J. Dadkhah and M. Niroomand, “Optimization methods of MPPT parameters for PV systems: Review, classification, and comparison,” in *Journal of Modern Power Systems and Clean Energy*, vol. 9, no. 2, pp. 225–236. 2021.
17. R. B. Bollipo, S. Mikkili, and P. K. Bonthagorla, “Hybrid, optimal, intelligent and classical PV MPPT techniques: A review,” in *CSEE Journal of Power and Energy Systems*, vol. 7, no. 1, pp. 9–33. 2021.
18. Y. Ohnuma, K. Orikawa, and J.-I. Itoh, “A single-phase current-source PV inverter with power decoupling capability using an active buffer,” in *IEEE Transactions on Industry Applications*, vol. 51, no. 1, pp. 531–538. 2015.

Chapter 2

Adaptive voltage regulation in inverter-based power systems

Anubrata Das

2.1 INTRODUCTION

The Government of India launched the National Solar Mission to encourage the development and deployment of solar photovoltaic (PV) technology across the country. Over the years, the mission's targets have been progressively revised upward, with the latest goal aiming for 100 GW of solar PV capacity by 2022 [1]. Of this target, 88% comprises grid-connected inverter-based resources (IBRs), while the remaining 12% consists of stand-alone PV systems [2].

Additionally, at COP26, India introduced the "One Sun One World One Grid (OSOWOG)" initiative, which envisions the creation of a transnational electricity grid. This grid aims to enable the transfer of solar energy from regions with abundant solar generation to areas with lower generation capacity [3], thereby supporting continuous operation of PV-based IBR.

Moreover, the Government of India has established the National Institute of Wind Energy (NIWE) to support research and development in the wind energy sector [4]. The Ministry of New and Renewable Energy (MNRE) has also set a target of installing 30 GW of offshore wind energy capacity by 2030 [5].

The growing deployment of renewable energy sources (RES) poses serious challenges to the network operator, particularly due to their intermittent power profile. Poor regulation of network voltage and reduced reliability of IBR are some of the few major challenges. Low-power IBR, particularly rooftop PV inverters, are often deployed in the distribution network, which is a combination of L and r, making the grid voltage dependent on active (P) and reactive (Q) power [6]. In such a network, the embedded PV inverters often introduce short-term or minute-scale voltage intermittency primarily because of the fast cloud dynamics. Short-term voltage intermittency often leads to undesired tripping of voltage-sensitive loads, such as medical appliances. Also, it may lead to a large number of tap operations of Load Tap Changer (LTC), thereby reducing their lifetime [7].

Power intermittency in RES also poses a serious threat to the reliability of IBR. Frequent wear and tear of the inverter switch because of varying

 DOI: 10.1201/9781003515807-2

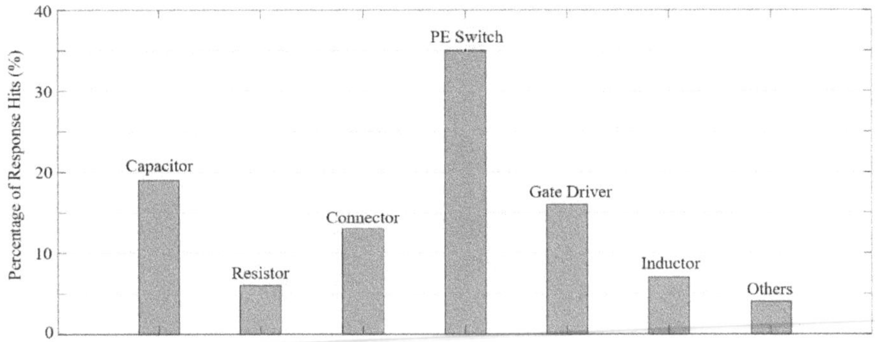

Figure 2.1 Fragile components in an inverter [8].

power profiles leads to poor reliability of IBRs. According to Yang et al. [8], 31% of the participants from different industries, including aerospace, motor drive, automation, and other utility manufacturers in the survey, identified the power semiconductor switches to be the most fragile component of the inverter, followed by the DC link capacitors and gate driver circuits, as shown in Figure 2.1. The two main factors that quantify the reliability or the lifetime of the IBR are (i) mean junction temperature (T_{jm}) and junction temperature ripple (ΔT_j). These quantities depend on the power processed by the IBR. Because of the power intermittency, the current through the switches varies, which results in the junction temperature swing. The problem is even more severe for the IBR installed in off-grid or microgrids. However, in this chapter, we will limit the discussion to the voltage regulation capability of IBR.

2.2 VOLTAGE REGULATION BY IBR

Traditionally, voltage regulation has been managed by dedicated custom power devices and centralized generating stations. However, with the rising share of IBRs at both transmission and distribution levels, this approach has evolved [13]. While grid support functionalities may differ slightly across countries, they are generally based on similar principles, with many adhering to the IEEE 1547-2018 standard [14]. In regions with high IBR penetration, grid codes typically enforce strict voltage limits, often within a ±10% range. Figure 2.2 illustrates the allowable operating voltage ranges for low-voltage (LV) and medium-voltage (MV) networks. Table 2.1 provides a comparative overview of the permissible voltage ranges defined by various national grid codes [9–12].

To mitigate voltage regulation issues, grid codes typically recommend the following methods: (i) fixed reactive power compensation, (ii) operation at a fixed power factor, (iii) voltage-dependent reactive power control (Volt-Var), and (iv) power factor control based on active power output. However, it is

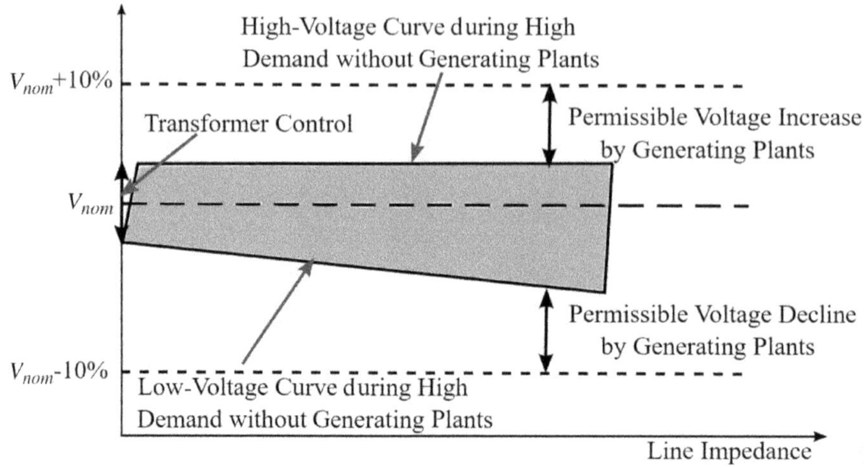

Figure 2.2 Operating voltage range for LV and MV networks in Germany [15].

Table 2.1 Supply voltage variation

CEA India [9]	*VDE-AR-N 4015 Germany [10]*	*RD 661/2007 Spain [11]*	*Arrete 2011 France [12]*
0.9 pu < V < 1.1 pu	0.8 pu < V < 1.1 pu	0.85 pu < V < 1.1 pu	0.9 pu < V < 1.1 pu

important to recognize that reactive power compensation via IBRs should not be restricted to only these methods. For distribution feeders supplying sensitive equipment, such as medical devices, voltage must be maintained within a tighter tolerance band of ±5%. Additionally, other power quality parameters—such as voltage flicker indices (P_{st} and P_{lt})—should remain within the specified upper limits outlined in Ref. [16]. Voltage flicker can cause noticeable variations in light intensity, leading to discomfort or annoyance. Therefore, maintaining P_{st} and P_{lt} within acceptable thresholds is equally critical. While current grid codes and standards do not fully address all these aspects, such requirements are likely to become more comprehensive with the continued expansion of IBR deployment.

While the intermittent nature of power from RES presents several challenges, it also offers an opportunity to utilize the unused capacity of IBRs for voltage regulation. The issue of voltage regulation—particularly the mitigation of overvoltage using IBRs—has been extensively studied in the existing literature [17–20]. Managing overvoltage is relatively straightforward, as reducing both active (P) and reactive power (Q) can effectively lower the voltage. On the other hand, addressing undervoltage conditions is more complex. To raise the voltage level, reactive power (Q) is increased up to the available capacity of the IBR, often requiring a reduction in active power (P) output to accommodate the additional Q. However, in systems where voltage is

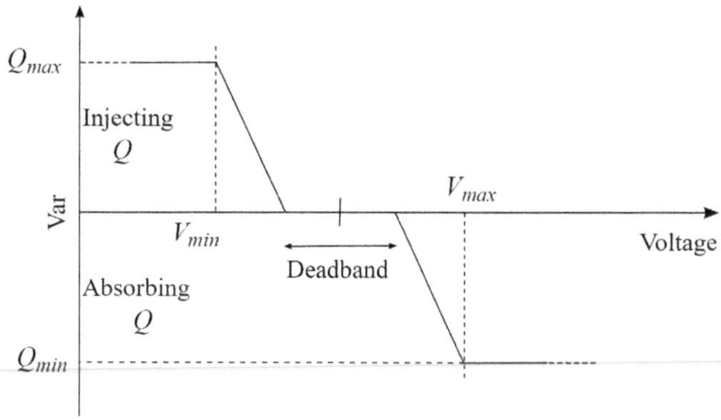

Figure 2.3 Volt-Var characteristic as per IEEE 1547-2018 [14].

influenced by both P and Q, increasing Q at the expense of P during a significant overvoltage event may exacerbate the problem. Despite this complexity, most voltage regulation strategies in the literature primarily rely on managing reactive power from IBRs. The most widely used method involves controlling Q based on a static Q–V (Volt-Var) characteristic [21, 22], a typical example of which is illustrated in Figure 2.3.

While the droop control method is easy to implement, its open-loop nature often results in voltage regulation errors. To overcome these limitations, several dynamic droop-based approaches have been proposed [23–25]. Singhal et al. [23] introduced an adaptive Volt-Var control method to achieve two key goals: (i) reduce steady-state error while enhancing control stability, and (ii) enable self-adjusting control parameters to adapt to changing network conditions. Jabr [24] enhances the traditional Volt-Var characteristic by incorporating an active power-dependent term, improving robustness against fluctuations in active power output from PV-based IBRs. Reza Malekpour and Pahwa [25] proposed a multiobjective control strategy, utilizing voltage sensitivity data to derive reactive power references through a modified droop control, aiming to maintain a smooth voltage profile with minimal power loss.

However, these methods typically focus solely on regulating reactive power (Q), which may not be sufficient in sub-transmission and distribution networks where the resistance-to-reactance (r/x) ratio exceeds 1. In such systems, relying only on Q adjustment often fails to achieve effective voltage regulation [26] and may lead to oversizing of the power converters [27]. A more effective approach involves the coordinated adjustment of both active (P) and reactive (Q) power. Modifying active power, however, is not straightforward—while reducing P is simple in renewable-based IBRs, increasing P requires an additional power reserve. This reserve can come either from energy storage systems [28–30] or from maintaining a power margin by curtailing active power output. Although the former is more precise, it involves additional maintenance for the battery energy storage system [31]. The latter, known as active power curtailment (APC), is a simpler and more cost-effective method to enhance the grid-friendliness of RES, though it results in underutilization of available capacity [32]. Several voltage regulation strategies in the literature utilize both *P* and *Q* from IBRs. A *P–Q–V* droop control scheme is presented in Ref. [33], where the active power of the IBR is modulated to counteract voltage drops caused by load demand. Ouyang et al. [34] implemented APC in a wind farm to align reactive power supply with system demand, thereby achieving voltage reference targets. Coordination among various techniques—such as voltage-dependent power factor, APC, and *Q–V* droop control—is explored in Ref. [35], where APC is activated only during overvoltage conditions. Ghosh et al. [36] employed short-term *PV* generation forecasts to determine appropriate levels of APC. Optimization-based APC strategies have also been proposed [37, 38]. For instance, Chamana and Chowdhury [37] present a zone-based Volt-Var optimization approach to reduce tap operations of upstream voltage regulators, while Li et al. [38] introduce a distributed optimization method that maintains grid voltage while minimizing line losses. Both approaches aim to minimize APC while enhancing voltage control.

All the previously discussed methods leverage the additional flexibility of adjusting active power (*P*), but none explicitly account for the actual relationship between the active (*P*) and the reactive power (*Q*) of the IBR and its corresponding node voltage. These approaches generally assume that this relationship is monotonic. In practice, however, the voltage exhibits a non-monotonic dependency on both *P* and *Q*. As a result, increasing *Q* by curtailing *P* beyond a certain threshold can paradoxically cause a drop in grid voltage, potentially leading to a brownout condition.

2.3 THE RELATION BETWEEN VOLTAGE AND *P* AND *Q* OF IBR

The equivalent circuit of the network as seen from the PV node is shown in Figure 2.4. In case of an undervoltage event, IBR increases its *Q* up to the kVA rating of the inverter to boost the voltage. However, further injection of *Q* at the expense of curtailing *P* has a detrimental effect on the voltage. The mathematical relation between voltage and *P*, *Q* of the inverter is derived in

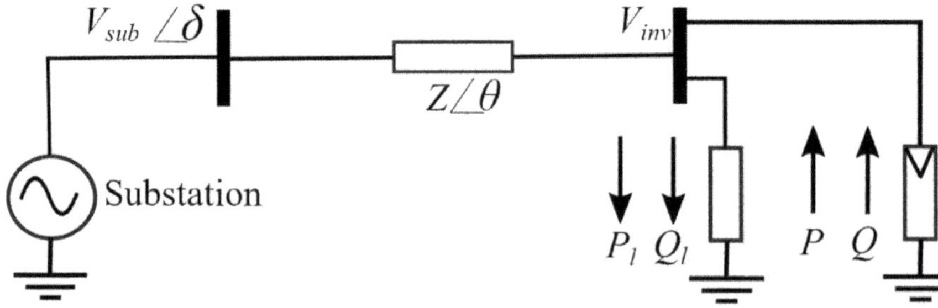

Figure 2.4 Single-node equivalent circuit of the power system as seen by the IBR.

the following, and it can be seen that only one maxima exists on the non-monotonic characteristics.

From the load flow analysis for the system shown in Figure 2.4, the following is obtained:

$$S = 3V_{inv}I^{*} \tag{2.1}$$

where S is the apparent power and I is the current flowing through the network. Replacing S in terms of P and Q and replacing I in terms of voltage and line impedance in Equation (2.1) gives

$$(P_l - P) + j(Q_l - Q) = 3V_{inv} \times \left[\frac{V_{sub}\angle\delta - V_{inv}\angle 0}{Z\angle\theta}\right]^{*} \tag{2.2}$$

Equating the real and imaginary parts of Equation (2.2),

$$\frac{P_l - P}{3} + \frac{V_{inv}^2}{Z}\cos\theta = \frac{V_{inv}V_{sub}}{Z}\cos(\theta - \delta) \tag{2.3}$$

$$\frac{Q_l - Q}{3} + \frac{V_{inv}^2}{Z}\sin\theta = \frac{V_{inv}V_{sub}}{z}\sin(\theta - \delta) \tag{2.4}$$

where P_l and Q_l are the load active and reactive power demand, and P and Q are the active and the reactive power supplied by the IBR. Variables δ and θ are the substation voltage and line impedance phase angle, respectively, considering the IBR node voltage to be the reference. V_{sub} and V_{inv} are the substation and inverter node voltage magnitudes. Z is the magnitude of the line impedance. Assuming the load power (P_l and Q_l) to be zero, as load power demand does not affect the result of the analysis, squaring and adding Equations (2.3) and (2.4) yields

$$V_{inv}^4 - \left[\frac{2P}{3}r + \frac{2Q}{3}x + V_{sub}^2\right]V_{inv}^2 + \frac{P^2 + Q^2}{9}Z^2 = 0 \tag{2.5}$$

In order to have the maximum voltage support from the IBR during a heavy undervoltage event, it is required to operate the IBR at its nominal rating, i.e., leveraging all the available P, Q capacity,

$$P^2 + Q^2 = S_{nom}^2 \tag{2.6}$$

Replacing the expression of P in Equation (2.5) gives

$$V_{inv}^4 - \left[\frac{2\sqrt{(S_{nom}^2 - Q^2)}}{3}r + \frac{2Q}{3}x + V_{sub}^2\right]V_{inv}^2 + \frac{S_{nom}^2}{9}Z^2 = 0 \tag{2.7}$$

Differentiating Equation (2.7) w.r.t Q yields

$$\frac{dV_{inv}}{dQ} = \frac{\left[\frac{2r}{3} \times \frac{Q}{P} - \frac{2}{3}x\right]}{4V_{inv}^3 - \left[\frac{2P}{3}r + \frac{2Q}{3}x + V_{sub}^2\right]} \tag{2.8}$$

Now for maximizing V_{inv}, equating $\frac{dV_{inv}}{dQ} = 0$ in Equation (2.8) gives only one solution:

$$\frac{P}{Q} = \frac{r}{x} \tag{2.9}$$

This shows that under a particular operating condition, there exists a single maximum point in the non-monotonic relation between V_{inv} and P and Q. In Figure 2.5, $V_{inv} - P/Q$ relation for different values of r/x ratio is shown. The line impedance is kept constant while varying the r/x ratio. Figure 2.5 shows that the voltage peak occurs when the relation given in Equation (2.9) holds true. Therefore, a conventional linear controller that disregards this non-monotonic relation risks voltage instability if it overtakes V_{peak} at any time and keeps on injecting Q by curtailing P.

It is worth noting that the r/x for an actual network is not a constant number. The line diagram shown in Figure 2.4 with constant resistance and inductance values represents the Thevenin equivalent network of the entire power system. The r and x values represent different loading and operating conditions, like the combination of lines, equipment, and power flows in the

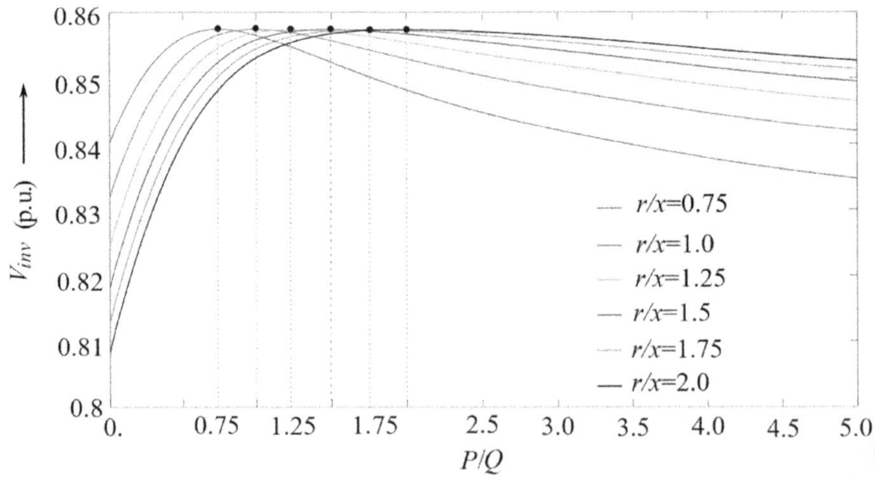

Figure 2.5 IBR node voltage for various *r/x* ratio.

network. Therefore, the r and x values are not constant, and therefore the optimal P/Q changes with loading condition or the network reconfiguration. Hence, to ensure the maximum voltage regulation, the continuous real-time estimation of the r/x ratio is required. However, this requires additional sensors or real-time data monitoring and communication of these data to the inverter controller, which might not always be feasible. Hence, in the following, a network-agnostic method is discussed, which can ensure maximum voltage regulation without using additional sensors or data measurement.

2.4 NETWORK AGNOSTIC VOLTAGE REGULATION METHOD

A network-agnostic voltage regulation approach, as presented in Refs. [39, 40], is briefly outlined here. This method introduces an adaptive algorithm to determine the scheduled or reference reactive power (Q_{sch}) that the IBR should inject in order to maintain the grid voltage at the inverter node equal to a predefined voltage reference (V_{ref}) or the maximum allowable voltage (V_{max}), whichever is lower.

The algorithm, illustrated in the flowchart in Figure 2.6, adjusts Q_{sch} incrementally using a fixed step size (ΔQ). The operation resembles a Perturb and Observe (P&O) algorithm commonly used for tracking the maximum point on a P–V curve. To better illustrate the functioning of the algorithm, the flowchart is further explained using two representative plots corresponding to different network scenarios.

- *Feasible Voltage Range ($V_{ref} < V_{max}$):* An illustrative plot is presented in Figure 2.7a, where each point on the curve represents a specific value of scheduled reactive power (Q_{sch})—or equivalently, a corresponding level of APC (ΔP), assuming the IBR operates at its full capacity at all times. Consider the initial operating point 1. At this point, the operating voltage (V_1) is less than the reference set point (V_{ref}), which prompts the Q_{sch} to increase in the step size of ΔQ. After a few several such steps, ΔQ, the operating point, transitions to point 2, where the operating voltage exceeds V_{ref}. Hence, at this point, the IBR begins to reduce Q_{sch}, thereby bringing the operating point back to V_{ref}. As a result, the operating point fluctuates around the vicinity of V_{ref}, maintaining the voltage regulation around the desired set point.
- *Infeasible Reference Voltage ($V_{max} < V_{ref}$):* This example is shown in Figure 2.7b. Similar to the previous case, the initial operating point begins at a voltage V_1, which is below the reference voltage V_{ref}. Consequently, the algorithm responds by increasing Q_{sch}, moving the operating point upward along the curve. However, unlike the earlier case, this time V_{max} is lower than V_{ref} due to specific operating conditions. As a result, the operating point surpasses V_{max} and enters in the

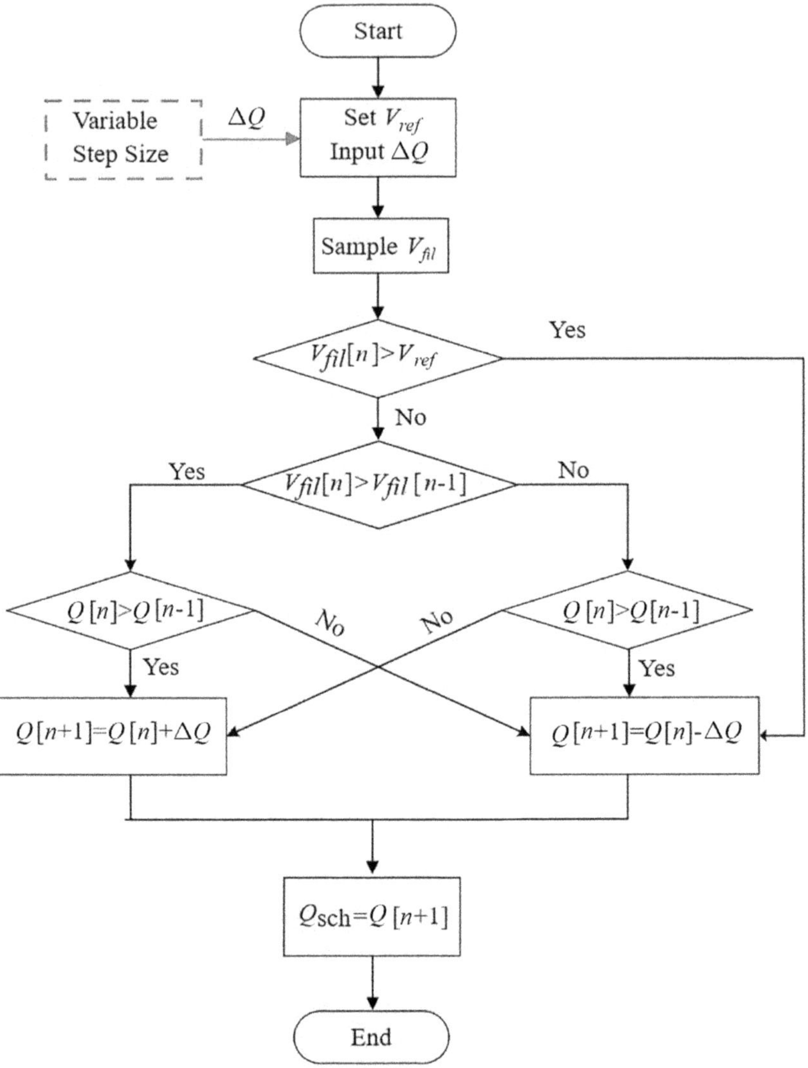

Figure 2.6 Flowchart of Adaptive PQ adjustment algorithm.

right half of the non-monotonic *P–V* curve, where voltage begins to decrease with increasing Q_{sch}. Detecting the decrease in the voltage drop in the last step, the algorithm reduces Q_{sch}, thereby shifting the operating point back to the right-hand side. Thus, the operating point would oscillate around V_{max}.

Hence, with the proposed algorithm, the Q_{sch} is adjusted such that the inverter voltage is regulated to V_{ref} when this is possible or to V_{max} when V_{ref} is not achievable. Thus, the operating principle of the proposed algorithm is

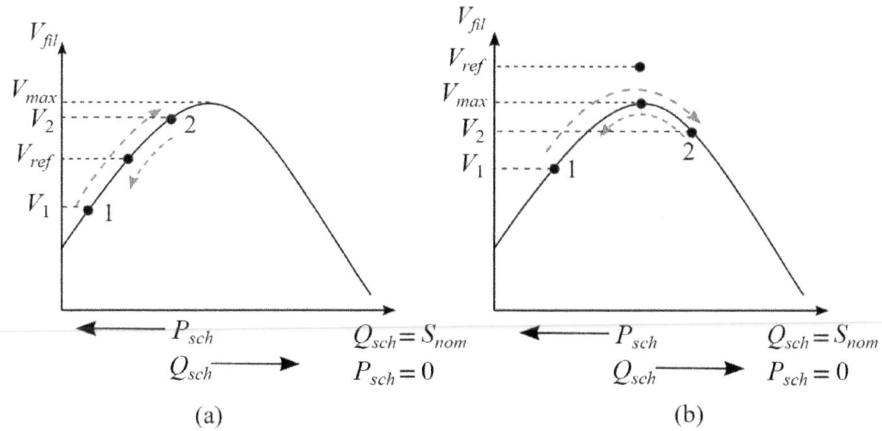

Figure 2.7 Indicative figure of feasible and infeasible reference voltage. (a) V_{ref} is feasible. (b) V_{ref} is infeasible.

very similar to the P&O-based search algorithm, where no prior information about the network or inputs are required. This makes the scheme easy to use and, like the P&O algorithm, keeps tracking the maxima at the constantly changing operating conditions.

2.5 VARIABLE STEP SIZE IMPLEMENTATION

Figure 2.7 shows that the slope of $V_{fil} - Q_{sch}$ relation strongly depends on the operating point. Therefore, the fixed step size ΔQ leads to steady-state voltage ripple, which may violate the voltage flicker requirement mandated by the IEEE 1453-2015 standard. This is the main drawback of the technique proposed in Ref. [39]. To address this, a variable step size algorithm is proposed in Ref. [40]. The proposed method manages the steady-state voltage ripple by distinguishing between two operating conditions—"steady state" and "transient." The steady-state condition refers to when the operating point has reached its target (V_{ref} or V_{max}) and oscillates around it, so the ripple needs to be regulated around an ϵ_2 limit. The transient condition corresponds to when the operating point traverses the $V_{fil} - Q_{sch}$ characteristic in search of its V_{ref} or V_{max}, in which case ΔQ should be large to make the transition fast.

The full picture is shown in Figure 2.8. The algorithm utilizes two running windows, W_1 and W_2, for the identification of the operating mode and voltage ripple calculation, respectively. W_1 stores the most recent Q_{sch} values to figure out whether it is in steady-state or transient mode. This mechanism is also explained with the plots of Figure 2.9a and b. When in transient mode, Q_{sch} is constantly increasing (or decreasing) and every value is higher (or lower) than the previous, thus the sign of each step will always be positive (or negative)

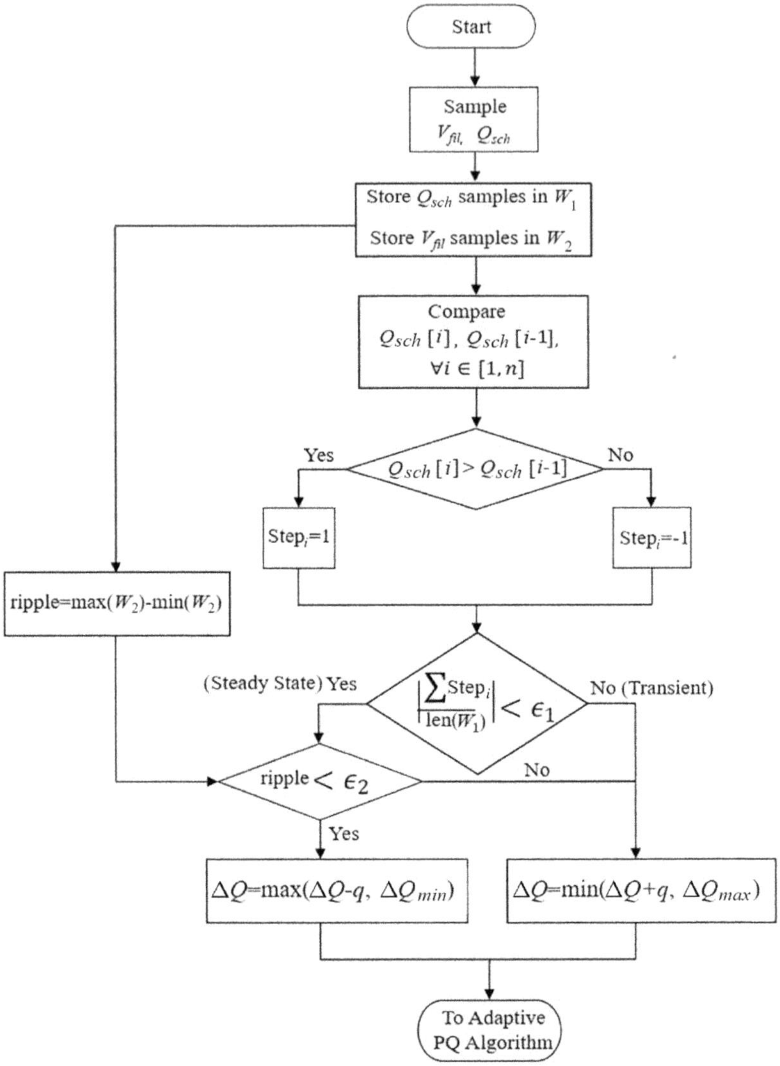

Figure 2.8 Flowchart of variable step size algorithm.

(see Figure 2.9a). On the contrary, in steady state, the step sign oscillates between positive and negative values, as Figure 2.9b shows. Therefore, the average of the step signs within the window should be close to +1 (or −1) in transient and close to 0 in steady state.

This is the premise of the central branch in Figure 2.8, in which the absolute step signs average is compared to a tolerance $\epsilon_1 = 0.5$ to infer the operating mode.

The operating mode decides how the ΔQ step is adjusted. When in transient mode, the step is always increased to reach the voltage target promptly,

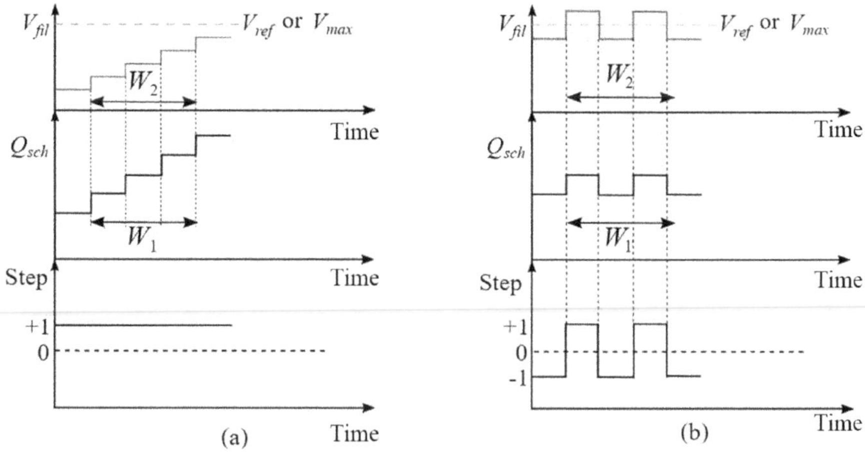

Figure 2.9 Indicative figures of steady-state and transient operating conditions. (a) V_{fil} and Q_{sch} are in transient state. (b) V_{fil} and Q_{sch} are in steady state.

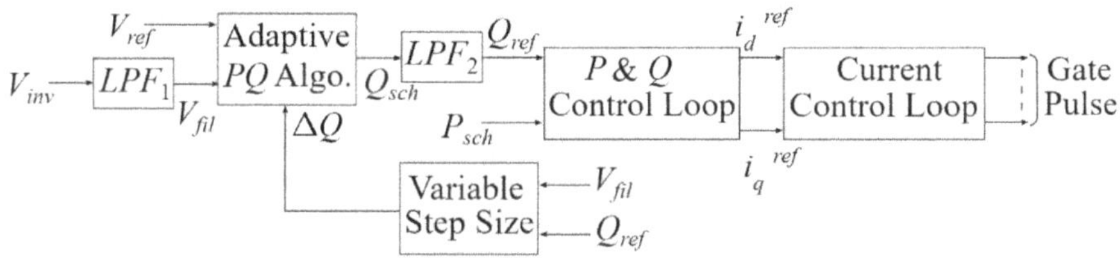

Figure 2.10 Complete control diagram.

whereas in steady-state operation, the step is regulated up or down to meet the grid voltage ripple requirements. This last task is performed with W_2, which keeps record of the recent V_{fil} samples and captures the ripple; this is then compared to the ripple target ϵ_2 to determine up- or downregulation of the ΔQ step. These steps adjustments are made with a resolution of q and abide by a lower ΔQ_{min} and upper ΔQ_{max} bound. The selection of all these parameters is discussed in the latter part of this section. The complete control diagram of the proposed scheme is shown in Figure 2.10.

2.6 CONTROL PARAMETER SELECTION

In the following, the selection of parameters for the discussed voltage regulation technique is discussed.

1. *LPF_1 Time Constant τ_1:* This low pass filter is used to reject the voltage measurement noise from V_{inv}, which is an RMS quantity. Therefore, in the steady state, V_{inv} value does not change. During a transient a change

in the order of few Hz is expected in V_{inv} value based on load driven dynamics. So, a cutoff frequency around 25–50 Hz is reasonable.

2. *LPF_2 Time Constant τ_2:* This low pass filter is used to smoothen out the step change of Q_{sch} according to the operator's requirements. A cutoff frequency on the order of 1–2 Hz is considered here sensible, but it could be lower for even smoother response.
3. *Execution Period T_s:* This refers to the execution frequency of the entire algorithm incorporating both Adaptive PQ and variable step size mechanisms. T_s should be sufficiently large for all dynamics to settle after any step. The network dynamics are very fast and can be ignored, while LPF2 is much slower than LPF1 and dominates the settling time. Therefore, T_s is selected as four times τ_2, based on the common empirical rule for first-order filters.
4. *Window Length of W_1:* W_1 is used to detect the operating mode, steady state or transient. For this purpose, it should hold an even number of samples greater or equal to 4; a lengthier window gives a more robust, albeit slower, estimation.
5. *Window Length of W_2:* This window is used for the grid voltage ripple calculation and requires at least two V_{fil} samples. It is recommended though to use an even number greater or equal to 4 for more reliable results.
6. *Tolerance ϵ_1:* This criterion is applied to the running average of step signs within W_1 to infer the operating mode. This average should be 1 in transients and 0 in steady state, so ε_1 is selected midway that range at 0.5.
7. *Tolerance ϵ_2:* This ripple target is application-dependent and is imposed by the operator to strike a balance between power quality and reliable operation of the algorithm. ϵ_2 should be decided at the planning stage depending on minimum flicker requirement. Hence, ϵ_2 should not be too high due to power quality impact (e.g., as per IEEE 1453–2015 standard [16]), not too low because the algorithm would struggle to perceive the voltage changes. In fact, the optimal value of ϵ_2 does not depend on the inverter control but on external parameters, primarily the noise level and network impedance. A value of 0.1–0.2% of ϵ_2 is deemed sufficient for reasonable noise condition.
8. *Maximum Step ΔQ_{max}:* The upper bound of the step size should be high enough to achieve the ϵ_2 ripple at the voltage peak V_{max}, but not too high to impede the step adjustment. Empirically, it is found that 10% of S_{nom} is a good value, but higher levels would work as well.
9. *Minimum Step ΔQ_{min}:* The lower bound is not very critical, but needs to be a positive value greater than 0. It is recommended to set this equal to the adjustment resolution q.
10. *Step Adjustment Resolution q:* The step size ranging from ΔQ_{min} to ΔQ_{max} should be split into five to ten levels to achieve a resolution that balances fine-tuning and adjustment speed.

It is noteworthy that the proposed algorithm is deterministic in nature. Therefore, the proposed algorithm does not require any additional computation requirement and can be very well implemented in any commercial digital signal processor (DSP).

2.7 SIMULATION RESULTS

The proposed technique is validated through simulations in MATLAB-Simulink for the system shown in Figure 2.11. This system is loosely based on UKGDS-95 bus radial distribution network where lines and loads are aggregated to improve the simulation time. The system parameters are given in Table 2.2 and the line and load parameters in Figure 2.11.

The grid voltage fluctuation is emulated by manipulating the loading profile of Load 4 as shown in Figure 2.12. It is worth mentioning that in a real case the load would vary more smoothly, but here it is deliberately assumed to change in an abrupt step manner to clearly evaluate the merits of the techniques under consideration.

In the following discussion, the voltage regulation performance of the proposed Adaptive PQ+ technique in Ref. [40] is compared with another three approaches: (i) the Conventional *Q* injection (Conv. *Q*) [14, 21], which utilizes only the reactive power capability of the inverter, (ii) the Conventional APC (Conv. PQ) [36, 41], which includes active power adjustment through APC but does not account for the non-monotonic relationship between power and

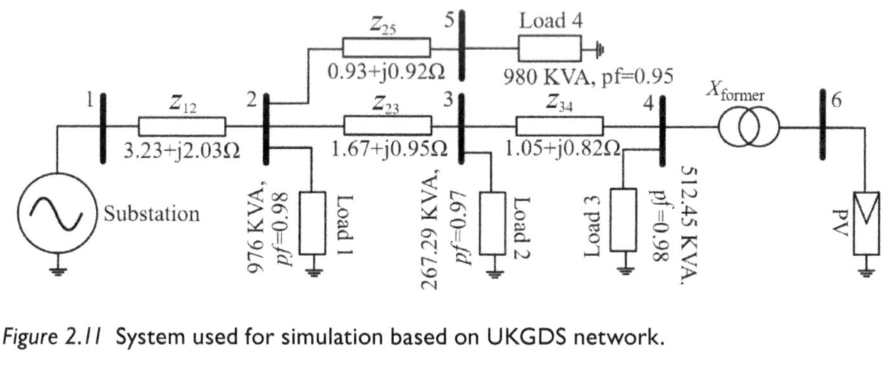

Figure 2.11 System used for simulation based on UKGDS network.

Table 2.2 Specifications of the simulated system

Symbol	*Parameter*	*Value*
S_{CCM}	Nominal rating of IBR	1.05 MVA
f_{nom}	Nominal frequency	50 Hz
V_{nom}	Nominal voltage	11 kV (RMS)
f_s	Switching frequency	10 kHz

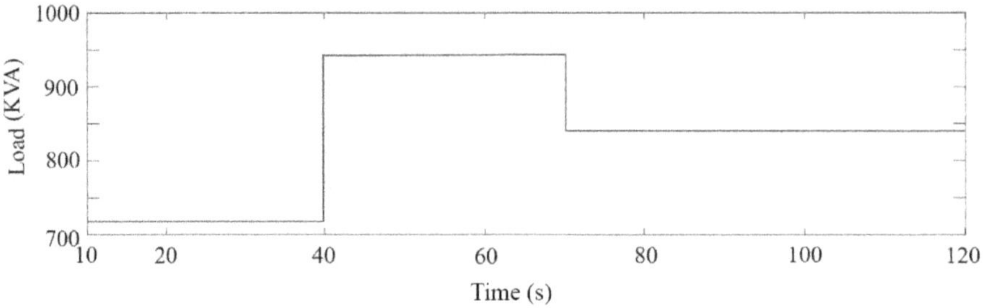

Figure 2.12 Simulation results—loading profile of Load 4.

voltage; and (iii) earlier version of Adaptive PQ algorithm [39], which lacks features such as adaptive step sizing and ripple regulation. The key parameters of the Adaptive PQ+ are given in Table 2.3.

Figure 2.13 illustrates the voltage and power profiles for all four voltage regulation methods, while Figure 2.14 provides a more detailed view of the control signals associated with the Adaptive PQ+ approach. In the first interval t_1, the Load 4 operates under normal conditions and all methods successfully regulate the filtered voltage V_{fil} to its reference value V_{ref} (as indicated by the black dotted line in Figure 2.13c. In this interval, the required reactive power Q_{pv} stays within the inverter's capability, eliminating the need for APC. Consequently, the active power output remains at its maximum power point P_{mpp} across all methods, as seen in Figure 2.13b. The two adaptive algorithms induce some ripple on the grid voltage, which is substantially less in Adaptive PQ+ compared to that given in Ref. [39] due to the adaptive step mechanism. This is more clearly shown in Figure 2.14c and d, where the voltage ripple is rapidly reduced to meet the ϵ_2 target in the former plot, as a result of the step size modification in the latter plot.

In the next interval t_2, a substantial load increase (see Figure 2.12) leads to a voltage sag in Figure 2.13c. Now, V_{ref} becomes infeasible, as V_{max} falls to 0.977 pu (pink dotted line). In response to this sag, Conv. Q performs suboptimal regulation reaching up to 0.972 pu (blue line), as the reactive power capacity does not suffice. Interestingly, the Conv. PQ method reaches and overtakes V_{max} in search of the infeasible V_{ref} target, and as a result it curtails

Table 2.3 Parameters of the Adaptive PQ+ algorithm (network agnostic variable step size)

Parameter	*Value*
LPF time constant	$\tau_1 = 40$ ms; $\tau_2 = 500$ ms
Execution period	$T_s = 2$ s
Window length	$W_1 = W_2 = 4$ samples
Tolerance	$\epsilon_1 = 0.5$, $\epsilon_2 = 0.15\%$
Step size	$\Delta Q_{min} = 20$ kVAr; $\Delta Q_{max} = 120$ KVAr, $q = 20$ kVAr

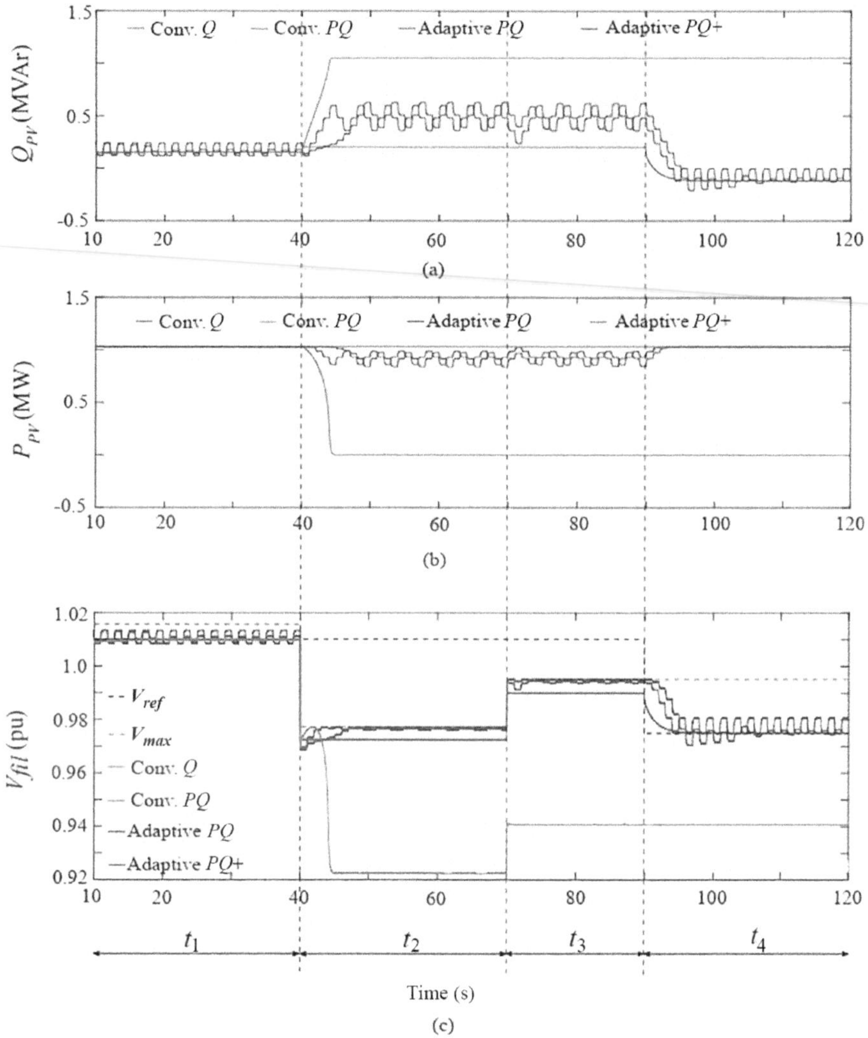

Figure 2.13 Simulation results. (a) Reactive power profile. (b) Active power profile. (c) IBR node voltage profile.

all available P_{pv} (red lines in Figure 2.13b and c). This yields not only the lowest voltage (0.922 pu), but also massive amounts of unnecessary APC. This highlights why it is important to take the non-monotonic relation between *P*, *Q*, and *V* into account when adjusting both power components in voltage regulation.

The two Adaptive PQ methods ensure operation around the critical *P*/*Q* ratio that yields V_{max} by curtailing some *P* and increasing *Q* (red and purple lines in Figure 2.13). This is in contrast to the conventional approaches that

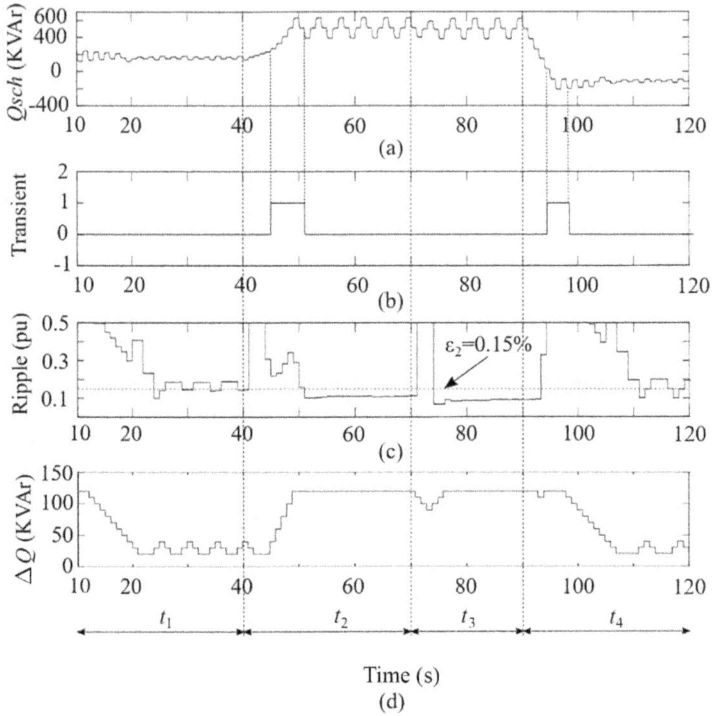

Figure 2.14 Control signals. (a) Reactive power profile. (b) Operating mode. (c) Grid voltage profile. (d) ΔQ step.

do not attain this value. Figure 2.14b shows when Adaptive PQ+ detects the transient condition, during which the ΔQ step increases rapidly in Figure 2.14d to regulate the ripple below ϵ_2 in Figure 2.14c. The steady-state voltage values and ripple are given for all methods and phases in Table 2.4.

In the subsequent interval t_3, a decrease in load results in an increase of V_{max} to 0.995 pu (indicated by the pink dotted line), which remains below the

Table 2.4 Comparison of voltage regulation performance

			Steady-state value of V_{fil} (pu)				*Steady-state voltage ripple (%)*	
Duration	*V_{ref} (pu)*	*V_{max} (pu)*	*a*	*b*	*c*	*d*	*c*	*d*
t_1	1.01	1.017	1.01	1.01	1.01	1.01	0.49	0.14
t_2	1.01	0.977	0.972	0.922	0.977	0.977	0.11	0.09
t_3	1.01	0.995	0.99	0.94	0.995	0.995	0.12	0.10
t_4	1.01	0.995	0.975	0.94	0.975	0.975	0.26	0.15

a: Conventional *Q* [10, 21]; *b*: Conventional PQ [36, 41]; *c*: Adaptive PQ [39]; *d*: Adaptive PQ+ [40].

reference voltage target. Consequently, the behavior of all four control methods mirrors that of the previous interval: Conv. Q continues to underperform due to limited available reactive power margin; Conv. PQ remains trapped in a state of $P_{pv} = 0$, $Q_{pv} = S_{nom}$; while the two adaptive methods effectively track the optimal active/reactive power ratio (see Figure 2.13). This phase highlights the robustness of the adaptive algorithms, demonstrating their ability to maintain alignment with the voltage peak even amid load variations. At this point, the voltage ripple observed in both adaptive approaches is comparable, as they are operating near the voltage peak and approaching the ϵ_2 ripple threshold.

In the final interval t_4, the voltage reference V_{ref} is lowered to 0.975 pu without any change in load conditions, making the new target feasible (see Figure 2.13c). During this phase, the required reactive power Q_{pv} remains within the inverter's reactive power capacity, allowing all control methods to reach the target without the need for APC. However, the Conv. PQ method remains stuck in the previous 100% curtailment state and fails to recover without manual intervention—highlighting once again the risks associated with ignoring the non-monotonic voltage–power relationship. The two adaptive algorithms respond to the updated reference and successfully track the new target, albeit with a slower convergence rate compared to Conv. Q. Notably, the final voltage ripple observed with the Adaptive PQ+ method is significantly lower than that of the method in Ref. [39], as seen in Figure 2.13c. This improvement is corroborated by Figure 2.14c and d, which demonstrates how the adaptive step size is reduced to maintain ripple within acceptable bounds.

The overall voltage regulation performance of the techniques discussed is summarized in Table 2.4. Based on the steady-state values of the filtered voltage V_{fil}, it is evident that both the methods presented in Ref. [39] and the proposed Adaptive PQ+ technique consistently achieve the desired voltage targets, maintaining either V_{ref} or V_{max}, whichever is lower. While the Conv. Q method is dependable, it does not fully utilize the inverter's apparent power (VA) capacity, leading to suboptimal performance in intervals t_2 and t_3. The Conv. PQ control approach, on the other hand, is susceptible to instability. It overlooks the non-monotonic relationship between voltage and power, and in some cases becomes trapped at $P_{pv} = 0$, resulting in a poor voltage profile. In terms of voltage ripple, the two conventional methods exhibit superior performance under simulation conditions, showing virtually no ripple due to their smooth linear control actions. However, among the adaptive strategies, the proposed Adaptive PQ+ algorithm significantly outperforms the earlier version [39], consistently maintaining voltage ripple within or below the 0.15% threshold—a reduction of two to three times compared to that given in Ref. [39], thanks to its variable step-size mechanism.

The key takeaway from this study is that the Adaptive PQ+ method effectively balances power quality with voltage regulation performance. It maximizes the utilization of the inverter's VA capacity while inducing only minimal and tightly controlled voltage ripple.

2.8 CONCLUSIONS

This study demonstrates that the conventional Q-based voltage regulation method is inherently limited by the inverter's reactive power capacity and may lead to inverter tripping under high-voltage conditions. On the other hand, conventional PQ-based linear control methods are not constrained by this limitation, but they introduce a significant risk of instability—particularly when the voltage set point becomes temporarily infeasible, a scenario that is plausible in networks with highly variable power flows. The proposed adaptive method addresses this gap by employing a hill-climbing strategy that dynamically adjusts control inputs. Simulation and experimental results using a hardware prototype confirm the effectiveness of this method. It optimally utilizes both active and reactive power capacities of the inverter, while successfully managing the non-monotonic relationship between power and voltage.

While the step-wise nature of the algorithm introduces some ripple and potential short-term voltage flicker (P_{st}), this is mitigated through the adaptive step-size mechanism introduced in this work. The observed flicker remains only marginally above the level of measurement noise in experimental tests, indicating a well-balanced trade-off between regulation performance and power quality.

Thanks to its network-agnostic and robust performance, the proposed approach is particularly well-suited for PV systems and other IBRs connected to medium- and low-voltage distribution networks. This research has approached the challenge from the standpoint of closed-loop voltage regulation, emphasizing practical applicability and system resilience.

REFERENCES

1. "Solar Energy," Ministry of New and Renewable Energy, Technical Report, 2021. Available at: https://mnre.gov.in/solar/current-status/ (accessed January 6, 2022).
2. "Jawaharlal Nehru National Solar Mission Phase II Policy Document (Draft)," Ministry of New and Renewable Energy, Technical Report, 2013. Available at: http://mnre.gov.in/ (accessed January 6, 2022).
3. "Explained: How India-UK 'Green Grids' Initiative Can Boost Transition to Clean Energy," *The Indian Express*, Technical Report, 2021. Available at: https://indianexpress.com/article/explained/cop26-india-uk-green-grids-project-clean-energy-7605849/ (accessed January 6, 2022).
4. "National Offshore Wind Energy Policy," Ministry of New and Renewable Energy, Technical Report, 2015. Available at: https://www.indianwindpower.com/pdf/Policies/National-Offshore-Wind-Energy-Policy.pdf (accessed January 8, 2022).
5. "Offshore Wind," Ministry of New and Renewable Energy, Technical Report, 2015. Available at: https://mnre.gov.in/wind/offshore-wind/ (accessed January 8, 2022).
6. L. Wang, R. Yan, and T. K. Saha, "Voltage regulation challenges with unbalanced PV integration in low voltage distribution systems and the corresponding solution," in *Applied Energy*, vol. 256, p. 113927, 2019. https://doi.org/10.1016/j.apenergy.2019.113927 (accessed January 12, 2022).

7. Y. Liu, J. Bebic, B. Kroposki, J. de Bedout, and W. Ren, "Distribution system voltage performance analysis for high-penetration PV," 2008 IEEE Energy 2030 Conference, pp. 1–8, 2008. doi: 10.1109/ENERGY.2008.4781069.
8. S. Yang, A. Bryant, P. Mawby, D. Xiang, L. Ran, and P. Tavner, "An industry-based survey of reliability in power electronic converters," in *IEEE Transactions on Industry Applications*, vol. 47, no. 3, pp. 1441–1451, 2011. doi: 10.1109/TIA.2011.2124436.
9. "Draft Second Amendment in CEA (Technical Standards for Connectivity to the Grid) Regulations," CEA Grid Management Division, Technical Report, 2013. Available at: https://cea.nic.in/old/reports/regulation/drafttechnicalstdgridregulations.pdf
10. "VDE-AR-N 4105 Generators Connected to the Low-Voltage Distribution Network. Technical Requirements for the Connection to and Parallel Operation with Low-Voltage Distribution Networks," VDE-AR-N4105, Technical Report, 2011. Available at: https://www.vde-verlag.de/standards/0105029/vde-ar-n-4105-anwendungsregel-2011-08.html
11. "10556 real decreto 661/2007, de 25 de mayo, por el que se regula la actividad de producción de energıa eléctrica en régimen especial," Royal Decree 661, Technical Report, 2007. Available at: https://www.boe.es/eli/es/rd/2007/05/25/661/con
12. "Arrêté du 23 avril 2008 relatif aux prescriptions techniques de conception et de fonctionnement pour le raccordement à un réseau public de distribution d'électricité en basse tension ou en moyenne tension d'une installation de production d'énergie électrique," ARRETE, Technical Report, 2011. Available at: https://www.legifrance.gouv.fr/loda/id/JORFTEXT000018698004/
13. E. Troester, "New German grid codes for connecting PV systems to the medium voltage power grid," in Proceedings of the 2nd International Workshop on Concentrating Photovoltaic Power Plants: Optical Design, Production, Grid Connection, pp. 9–10, 2009.
14. "IEEE Standard for Interconnection and Interoperability of Distributed Energy Resources with Associated Electric Power Systems Interfaces," IEEE Standard 1547-2018 (Revision of IEEE Standard 1547-2003), pp. 1–138, 2018. doi: 10.1109/IEEESTD.2018.8332112.
15. "Technische richtlinie erzeugungsanlagen am mittelspannungsnetz," BDEW, Technical Report, 2008. Available at: https://www.bdew.de/media/documents/20080601BDEW-Mittelspannungsrichtlinie.pdf
16. "IEEE Recommended Practice for the Analysis of Fluctuating Installations on PowerSystems," IEEEStandard1453-2015(RevisionofIEEEStandard1453-2011), pp. 1–74, 2015. doi: 10.1109/IEEESTD.2015.7317469.
17. A. Samadi, R. Eriksson, L. Söder, B. G. Rawn, and J. C. Boemer, "Coordinated active power-dependent voltage regulation in distribution grids with PV systems," in *IEEE Transactions on Power Delivery*, vol. 29, no. 3, pp. 1454–1464, 2014. doi: 10.1109/TP-WRD.2014.2298614.
18. F. Olivier, P. Aristidou, D. Ernst, and T. Van Cutsem, "Active management of low-voltage networks for mitigating overvoltages due to photovoltaic units," in *IEEE Transactions on Smart Grid*, vol. 7, no. 2, pp. 926–936, 2016. doi: 10.1109/TSG.2015.2410171.
19. A. Safayet, P. Fajri, and I. Husain, "Reactive power management for overvoltage prevention at high PV penetration in a low-voltage distribution system," in *IEEE Transactions on Industry Applications*, vol. 53, no. 6, pp. 5786–5794, 2017. doi: 10.1109/TIA.2017.2741925.
20. S. Weckx, C. Gonzalez, and J. Driesen, "Combined central and local active and reactive power control of PV inverters," in *IEEE Transactions on Sustainable Energy*, vol. 5, no. 3, pp. 776–784, 2014. doi: 10.1109/TSTE.2014.2300934.

21. P. Jahangiri and D. Aliprantis, "Distributed Volt/Var control by PV inverters," in 2014 IEEE PES General Meeting—Conference Exposition, pp. 1–1, 2014. doi: 10.1109/PESGM.2014.6939383.
22. C. Zhang, Y. Xu, Z. Dong, and J. Ravishankar, "Three-stage robust inverter-based voltage/var control for distribution networks with high-level PV," in *IEEE Transactions on Smart Grid*, vol. 10, no. 1, pp. 782–793, 2019. doi: 10.1109/TSG.2017.2752234.
23. A. Singhal, V. Ajjarapu, J. Fuller, and J. Hansen, "Real-time local Volt/Var control under external disturbances with high PV penetration," in *IEEE Transactions on Smart Grid*, vol. 10, no. 4, pp. 3849–3859, 2019. doi: 10.1109/TSG.2018.2840965.
24. R. A. Jabr, "Robust Volt/Var control with photovoltaics," in *IEEE Transactions on Power Systems*, vol. 34, no. 3, pp. 2401–2408, 2019. doi: 10.1109/TPWRS.2018.2890767.
25. A. Reza Malekpour and A. Pahwa, "A dynamic operational scheme for residential PV smart inverters," in *IEEE Transactions on Smart Grid*, vol. 8, no. 5, pp. 2258–2267, 2017. doi: 10.1109/TSG.2016.2521367.
26. T. Tewari, A. Mohapatra, and S. Anand, "Coordinated control of OLTC and energy storage for voltage regulation in distribution network with high PV penetration," in *IEEE Transactions on Sustainable Energy*, vol. 12, no. 1, pp. 262–272, 2021. doi: 10.1109/TSTE.2020.2991017.
27. E. Demirok, P. C. González, K. H. B. Frederiksen, D. Sera, P. Rodriguez, and R. Teodorescu, "Local reactive power control methods for overvoltage prevention of distributed solar inverters in low-voltage grids," in *IEEE Journal of Photovoltaics*, vol. 1, no. 2, pp. 174–182, 2011. doi: 10.1109/JPHOTOV.2011.2174821.
28. L. Wang, F. Bai, R. Yan, and T. K. Saha, "Real-time coordinated voltage control of PV inverters and energy storage for weak networks with high PV penetration," in *IEEE Transactions on Power Systems*, vol. 33, no. 3, pp. 3383–3395, 2018. doi: 10.1109/TPWRS.2018.2789897.
29. M. Zeraati, M. E. Hamedani Golshan, and J. M. Guerrero, "Distributed control of battery energy storage systems for voltage regulation in distribution networks with high PV penetration," in *IEEE Transactions on Smart Grid*, vol. 9, no. 4, pp. 3582–3593, 2018. doi: 10.1109/TSG.2016.2636217.
30. M. N. Kabir, Y. Mishra, G. Ledwich, Z. Y. Dong, and K. P. Wong, "Coordinated control of grid-connected photovoltaic reactive power and battery energy storage systems to improve the voltage profile of a residential distribution feeder," in *IEEE Transactions on Industrial Informatics*, vol. 10, no. 2, pp. 967–977, 2014. doi: 10.1109/TII.2014.2299336.
31. M. Zillmann, R. Yan, and T. K. Saha, "Regulation of distribution network voltage using dispersed battery storage systems: A case study of a rural network," 2011 IEEE Power and Energy Society General Meeting, pp. 1–8, 2011. doi: 10.1109/PES.2011.6039094.
32. E. Batzelis, Z. H. Rather, J. Barton, B. R. Naidu, B. Wu, U. N. Firdous et al., "Solar Integration in the UK and India: Technical Barriers and Future Directions," JVCEC, Technical Report, 2021. Available at: https://doi.org/10.17028/rd.lboro.14453133.v1
33. A. Moawwad, V. Khadkikar, and J. L. Kirtley, "A New P–Q–V Droop Control Method for an Interline Photovoltaic (I-PV) Power System," in *IEEE Transactions on Power Delivery*, vol. 28, no. 2, pp. 658–668, 2013. doi: 10.1109/TPWRD.2013.2242906.
34. J. Ouyang, T. Tang, J. Yao, and M. Li, "Active voltage control for DFIG-based wind farm integrated power system by coordinating active and reactive powers under wind speed variations," in *IEEE Transactions on Energy Conversion*, vol. 34, no. 3, pp. 1504–1511, 2019. doi: 10.1109/TEC.2019.2905673.

35. S. Pukhrem, M. Basu, M. F. Conlon, and K. Sunderland, "Enhanced network voltage management techniques under the proliferation of rooftop solar PV installation in low-voltage distribution network," in *IEEE Journal of Emerging and Selected Topics in Power Electronics*, vol. 5, no. 2, pp. 681–694, 2017. doi: 10.1109/JESTPE.2016.2614986.
36. S. Ghosh, S. Rahman, and M. Pipattanasomporn, "Distribution voltage regulation through active power curtailment with PV inverters and solar generation forecasts," in *IEEE Transactions on Sustainable Energy*, vol. 8, no. 1, pp. 13–22, 2017. doi: 10.1109/TSTE.2016.2577559.
37. M. Chamana and B. H. Chowdhury, "Optimal voltage regulation of distribution networks with cascaded voltage regulators in the presence of high PV penetration," in *IEEE Transactions on Sustainable Energy*, vol. 9, no. 3, pp. 1427–1436, 2018. doi: 10.1109/TSTE.2017.2788869.
38. J. Li, Z. Xu, J. Zhao, and C. Zhang, "Distributed online voltage control in active distribution networks considering PV curtailment," in *IEEE Transactions on Industrial Informatics*, vol. 15, no. 10, pp. 5519–5530, 2019. doi: 10.1109/TII.2019.2903888.
39. A. Das, E. Batzelis, S. Anand, and S. R. Sahoo, "An adaptive P–Q management technique for grid voltage regulation using PV inverter," 2020 21st National Power Systems Conference (NPSC), pp. 1–6, 2020. doi: 10.1109/NPSC49263.2020.9331760.
40. A. Das, E. Batzelis, S. Anand, and S. R. Sahoo, "Network-agnostic Adaptive PQ adjustment control for grid voltage regulation in PV systems," in *IEEE Transactions on Industry Applications*, vol. 58, no. 5, pp. 5792–5804, 2022. doi: 10.1109/TIA.2022.3180280.
41. R. Tonkoski and L. A. Lopes, "Impact of active power curtailment on overvoltage prevention and energy production of PV inverters connected to low voltage residential feeders," in *Renewable Energy*, vol. 36, no. 12, pp. 3566–3574, 2011. doi: https://doi.org/10.1016/j.renene.2011.05.031

Chapter 3

The neutral pointless H-type (NPL-H) three-level dual inverter topology for renewable energy integration

Dharavath, Bhukya Hari Krishna, and Udaykumar R. Yaragatti

3.1 INTRODUCTION

3.1.1 Background

An inverter is a power electronic device that transforms DC input power into AC output power at a desired voltage and known frequency. An inverter that generates two levels of voltages at the output is known as two-level inverter, as seen in Figure 3.1. Here the voltage switches between two levels +/–V according to the switching pulses. A two-level conventional inverter has higher switching losses at high switching frequency along with device power rating constraints for high-power applications. Besides, they also face harmonic distortion, EMI, and voltage stress. To overcome these issues of a conventional inverter, the concept of multilevel inverter (MLI) was introduced. An MLI generates a sinusoidal waveform with many levels of DC voltage as its input. The more the quality of levels, the smoother the sine wave generated. A higher

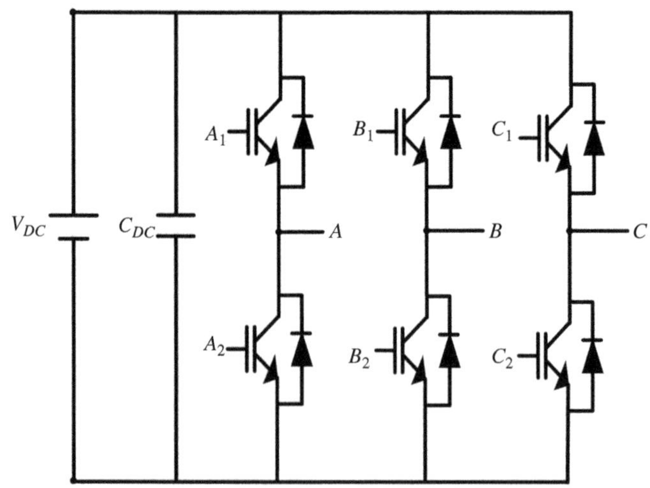

Figure 3.1 Conventional H-bridge inverter circuit.

DOI: 10.1201/9781003515807-3

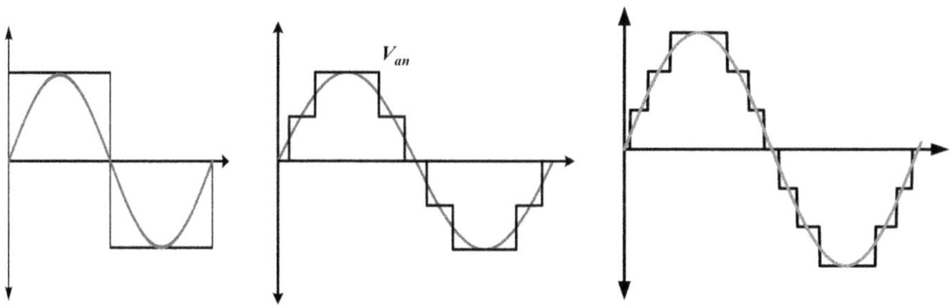

Figure 3.2 Three-, five-, and seven-level inverter output waveforms.

power rating is possible for an MLI with high voltage levels. Device power rating is also reduced, although at the cost of increased number of switching devices. Figure 3.2 shows the waveforms of a three-, five-, and seven-level MLI topology. It is evident that the waveforms approach a sinusoidal waveform as the voltage levels rise [1].

The output waveform has a lower rate of change of voltage and less harmonic distortion in contrast to an inverter with two levels. In an MLI topology, the circuit design is more complex as the voltage level increases along with a complex switching controller. In summary, MLIs have lower THD, lower switching losses, lower EMI, and higher power quality and they are suitable to operate at higher switching frequency and higher power applications.

3.1.2 Literature review

In recent years, MLIs have drawn a lot of attention because of their capacity to increase power efficiency and quality across a range of applications. In contrast to the conventional two-level inverters, MLIs are capable of producing output voltages with multiple levels and reducing harmonic distortion and electromagnetic interference (EMI) [2].

3.1.3 Topologies

Several topologies of MLIs have been developed, and most widely known topologies are described below:

Neutral Point Clamped (NPC): The NPC inverter clamps the DC bus voltage with diodes to produce a variety of output voltage levels. Every phase leg is linked to a number of capacitors, and the diodes control the voltage levels by clamping them to specific points on the DC bus. An alternative name for this architecture is the diode clamped inverter. The primary benefits of this topology is its ability to produce multiple voltage levels, which helps in reducing harmonic distortion. However, numerous capacitors and diodes are needed. It may raise the cost and complicate the design.

Flying Capacitor Inverter: The flying capacitor inverter uses capacitors that "fly" between different voltage levels to generate the output voltage. Each phase leg consists of switching devices and capacitors that are charged and discharged to create the desired voltage levels. This topology is advantageous because it does not require additional voltage sources, and the capacitors can balance the voltage levels naturally. However, the voltage balancing of the capacitors can be challenging, especially under dynamic load conditions.

Cascaded H-Bridge Inverter: The cascaded H-bridge inverter consists of a series of H-bridge cells connected in series. Each H-bridge cell is powered by its own DC source, and the cells are cascaded to generate the multilevel output voltage. This topology is modular and scalable, making it suitable for high-voltage applications. The main advantage of this topology is its simplicity and ease of control, as each H-bridge cell can be controlled independently. However, it requires multiple DC sources, which can increase the overall cost and complexity of the system.

T-Type Multilevel Inverter (T-MLI) [3, 4]: The T-MLI is a type of multilevel inverter that employs a combination of bidirectional and unidirectional switches to generate multiple voltage levels. Each phase leg consists of one bidirectional switch and two unidirectional switches. This configuration allows the inverter to produce a three-level output voltage by selecting different switches to be turned on. The T-MLI is known for its simplicity and reduced switch count compared to other multilevel topologies, making it an efficient choice for medium- and high-power applications.

3.1.4 Control techniques

Effective control techniques are crucial for the performance of MLIs [5]. Commonly used control techniques for the MLI are as follows:

Pulse Width Modulation (PWM): PWM is a popular method for managing multilayer inverters' output voltage and frequency. In order to produce a sequence of voltage pulses, it entails turning the inverter's power components on and off at a high frequency. It is possible to regulate the average voltage output by altering the pulses' width. PWM helps in reducing harmonic distortion and enhancing the output waveform's quality.

Space Vector Modulation (SVM): It is a sophisticated PWM method that lowers switching losses and improves DC link voltage usage. It represents the inverter's output voltage as a vector in a two-dimensional space, and the control algorithm determines the optimal switching states to achieve the desired voltage vector. SVM is known for its ability to produce a more sinusoidal output and reduce total harmonic distortion (THD).

Selective Harmonic Elimination (SHE) [6]: SHE targets specific harmonics for elimination, improving the output waveform quality. With this method, the ideal switching angles that remove certain harmonics are found by solving nonlinear equations. SHE is particularly useful in applications where harmonic performance is critical, such as in high-power drives and renewable energy systems.

Third Harmonic Injection–Pulse Width Modulation (THI-PWM): THI-PWM enhances the performance of traditional PWM by injecting a third harmonic component into the reference signal. As a result, the DC bus voltage is better utilized and the voltage amplitude increases. Comparing THIPWM to conventional PWM techniques, the third harmonic can result in reduced THD and better output waveform quality.

3.1.5 Applications

MLIs have found applications in various fields due to their ability to produce high-quality waveforms, improve power efficiency, and reduce harmonic distortion. Here are some key applications:

Solar Photovoltaic Systems: MLIs are used to convert the DC output from solar panels 4 into AC power with high efficiency and low harmonic distortion. This is crucial for integrating solar power into the grid or for off-grid applications.

Wind Energy Systems: In wind turbines, MLIs convert the variable frequency and amplitude output from the generator into stable AC power, ensuring efficient power transfer and quality.

Electric Motor Drives: MLIs are widely used in industrial applications to drive high-power electric motors. They offer better control over motor speed and torque while reducing losses and improving efficiency.

Flexible AC Transmission Systems (FACTS): MLIs are employed in FACTS devices to enhance the stability and capacity of power transmission networks. They help in controlling voltage, phase angle, and impedance, thus improving the overall performance of the power grid.

Electric Vehicles (EVs): MLIs are used in electric and hybrid vehicles for traction drive applications. They improve the efficiency and performance of the electric drive system while reducing noise and EMI.

Uninterruptible Power Supplies (UPS): MLIs are used in UPS systems to provide reliable and high-quality power to critical loads during power outages. Their ability to produce high-quality AC output with low harmonic content makes them ideal for sensitive electronic equipment.

Harmonic Compensation: MLIs are employed in active power filters to compensate for harmonics and reactive power in power systems. This helps in improving the overall power quality and reducing losses in the system.

Voltage Regulation: MLIs are used in dynamic voltage restorers (DVRs) to protect sensitive loads from voltage sags and swells, ensuring stable and reliable power supply.

Grid Integration of Distributed Generation: MLIs play a significant role in the integration of distributed generation sources such as microgrids and small-scale renewable energy systems into the main power grid. They ensure smooth and efficient power transfer between the sources and the grid.

3.2 NPL-H THREE-LEVEL DUAL INVERTER TOPOLOGY

3.2.1 Introduction

The NPL-H three-level dual inverter topology [7] is an advanced multilevel inverter design that combines the benefits of NPL and H-type configurations. This topology is designed to improve efficiency, reduce component count, and enhance performance in various applications. The NPL-H three-level dual inverter can generate three voltage levels. It enhances the output waveform's quality and lessens harmonic distortion. This topology features two inverters working in parallel, allowing for independent control of dual-phase outputs. This is particularly useful in applications requiring dual-phase power, such as advanced motor drive systems and UPS. By optimizing the switch configuration, the NPL-H three-level dual inverter reduces the number of required switches, leading to lower manufacturing and maintenance costs. The dual inverter configuration supports a wide range of output frequencies, making it suitable for applications with varying speed requirements. This chapter explains the circuit diagram and switching modes in an NPL-H three-level dual inverter topology. It is said that the inverter's switching pulses are produced using a sinusoidal PWM approach. which will be employed in the simulation done in Section 3.3.2.

3.2.2 Circuit description

The NPL-H topology is shown in Figure 3.3. It comprises three interphase commutation legs and six half-bridge legs. The construction is comparable to that of a two-level three-phase inverter, with the interphase legs serving as a pathway for the generation of a three-level output voltage. In total, 18 switches are used in the inverter topology. The switches in the interphase legs that are in charge of producing a three-level output voltage waveform are switches X5 and X6 (where X might be A, B, or C). Here a single capacitance is used for the multilevel inverter topology and hence the problems of neutral point balancing are nonexistent unlike other multilevel inverter topologies such as NPC or T-MLIs [8]. This also helps in reducing the voltage ripple across the capacitor.

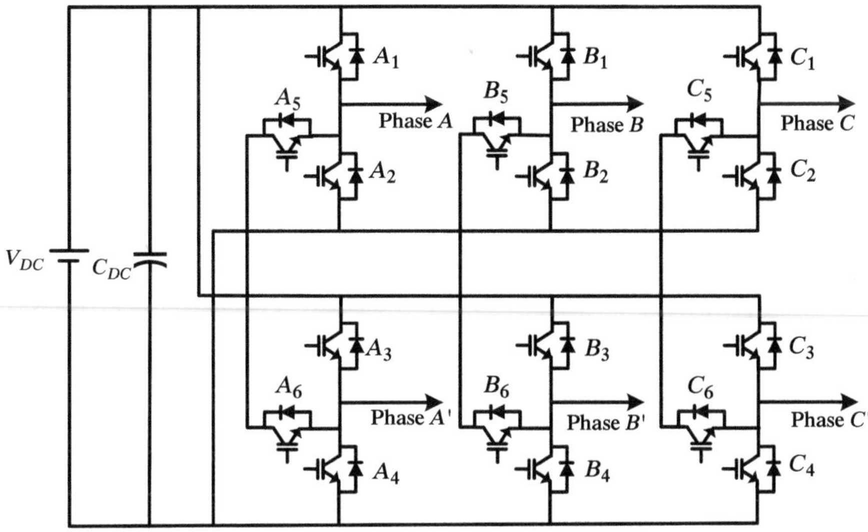

Figure 3.3 NPL-H topology.

A three-level voltage lowers voltage harmonics by slowing down the rate of voltage change. Each half-bridge's midpoint voltage with respect to the ground can be either V_{DC}, $V_{DC}/2$, or 0. Thus, V_{DC}, $V_{DC}/2$, and 0 will represent the line-to-line voltage levels $V_{DC}/2$ and V_{DC}. There are three different switching states for one leg of the inverter. The state where the switches X_1, X_4, and X_5 are on is called the P state (where X can be A, B, or C). In this state, the upper-phase terminal is connected to $+V_{DC}$ and the lower-phase terminal is connected to 0. The state where both interphase switches X_5 and X_6 is on is called the O state. Here both the terminals are essentially shorted and for a balanced load, the output voltage is $+ V_{DC}/2$ for both upper- and lower-phase terminals. State where switches X_2, X_3, and X_6 are on is called the N state. In this case, the lower-phase terminal is connected to $+ V_{DC}$ and the upper-phase terminal to 0.

When all the three phases of P, O, and N are taken into account, the inverter can have a total of 27 switching states. By alternating between these states, a range of line-to-line voltages with values of $+V_{DC}$, $+V_{DC}/2$, 0, $-V_{DC}/2$, and $-V_{DC}$ are obtained. Phase A's switching table is provided in Table 3.1. The

Table 3.1 Switching table for phase A

Switching mode	A_1	A_2	A_3	A_4	A_5	A_6	*VA*	*VA'*
P	On	Off	Off	On	On	Off	V_{DC}	0
O	Off	Off	Off	Off	On	On	$V_{DC}/2$	$V_{DC}/2$
N	Off	On	On	Off	Off	On	0	V_{DC}

same switching mode applies to phases *B* and *C* as well and the combination yields 27 different switching modes.

3.2.3 Modulation scheme

For the initial study, sinusoidal pulse width modulation (SPWM) scheme was implemented. The modulation and carrier waveforms for SPWM is shown in Figure 3.4 and gate pulses for phase *A* switches are shown in Figure 3.5. Actual switching frequency used for simulation is not shown in the figure for illustration purposes. The required PWM pulses are produced by carrier wave 1 during the modulation signal's positive half-cycle, while carrier wave 2 produces the PWM pulses during the modulation signal's negative half-cycle. The generated PWM pulses are directed to the corresponding gate inputs of the switches according to the switching logic illustrated in Section 3.2.2.

As shown in Figure 3.5, the gates GA1 and GA4 have the same signal and so do GA2 and GA3. GA6 has the complement signal of GA1 and GA5, whereas GA5 has the complement signal of GA2. The inverter produces two sets of three-phase output voltages, with a 180° phase difference between the two sets.

3.2.4 SVPWM implementation

3.2.4.1 Introduction

Space vector pulse width modulation (SVPWM) is an advanced PWM technique used to control three-phase inverters efficiently. Unlike traditional sinusoidal PWM, SVPWM utilizes space vector theory to maximize the DC bus voltage utilization, thereby reducing harmonics and improving motor performance [9].

In three-level SVPWM, the inverter topology uses three distinct voltage levels: positive (*P*), neutral (*O*), and negative (*N*), instead of the conventional two-level inverter that only switches between positive and negative. This additional level improves voltage waveform quality, reduces THD, and enhances control flexibility. The three-level SVPWM divides the space vector plane into six sectors, each containing multiple vector regions. By selecting appropriate switching states based on reference vector position, SVPWM optimizes voltage utilization and minimizes switching losses. In this project, a three-level SVPWM implementation is done on Simulink for the NPL-H dual inverter topology and its output waveforms are observed and compared with the SPWM implementation.

3.2.4.2 Principle of three-level SVPWM

SVPWM is one of the preferred real-time modulation techniques and is widely used for voltage-source inverters control. The SVM approach is represented by the three states (*P*), (*O*), and (*N*) in Table 3.1. The inverter may transition

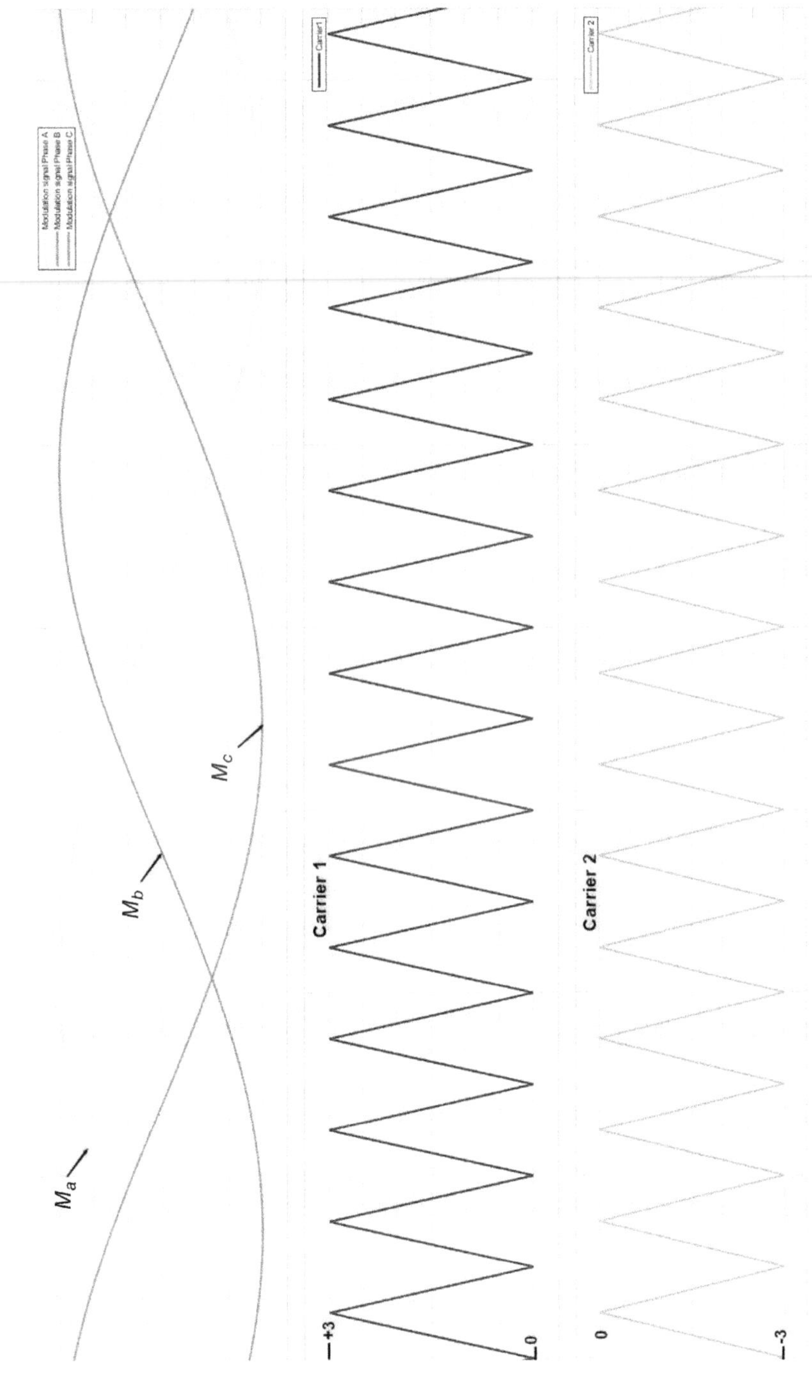

Figure 3.4 Modulation signal and carrier.

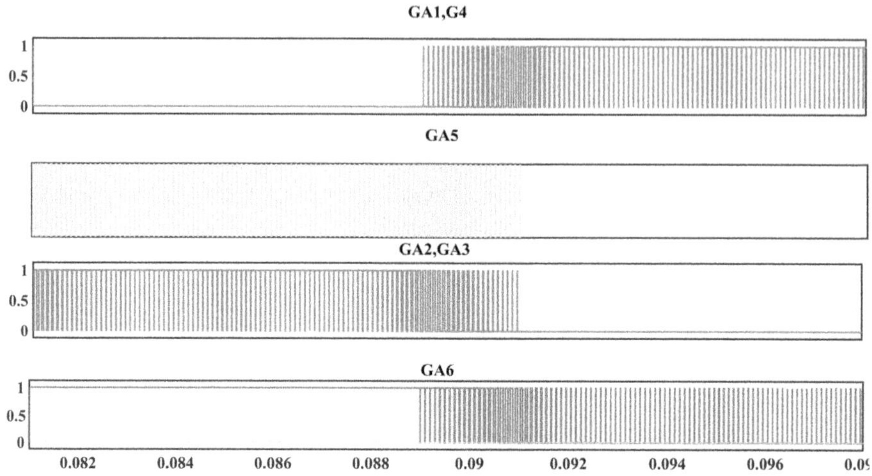

Figure 3.5 Gate pulses for phase *A*.

between 27 different states while three legs are operating. Three 0 vectors (PPP, OOO, and NNN) and 24 active vectors are located in the hexagon's center. As seen in Figure 3.6, the hexagon's area can be split into six sectors (I–VI), each of which has four regions (1–4), for a total of 24 regions of activity.

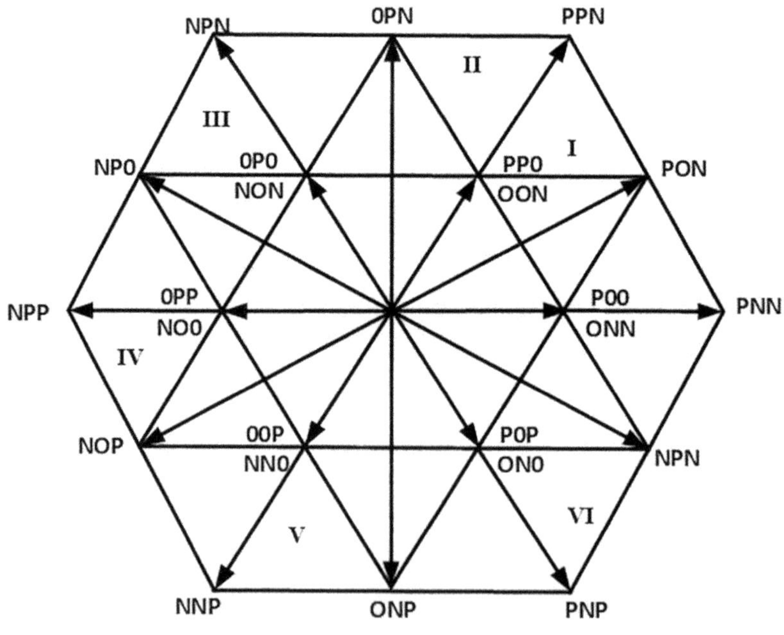

Figure 3.6 Switching states for three-level SVPWM.

The three-phase sine wave's instantaneous voltage is determined by

$$V_a = V_m \sin(\omega t)$$

$$V_b = V_m \sin\left(\omega t - \frac{\pi}{3}\right)$$

$$V_c = V_m \sin\left(\omega t + \frac{\pi}{3}\right)$$

Clark's transformation can be used to determine the spinning vector's magnitude and angle:

$$\begin{bmatrix} V_\alpha \\ V_\beta \end{bmatrix} = \begin{bmatrix} \frac{2}{3} & -\frac{1}{3} & -\frac{1}{3} \\ 0 & \frac{1}{\sqrt{3}} & -\frac{1}{\sqrt{3}} \end{bmatrix} \begin{bmatrix} V_a \\ V_b \\ V_c \end{bmatrix}$$

$$\vec{V}_{ref} = V_\alpha + jV_\beta$$

$$\left|\vec{V}_{ref}\right| = \sqrt{V_\alpha^{\ 2} + V_\beta^{\ 2}}$$

$$\theta = \tan^{-1}\left(\frac{V_\beta}{V_\alpha}\right)$$

The sector to which V_{ref} belongs is computed using θ, and is determined as follows:

$$\begin{cases} 0^\circ \le \theta < 60^\circ, & V_{ref} \text{ is in Sector I} \\ 60^\circ \le \theta < 120^\circ, & V_{ref} \text{ is in Sector II} \\ 120^\circ \le \theta < 180^\circ, & V_{ref} \text{ is in Sector III} \\ 180^\circ \le \theta < 240^\circ, & V_{ref} \text{ is in Sector IV} \\ 240^\circ \le \theta < 300^\circ, & V_{ref} \text{ is in Sector V} \\ 300^\circ \le \theta < 360^\circ, & V_{ref} \text{ is in Sector VI} \end{cases}$$

Following the selection of the sector containing the reference vector, it is necessary to determine the 26 regions in which the vector is placed, which is

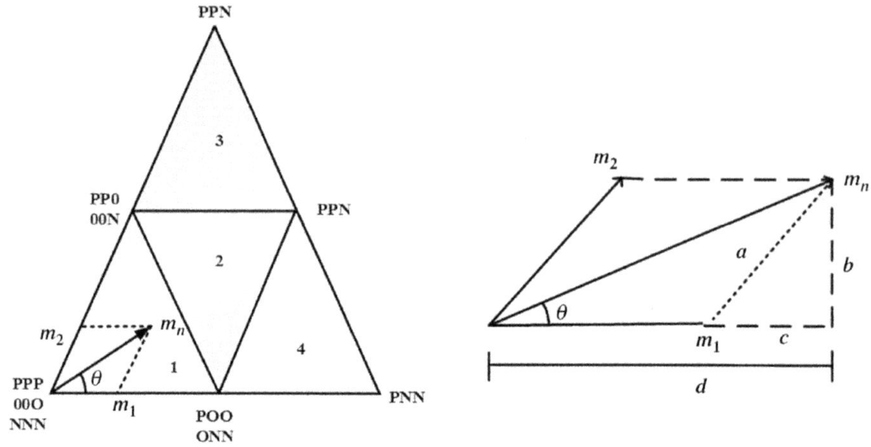

Figure 3.7 Space vector diagram for m_1 and m_2.

determined by m_1 and m_2, as illustrated in Figure 3.7. The following formula can be used to get the parameters m_1 and m_2 [10]:

$$a = m_2 = \frac{b}{\sin\left(\frac{\pi}{3}\right)} = \frac{2}{\sqrt{3}} \cdot b = \frac{2}{\sqrt{3}} \cdot \sin\theta$$

$$m_1 = m_n \cos(\theta)\left(\frac{2}{\sqrt{3}} \cdot m_n \sin\theta\right)\cos\left(\frac{\pi}{3}\right)$$

$$m_1 = m_n\left(\cos\theta - \frac{\sin\theta}{\sqrt{3}}\right)$$

The vector region is then determined using the subsequent tests:

$$\begin{cases} m_1, m_2, (m_1 + m_2) < 0.5, & ;\ V_{ref} \text{ is in Region 1} \\ m_1 > 0.5, & ;\ V_{ref} \text{ is in Region 4} \\ m_2 > 0.5, & ;\ V_{ref} \text{ is in Region 3} \\ m_1, m_2 < 0.5 \ and (m_1 + m_2) > 0.5, & ;\ V_{ref} \text{ is in Region 2} \end{cases}$$

3.2.4.2.1 Dwell time calculation region-wise

Once the sector and region are identified, the next step is to calculate the dwell times of each voltage vectors in each region (Figure 3.8). This dwell time calculation is necessary to calculate the duration for which each switch is turned off or on in the inverter.

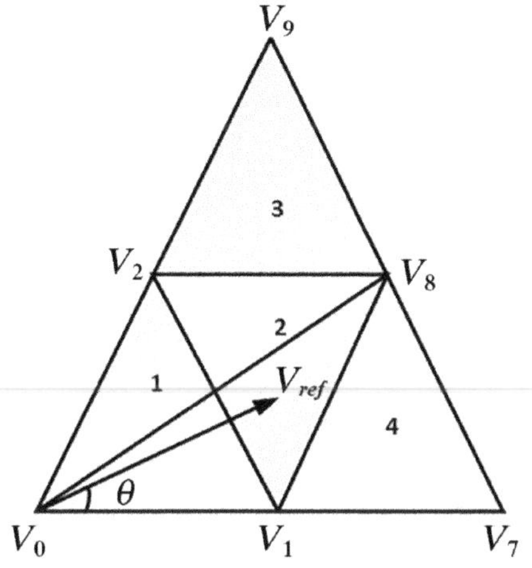

Figure 3.8 Vectors in Sector I.

For Sector I, *dwell times in Region 1* are calculated as follows:

V_0, V_1, and V_2 are the three voltage vectors in Region 1. The summary of vectors and their magnitudes are summarized in Table 3.2.

During a sampling period T_s, the equation for voltage vector on time is as follows:

$$V_{ref}T_s = V_0t_1 + V_1t_2 + V_2t_3$$

$$T_s = t_1 + t_2 + t_3$$

$$V_{ref}\cos\theta \cdot T_s = \frac{V_{DC}}{2}t_2 + \frac{V_{DC}}{2}\cos 60^\circ \cdot t_3$$

$$V_{ref}\sin\theta \cdot T_s = \frac{V_{DC}}{2}\sin 60^\circ \cdot t_3$$

Table 3.2 Switching states and associated vectors

Switching states	V_{an}	V_{bn}	V_{cn}	*Vector*	$\alpha + j\beta$
POO (V_1)	$V_{DC}/3$	$-V_{DC}/6$	$-V_{DC}/6$	$V_{DC}/2\angle 0^\circ$	$V_{DC}/2 + j0$
ONN (V_1)	$V_{DC}/3$	$-V_{DC}/6$	$-V_{DC}/6$	$V_{DC}/2\angle 0^\circ$	$V_{DC}/2 + j0$
PNN (V_7)	$2V_{DC}/3$	$-V_{DC}/3$	$-V_{DC}/3$	$V_{DC}\angle 0^\circ$	$V_{DC} + j0$
PON (V_8)	$V_{DC}/2$	0	$-V_{DC}/2$	$\sqrt{3}V_{DC}/2\angle 30^\circ$	$\left(3/4 + j\sqrt{3}/4\right)V_{DC}$
PPO (V_2)	$V_{DC}/2$	$-V_{DC}/2$	0	$V_{DC}/2\angle 60^\circ$	$\left(1/4 + j\sqrt{3}/4\right)V_{DC}$
OON (V_2)	0	0	$-V_{DC}/2$	$V_{DC}/2\angle 60^\circ$	$\left(1/4 + j\sqrt{3}/4\right)V_{DC}$
PPN (V_9)	$V_{DC}/2$	$-V_{DC}/2$	$-V_{DC}/2$	$V_{DC}\angle 60^\circ$	$\left(1/2 + j\sqrt{3}/4\right)V_{DC}$
NNN (V_0)	0	0	0	0	$0 + j0$
OOO (V_0)	0	0	0	0	$0 + j0$
PPP (V_0)	0	0	0	0	$0 + j0$

Solving for t_1, t_2, t_3:

$$t_1 = T_s - \frac{4}{3} m_a T_s \sin\left(\frac{\pi}{3} + \theta\right) \tag{3.1}$$

$$t_2 = \frac{4}{3} m_a T_s \sin\left(\frac{\pi}{3} - \theta\right) \tag{3.2}$$

$$t_3 = T_s - \frac{4}{3} m_a T_s \sin\theta \tag{3.3}$$

where

$m_a = \dfrac{\sqrt{3}V_{ref}}{V_{DC}}$ is the modulation index.

Dwell times in Region 2:
V_1, V_2, and V_8 are the three voltage vectors in Region 2:

$$V_{ref}T_s = V_1 t_4 + V_2 t_5 + V_8 t_6$$

$$T_s = t_4 + t_5 + t_6$$

Also,

$$V_{ref} \sin\theta T_s = V_1 t_4 + V_2 t_5 + V_8 t_6$$

$$V_{ref} \cos\theta T_s = V_1 t_4 + V_2 \cos 60° t_5 + V_8 \cos 60° t_6$$

Solving for t_4, t_5, t_6:

$$t_4 = T_s - \frac{4}{3} m_a T_s \sin\theta \tag{3.4}$$

$$t_5 = T_s - \frac{4}{3} m_a T_s \sin\left(\frac{\pi}{3} - \theta\right) \tag{3.5}$$

$$t_6 = \frac{4}{3} m_a T_s \sin\left(\frac{\pi}{3} + \theta\right) - T_s \tag{3.6}$$

Dwell Times in Region 3:
V_2, V_8, and V_9 are the three voltage vectors in Region 3:

$$V_{ref}T_s = V_2 t_7 + V_8 t_8 + V_9 t_9$$

$$T_s = t_7 + t_8 + t_9$$

Also,

$$V_{ref}\sin\theta T_s = V_2 \sin 60°t_7 + V_8 \sin 30°t_7 + V_9 \sin 60°t_7$$

$$V_{ref}\cos\theta T_s = V_2 \cos 60°t_7 + V_8 \cos 30°t_8 + V_9 \cos 60°t_9$$

Solving for t_7, t_8, t_9:

$$t_7 = 2T_s - \frac{4}{3}m_aT_s \sin\left(\frac{\pi}{3}+\theta\right) \tag{3.7}$$

$$t_8 = \frac{4}{3}m_aT_s \sin\left(\frac{\pi}{3}-\theta\right) \tag{3.8}$$

$$t_9 = \frac{4}{3}m_aT_s \sin\theta - T_s \tag{3.9}$$

Dwell Times in Region 4:
V_1, V_7, and V_8 are the three voltage vectors in Region 4:

$$V_{ref}T_s = V_1t_{10} + V_7t_{11} + V_8t_{12}$$

$$T_s = t_{10} + t_{11} + t_{12}$$

Also,

$$V_{ref}\sin\theta T_s = V_8 \sin 30°t_{12}$$

$$V_{ref}\cos\theta T_s = V_1t_{10} + V_7t_{11} + V_8 \cos 30°t_{12}$$

Solving for t_{10}, t_{11}, t_{12}:

$$t_{10} = 2T_s - \frac{4}{3}m_aT_s \sin\left(\frac{\pi}{3}+\theta\right) \tag{3.10}$$

$$t_{11} = \frac{4}{3}m_aT_s \sin\left(\frac{\pi}{3}-\theta\right) - T_s \tag{3.11}$$

$$t_{12} = \frac{4}{3}m_aT_s \sin\theta \tag{3.12}$$

Dwell time calculated for the regions are summarized in Table 3.3.

3.2.4.2.2 Switching sequence and switching time calculation for NPL-H inverter

Switching times for individual switches in each phase are calculated based on the switching sequences. Switching sequence is selected such that there is

Table 3.3 Voltage vectors and dwell times for different regions

Region	*Voltage vectors*	*Dwell times*
Region 1	V_0	$t_1 = T_s - \frac{4}{3} m_a T_s \sin\left(\frac{\pi}{3} + \theta\right)$
	V_1	$t_2 = \frac{4}{3} m_a T_s \sin\left(\frac{\pi}{3} - \theta\right)$
	V_2	$t_3 = T_s - \frac{4}{3} m_a T_s \sin\theta$
Region 2	V_1	$t_4 = T_s - \frac{4}{3} m_a T_s \sin\theta$
	V_2	$t_5 = T_s - \frac{4}{3} m_a T_s \sin\left(\frac{\pi}{3} - \theta\right)$
	V_8	$t_6 = \frac{4}{3} m_a T_s \sin\left(\frac{\pi}{3} + \theta\right) - T_s$
Region 3	V_2	$t_7 = 2T_s - \frac{4}{3} m_a T_s \sin\left(\frac{\pi}{3} + \theta\right)$
	V_8	$t_8 = \frac{4}{3} m_a T_s \sin\left(\frac{\pi}{3} - \theta\right)$
	V_9	$t_9 = \frac{4}{3} m_a T_s \sin\theta - T_s$
Region 4	V_1	$t_{10} = 2T_s - \frac{4}{3} m_a T_s \sin\left(\frac{\pi}{3} + \theta\right)$
	V_7	$t_{11} = \frac{4}{3} m_a T_s \sin\left(\frac{\pi}{3} - \theta\right) - T_s$
	V_8	$t_{12} = \frac{4}{3} m_a T_s \sin\theta$

minimum amount of switching between individual states. For Sector I, the switching sequence for each region are summarized in Table 3.4.

In Sector 1, the sequence of switching for Region 1 is as follows:

$$\text{PPO} \rightarrow \text{POO} \rightarrow \text{OOO} \rightarrow \text{OON} \rightarrow \text{ONN}$$

There are 18 switches in the NPL-H inverter topology. Switching table for phase *A* is 32 as mentioned in Table 3.2, where

- Switches A_1, A_4, and A_5 will be *on* for *P*.
- Switches A_5 and A_6 will be *on* for O.
- Switches A_2, A_3, and A_6 will be *on* for *N*.

Similarly, the switching occurs for phases *B* and *C* using switches B_1 to B_6 and C_1 to C_6, corresponding to the sequence of switching mentioned in the table.

Table 3.4 Sequence of switching in Sector I

Region	*Switching state*	*Switching time*	*Vector*
Region 1	PPO	$t_3/2$	V_2
	POO	$t_2/2$	V_1
	OOO	t_1	V_0
	OON	$t_3/2$	V_2
	ONN	$t_2/2$	V_1
Region 2	PPO	$t_5/2$	V_2
	POO	$t_4/2$	V_1
	PON	t_6	V_8
	OON	$t_5/2$	V_2
	ONN	$t_4/2$	V_1
Region 3	PPO	$t_7/2$	V_2
	PPN	t_9	V_9
	PON	t_8	V_8
	OON	$t_7/2$	V_2
Region 4	POO	$t_{10}/2$	V_1
	PON	t_{12}	V_8
	PNN	t_{11}	V_7
	ONN	$t_{10}/2$	V_1

3.2.4.2.2.1 ON TIME CALCULATIONS

Considering the switching sequence, the *on* times for individual switches are calculated:

Phase A

- Switches A_1 and A_4: *On* for a time period of $(t_3 + t_2)/2$.
- Switch A_5: *On* for $t_1 + t_2 + t_3$.
- Switch A_6: *On* for $t_1 + (t_2 + t_3)/2$.
- Switches A_2 and A_3 remain *off* during the time period.

Phase B

- Switches B_1 and B_4: *On* for $t_3/2$.
- Switch B_5: *On* for $t_3 + (t_1 + t_2)/2$.
- Switch B_6: *On* for $(t_1 + t_2 + t_3)/2$.
- Switches B_2 and B_3: *On* for $t_2/2$.

Phase C

- Switches C_1 and C_4: Always *off*.
- Switches C_2 and C_3: *On* for $(t_2 + t_3)/2$.
- Switch C_5: *On* for $(t_3 + t_2)/2$.
- Switch C_6: *On* for $t_1 + t_2 + t_3$.

The switching sequence for *Region 2* is as follows:

PPO → POO → PON → OON → ONN

Phase A

- Switches A_1 and A_4 will be *on* for $(t_4 + t_5)/2 + t_6$.
- Switch A_5 will be *on* for $t_4 + t_5 + t_6$.
- Switch A_6 will be *on* for $t_5/2 + t_4/2$.
- Switches A_2 and A_3 will remain *off* continuously.

Phase B

- Switches B_1 and B_4 will be *on* for $t_5/2$.
- Switch B_5 will be *on* for $t_4/2 + t_5 + t_6$.
- Switch B_6 will be *on* for $t_4 + t_5/2 + t_6$.
- Switches B_2 and B_3 will be *on* for $t_4/2$.

Phase C

- Switches C_1 and C_4 will remain *off* continuously.
- Switch C_5 will be *on* for $(t_4 + t_5)/2$.
- Switches C_2 and C_3 will be *on* for $(t_4 + t_5)/2 + t_6$.
- Switch C_6 will be *on* for $t_4 + t_5 + t_6$.

The switching sequence for *Region 3* is as follows:

PPO → PPN → PON → OON

Phase A

- Switches A_1 and A_4 will be *on* for $t_7/2 + t_8 + t_9$.
- Switch A_5 will be *on* for $t_7 + t_8 + t_9$.
- Switch A_6 will be on for $t_7/2$.
- Switches A_2 and A_3 will remain *off* continuously.

Phase B

- Switches B_1 and B_4 will be *on* for $t_7/2 + t_9$.
- Switch B_5 will be *on* for $t_7 + t_8 + t_9$.
- Switch B_6 will be *on* for $t_7/2 + t_8$.
- Switches B_2 and B_3 will be *off* continuously.

Phase C

- Switches C_1 and C_4 will remain *off* continuously.
- Switch C_5 will be *on* for $t_7/2$.
- Switches C_2 and C_3 will be *on* for $t_7/2 + t_8 + t_9$.
- Switch C_6 will be *on* for $t_7 + t_8 + t_9$.

The switching sequence for *Region 4* is as follows:

POO → PON → PNN → OON

Phase A

- Switches A_1 and A_4 will be *on* for $t_{10}/2 + t_{11} + t_{12}$.
- Switch A_5 will be *on* for $t_{10} + t_{11} + t_{12}$.
- Switch A_6 will be *on* for $t_{10}/2$.
- Switches A_2 and A_3 will remain *off* continuously.

Phase B

- Switches B_1 and B_4 will be off continuously.
- Switch B_5 will be *on* for $t_{10}/2 + t_{11}$.
- Switch B_6 will be *on* for $t_{10} + t_{11} + t_{12}$.
- Switches B_2 and B_3 will be *on* for $t_{10}/2 + t_{11}$.

Phase C

- Switches C_1 and C_4 will remain *off* continuously.
- Switch C_5 will be *on* for $t_{10}/2$.
- Switches C_2 and C_3 will be *on* for $t_{10}/2 + t_{11} + t_{12}$.
- Switch C_6 will be *on* for $t_{10} + t_{11} + t_{12}$.

Summary for switching times for Sector I region-wise is represented in Table 3.5.

Table 3.5 Dwell times for each switch region-wise: Sector I

Switch	*Region 1*	*Region 2*	*Region 3*	*Region 4*
A_1	$(t_2 + t_3)/2$	$t_6 + (t_4 + t_5)/2$	$t_7/2 + t_8 + t_9$	$t_{11} + t_{12} + t_{10}/2$
A_2	0	0	0	0
A_3	0	0	0	0
A_4	$(t_2 + t_3)/2$	$t_6 + (t_4 + t_5)/2$	$t_7/2 + t_8 + t_9$	$t_{11} + t_{12} + t_{10}/2$
A_5	$t_1 + t_2 + t_3$	$t_4 + t_5 + t_6$	$t_7 + t_8 + t_9$	$t_{10} + t_{11} + t_{12}$
A_6	$t_1 + (t_2 + t_3)/2$	$(t_4 + t_5)/2$	$t_7/2$	$t_{10}/2$
B_1	$t_3/2$	$t_5/2$	$t_7/2 + t_9$	0
B_2	$t_2/2$	$t_4/2$	0	$t_{11} + t_{10}/2$
B_3	$t_2/2$	$t_4/2$	0	$t_{11} + t_{10}/2$
B_4	$t_3/2$	$t_5/2$	$t_7/2 + t_9$	0
B_5	$t_1 + t_3 + t_2/2$	$t_4/2 + t_5 + t_6$	$t_7 + t_8 + t_9$	$t_{12} + t_{10}/2$
B_6	$t_1 + t_2 + t_3/2$	$t_4/2 + t_5 + t_6$	$t_7/2 + t_8$	$t_{10} + t_{11} + t_{12}$
C_1	0	0	0	0
C_2	$(t_2 + t_3)/2$	$t_6 + (t_4 + t_5)/2$	$t_7/2 + t_8 + t_9$	$t_{11} + t_{12} + t_{10}/2$
C_3	$(t_2 + t_3)/2$	$t_6 + (t_4 + t_5)/2$	$t_7/2 + t_8 + t_9$	$t_{11} + t_{12} + t_{10}/2$
C_4	0	0	0	0
C_5	$t_1 + (t_2 + t_3)/2$	$(t_4 + t_5)/2$	$t_7/2$	$t_{10}/2$
C_6	$t_1 + t_2 + t_3$	$t_4 + t_5 + t_6$	$t_7 + t_8 + t_9$	$t_{10} + t_{11} + t_{12}$

Table 3.6 Region-wise switching sequence across sectors

Sector	*Region 1*	*Region 2*	*Region 3*	*Region 4*
Sector I	PPO, POO, OOO, OON, ONN	PPO, POO, PON, OON, ONN	PPO, PPN, PON, OON	POO, PON, PNN, ONN
Sector II	PPO, OPO, OOO, OON, NON	PPO, OPO, OPN, OON, NON	OPO, OPN, NPN, NON	PPO, PPN, OPN, OON
Sector III	OPP, OPO, OOO, NOO, NON	OPP, OPO, NPO, NOO, NON	NOO, NPO, NPP, OPP	NON, NPN, NPO, OPO
Sector IV	OPP, OOP, OOO, NOO, NNO	OPP, OOP, NOP, NOO, NNO	OOP, NOP, NNP, NNO	OPP, NPP, NOP, NOO
Sector V	POP, OOP, OOO, ONO, NNO	POP, OOP, ONP, ONO, NNO	ONO, ONP, PNP, POP	OOP, ONP, NNP, NNO
Sector VI	POP, POO, OOO, ONO, ONN	POP, POO, PNO, ONO, ONN	POO, PNO, PNN, ONN	ONO, PNO, PNP, POP

3.2.4.2.3 Switching times for Sector II to Sector VI

For calculating the switching times for Sector II to Sector VI, a simplified algorithm is followed, as explained in Ref. [10], where the dwell times calculated in Sector I is used to calculate the switching times for Sectors II–VI. In this algorithm, whenever the reference vector is in any sector other than Sector I, the reference vector is rotated by an angle of $n\pi/3$, where n is the value from Sector II to Sector VI. By this method, the dwell time calculation is simplified as the dwell times calculated in Sector I is used to calculate the *on* times for switches when the reference vector is in other regions as well.

Once the vector is rotated, switching states in the rotated sector should be replaced with the switching states in Sector I but the dwell time calculations remain the same, hence simplifying the calculations of switching times. Once the switching states and switching sequence are identified, the corresponding states are used to calculate the final switching times within each sector. Table 3.5 shows a summary of switching sequence for all sectors region-wise. Using this switching sequence, the *on* times for each switch are calculated in the same way as done for Sector I, as elaborated in the previous section. Finally, the switching times for each switch for each sector are summarized in Tables 3.7–3.11.

3.2.5 Simulink implementation of three-level SVPWM

3.2.5.1 Simulink design for NPL-H inverter

Figure 3.9 demonstrates the Simulink implementation for the three-level SVPWM using the algorithm discussed. In Ref. [11], an implementation of two-level SVPWM was done using Simulink environment. Based on this, the implementation concept was extended for a three-level SVPWM. The simulation block generates the switching pulses necessary to trigger the switches of the NPL-H inverter.

Table 3.7 Dwell times for each switch region-wise: Sector II

Switch	*Region 1*	*Region 2*	*Region 3*	*Region 4*
A_1	$t_2/2$	$t_4/2$	0	$t_{10}/2 + t_{11}$
A_2	$t_3/2$	$t_5/2$	$t_7/2 + t_9$	0
A_3	$t_3/2$	$t_5/2$	$t_7/2 + t_9$	0
A_4	$t_2/2$	$t_4/2$	0	$t_{10}/2 + t_{11}$
A_5	$t_1 + t_2 + t_3/2$	$t_4 + t_5/2 + t_6$	$t_7/2 + t_8$	$t_{10} + t_{11} + t_{12}$
A_6	$t_1 + t_2/2 + t_3$	$t_4/2 + t_5 + t_6$	$t_7 + t_8 + t_9$	$t_{10}/2 + t_{12}$
B_1	$(t_2 + t_3)/2$	$(t_4 + t_5)/2 + t_6$	$t_7/2 + t_8 + t_9$	$t_{10}/2 + t_{11} + t_{12}$
B_2	0	0	0	0
B_3	0	0	0	0
B_4	$(t_2 + t_3)/2$	$(t_4 + t_5)/2 + t_6$	$t_7/2 + t_8 + t_9$	$t_{10}/2 + t_{11} + t_{12}$
B_5	$t_1 + t_2 + t_3$	$t_4 + t_5 + t_6$	$t_7 + t_8 + t_9$	$t_{10}/2 + t_{11} + t_{12}$
B_6	$t_1 + (t_2 + t_3)/2$	$(t_4 + t_5)/2$	$t_7/2$	$t_{10}/2$
C_1	0	0	0	0
C_2	$(t_2 + t_3)/2$	$t_6 + (t_4 + t_5)/2$	$t_7/2 + t_8 + t_9$	$t_{11} + t_{12} + t_{10}/2$
C_3	$(t_2 + t_3)/2$	$t_6 + (t_4 + t_5)/2$	$t_7/2 + t_8 + t_9$	$t_{11} + t_{12} + t_{10}/2$
C_4	0	0	0	0
C_5	$t_1 + (t_2 + t_3)/2$	$(t_4 + t_5)/2$	$t_7/2$	$t_{10}/2$
C_6	$t_1 + t_2 + t_3$	$t_4 + t_5 + t_6$	$t_7 + t_8 + t_9$	$t_{10} + t_{11} + t_{12}$

Table 3.8 Dwell times for each switch region-wise: Sector III

Switch	*Region 1*	*Region 2*	*Region 3*	*Region 4*
A_1	0	0	0	0
A_2	$(t_2 + t_3)/2$	$t_6 + (t_4 + t_5)/2$	$t_7/2 + t_8 + t_9$	$t_{11} + t_{12} + t_{10}/2$
A_3	$(t_2 + t_3)/2$	$t_6 + (t_4 + t_5)/2$	$t_7/2 + t_8 + t_9$	$t_{11} + t_{12} + t_{10}/2$
A_4	0	0	0	0
A_5	$t_1 + (t_2 + t_3)/2$	$(t_4 + t_5)/2$	$t_7/2$	$t_{10}/2$
A_6	$t_1 + t_2 + t_3$	$t_4 + t_5 + t_6$	$t_7 + t_8 + t_9$	$t_{10} + t_{11} + t_{12}$
B_1	$(t_2 + t_3)/2$	$t_6 + (t_4 + t_5)/2$	$t_7/2 + t_8 + t_9$	$t_{11} + t_{12} + t_{10}/2$
B_2	0	0	0	0
B_3	0	0	0	0
B_4	$(t_2 + t_3)/2$	$t_6 + (t_4 + t_5)/2$	$t_7/2 + t_8 + t_9$	$t_{11} + t_{12} + t_{10}/2$
B_5	$t_1 + t_2 + t_3$	$t_4 + t_5 + t_6$	$t_7 + t_8 + t_9$	$t_{10} + t_{11} + t_{12}$
B_6	$t_1 + (t_2 + t_3)/2$	$(t_4 + t_5)/2$	$t_7/2$	$t_{10}/2$
C_1	$t_3/2$	$t_5/2$	$t_7/2 + t_9$	0
C_2	$t_2/2$	$t_4/2$	0	$t_{11} + t_{10}/2$
C_3	$t_2/2$	$t_4/2$	0	$t_{11} + t_{10}/2$
C_4	$t_3/2$	$t_5/2$	$t_7/2 + t_9$	0
C_5	$t_1 + t_3 + t_2/2$	$t_4/2 + t_5 + t_6$	$t_7 + t_8 + t_9$	$t_{12} + t_{10}/2$
C_6	$t_1 + t_2 + t_3/2$	$t_4/2 + t_5 + t_6$	$t_7/2 + t_8$	$t_{10} + t_{11} + t_{12}$

Table 3.9 Dwell times for each switch region-wise: Sector IV

Switch	*Region 1*	*Region 2*	*Region 3*	*Region 4*
A_1	0	0	0	0
A_2	$(t_2 + t_3)/2$	$t_6 + (t_4 + t_5)/2$	$t_7/2 + t_8 + t_9$	$t_{11} + t_{12} + t_{10}/2$
A_3	$(t_2 + t_3)/2$	$t_6 + (t_4 + t_5)/2$	$t_7/2 + t_8 + t_9$	$t_{11} + t_{12} + t_{10}/2$
A_4	0	0	0	0
A_5	$t_1 + (t_2 + t_3)/2$	$(t_4 + t_5)/2$	$t_7/2$	$t_{10}/2$
A_6	$t_1 + t_2 + t_3$	$t_4 + t_5 + t_6$	$t_7 + t_8 + t_9$	$t_{10} + t_{11} + t_{12}$
B_1	$t_2/2$	$t_4/2$	0	$t_{10}/2 + t_{11}$
B_2	$t_3/2$	$t_5/2$	$t_7/2 + t_9$	0
B_3	$t_3/2$	$t_5/2$	$t_7/2 + t_9$	0
B_4	$t_2/2$	$t_4/2$	0	$t_{10}/2 + t_{11}$
B_5	$t_1 + t_2 + t_3/2$	$t_4 + t_5/2 + t_6$	$t_7/2 + t_8$	$t_{10} + t_{11} + t_{12}$
B_6	$t_1 + t_2/2 + t_3$	$t_4/2 + t_5 + t_6$	$t_7 + t_8 + t_9$	$t_{10}/2 + t_{12}$
C_1	$(t_2 + t_3)/2$	$(t_4 + t_5)/2 + t_6$	$t_7/2 + t_8 + t_9$	$t_{10}/2 + t_{11} + t_{12}$
C_2	0	0	0	0
C_3	0	0	0	0
C_4	$(t_2 + t_3)/2$	$(t_4 + t_5)/2 + t_6$	$t_7/2 + t_8 + t_9$	$t_{10}/2 + t_{11} + t_{12}$
C_5	$t_1 + t_2 + t_3$	$t_4 + t_5 + t_6$	$t_7 + t_8 + t_9$	$t_{10} + t_{11} + t_{12}$
C_6	$t_1 + (t_2 + t_3)/2$	$(t_4 + t_5)/2$	$t_7/2$	$t_{10}/2$

Table 3.10 Dwell times for each switch region-wise: Sector V

Switch	*Region 1*	*Region 2*	*Region 3*	*Region 4*
A_1	$(t_2 + t_3)/2$	$(t_4 + t_5)/2 + t_6$	$t_7/2 + t_8 + t_9$	$t_{10}/2 + t_{11} + t_{12}$
A_2	0	0	0	0
A_3	0	0	0	0
A_4	$(t_2 + t_3)/2$	$(t_4 + t_5)/2 + t_6$	$t_7/2 + t_8 + t_9$	$t_{10}/2 + t_{11} + t_{12}$
A_5	$t_1 + t_2 + t_3$	$t_4 + t_5 + t_6$	$t_7 + t_8 + t_9$	$t_{10} + t_{11} + t_{12}$
A_6	$t_1 + (t_2 + t_3)/2$	$(t_4 + t_5)/2$	$t_7/2$	$t_{10}/2$
B_1	0	0	0	0
B_2	$(t_2 + t_3)/2$	$(t_4 + t_5)/2 + t_6$	$t_7/2 + t_8 + t_9$	$t_{10}/2 + t_{11} + t_{12}$
B_3	$(t_2 + t_3)/2$	$(t_4 + t_5)/2 + t_6$	$t_7/2 + t_8 + t_9$	$t_{10}/2 + t_{11} + t_{12}$
B_4	0	0	0	0
B_5	$t_1 + (t_2 + t_3)/2$	$(t_4 + t_5)/2$	$t_7/2$	$t_{10}/2$
B_6	$t_1 + t_2 + t_3$	$t_4 + t_5 + t_6$	$t_7 + t_8 + t_9$	$t_{10} + t_{11} + t_{12}$
C_1	$(t_2 + t_3)/2$	$(t_4 + t_5)/2 + t_6$	$t_7/2 + t_8 + t_9$	$t_{10}/2 + t_{11} + t_{12}$
C_2	$t_3/2$	$t_5/2$	$t_7/2 + t_9$	0
C_3	$t_3/2$	$t_5/2$	$t_7/2 + t_9$	0
C_4	$(t_2 + t_3)/2$	$(t_4 + t_5)/2 + t_6$	$t_7/2 + t_8 + t_9$	$t_{10}/2 + t_{11} + t_{12}$
C_5	$t_1 + t_2 + t_3/2$	$t_4 + t_5/2 + t_6$	$t_7/2 + t_8$	$t_{10} + t_{11} + t_{12}$
C_6	$t_1 + t_2/2 + t_3$	$t_4/2 + t_5 + t_6$	$t_7 + t_8 + t_9$	$t_{10}/2 + t_{12}$

Table 3.11 Dwell times for each switch region-wise: Sector VI

Switch	*Region 1*	*Region 2*	*Region 3*	*Region 4*
A_1	$t_3/2$	$t_5/2$	$t_7/2 + t_9$	0
A_2	$t_2/2$	$t_4/2$	0	$t_{10}/2 + t_{11}$
A_3	$t_2/2$	$t_4/2$	0	$t_{10}/2 + t_{11}$
A_4	$t_3/2$	$t_5/2$	$t_7/2 + t_9$	0
A_5	$t_1 + t_2/2 + t_3$	$t_4/2 + t_5 + t_6$	$t_7 + t_8 + t_9$	$t_{10}/2 + t_{12}$
A_6	$t_1 + t_2 + t_3/2$	$t_4 + t_5/2 + t_6$	$t_7/2 + t_8$	$t_{10} + t_{11} + t_{12}$
B_1	0	0	0	0
B_2	$(t_2 + t_3)/2$	$(t_4 + t_5)/2 + t_6$	$t_7/2 + t_8 + t_9$	$t_{10}/2 + t_{11} + t_{12}$
B_3	$(t_2 + t_3)/2$	$(t_4 + t_5)/2 + t_6$	$t_7/2 + t_8 + t_9$	$t_{10}/2 + t_{11} + t_{12}$
B_4	0	0	0	0
B_5	$t_1 + (t_2 + t_3)/2$	$(t_4 + t_5)/2$	$t_7/2$	$t_{10}/2$
B_6	$t_1 + t_2 + t_3$	$t_4 + t_5 + t_6$	$t_7 + t_8 + t_9$	$t_{10} + t_{11} + t_{12}$
C_1	$t_2/2$	$t_4/2$	0	$t_{10}/2 + t_{11}$
C_2	0	0	0	0
C_3	0	0	0	0
C_4	$t_2/2$	$t_4/2$	0	$t_{10}/2 + t_{11}$
C_5	$t_1 + t_2 + t_3$	$t_4 + t_5 + t_6$	$t_7 + t_8 + t_9$	$t_{10} + t_{11} + t_{12}$
C_6	$t_1 + (t_2 + t_3)/2$	$(t_4 + t_5)/2$	$t_7/2$	$t_{10}/2$

The implementation is divided into four parts. The first part is to determine a magnitude and an angle of the reference voltage vector. Three-phase sine wave signal generators are used to generate the reference voltage vector in the simulation. From the angle, the current sector is identified as well as the modified angle value calculated after rotating the vector to Sector I. Modified angle value ranges from 0 to $\pi/3$, and this angle value helps in determining the region in which the vector is currently at in Sector I, which is the second part of the implementation. The next part is to calculate the switching times based on the dwell time calculations done in the previous sections. Based on the sector and region, the corresponding switching times are calculated and the SVPWM-modulated signal is obtained as the output from the *Switching Time Calculation* block, as shown in Figure 3.10. Finally, the modulated signal is contrasted with a repeating sequence to produce the necessary PWM pulses for each individual switches. Normalized switch on times for switches A_1, A_2, A_3, and A_4 was obtained, as shown in Figure 3.11. In this instance, A_1 and A_4 share the same switching signal, as do A_2 and A_3. A_5 has a complementary switching with respect to A_1, and A_6 has a complementary switching with respect to A_2. The same applies to switches in phases B and C as well.

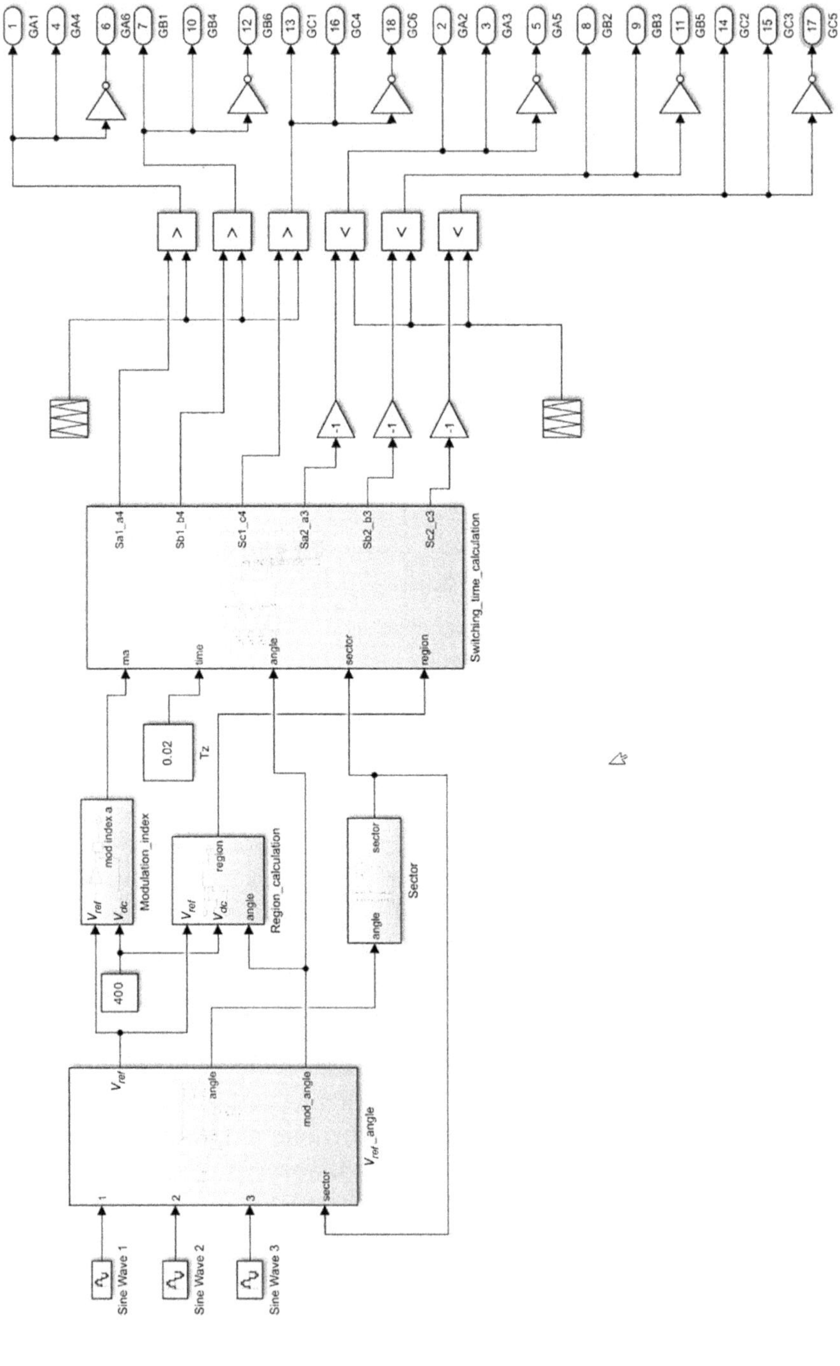

Figure 3.9 Simulink implementation of three-level SVPWM.

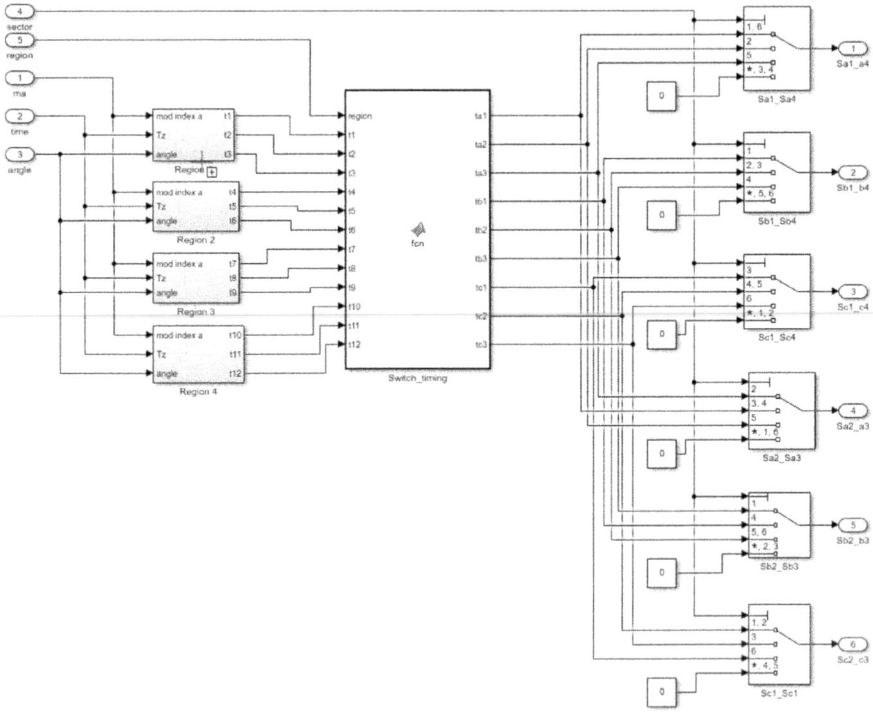

Figure 3.10 Switching time calculation in Simulink.

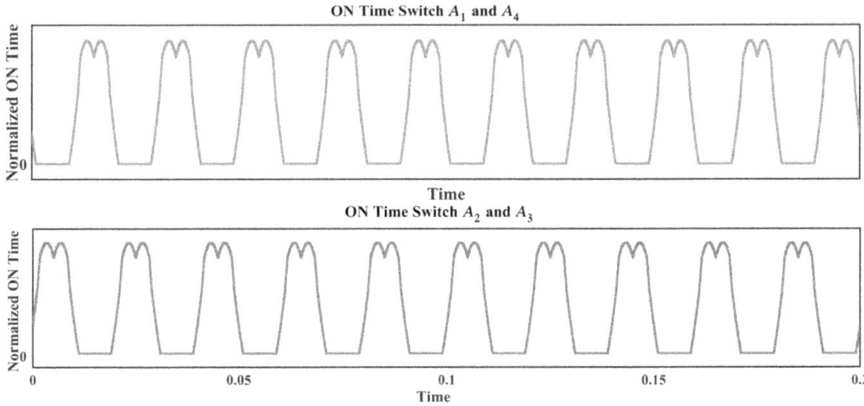

Figure 3.11 Simulink result of ON-time for switches A1 to A4.

3.2.5.2 Results

NPL-H three-level dual inverter topology was simulated in Simulink environment using the SVPWM algorithm discussed above based on the specifications mentioned in Table 3.12.

Using SVPWM modulation, the NPL-H inverter is operated and the current and voltage waveforms were observed, as seen in Figures 3.11 and 3.12. DC link capacitor voltage ripple of about 4.5 V was observed from the waveform as shown in Figure 3.13. This voltage ripple is about 1%, which indicates a very good performance of the inverter (Figure 3.14).

Table 3.12 Simulation specifications

Parameter	*Value*
V_{DC}	450 V
f_{sig}	50 Hz
f_{sw}	10 kHz
C_{DC}	100 μF
Modulation Index m_a	0.99

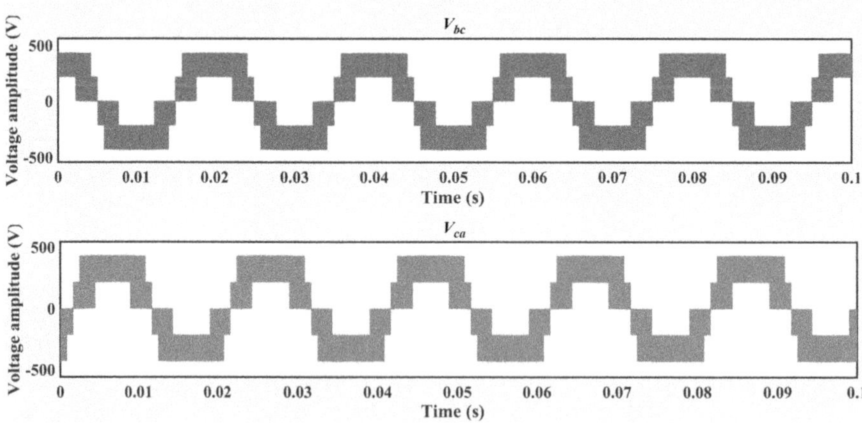

Figure 3.12 Output voltage waveform using three-level SVPWM.

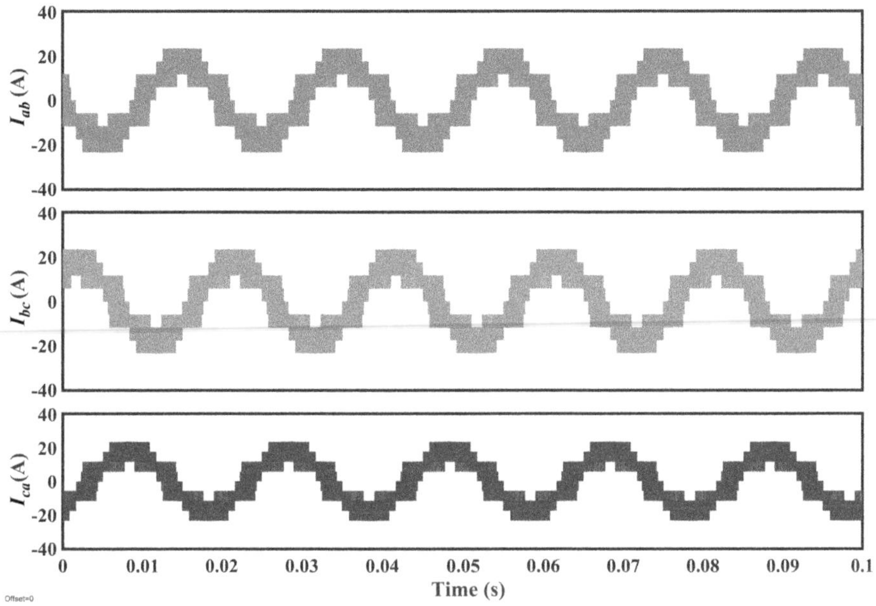

Figure 3.13 Output current waveform using three-level SVPWM.

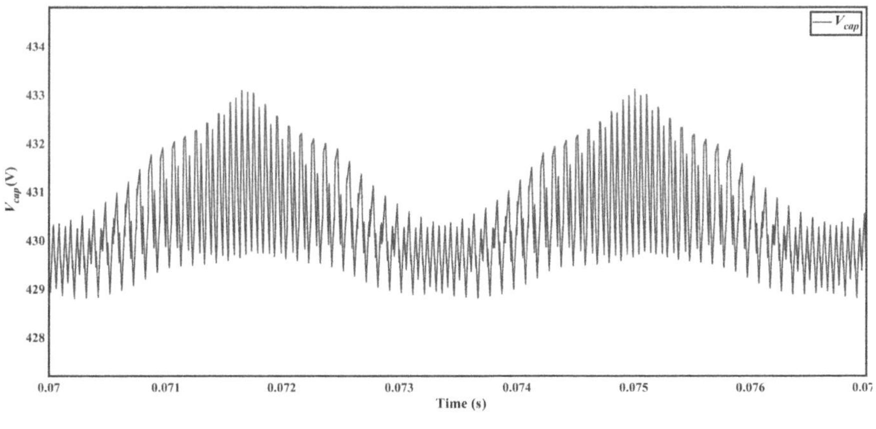

Figure 3.14 DC link capacitor voltage ripple during SVPWM.

THD value of the voltage waveform generated was observed to be 34.91%, which is slightly better than that of sinusoidal PWM, as shown in Figure 3.15. Fundamental voltage component was observed to be having a rms voltage of 302.9 V (peak of 428.4 V). This suggests that the SVPWM's DC link bus utilization is significantly higher than that of the sinusoidal PWM, which is also its greatest comparative advantage.

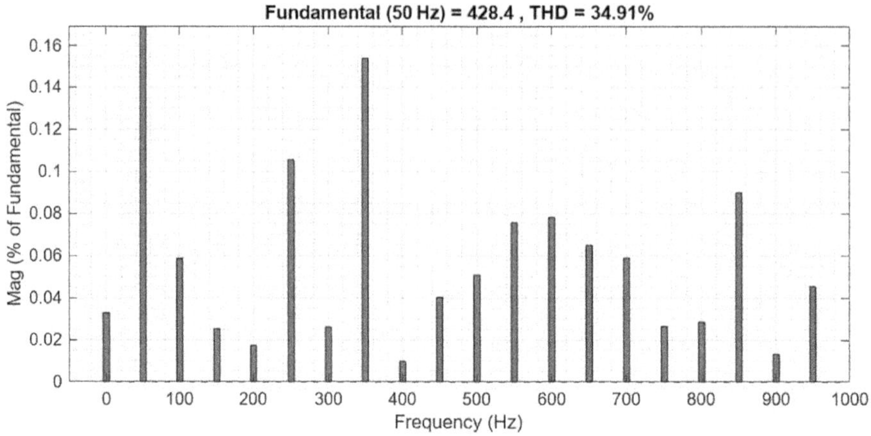

Figure 3.15 THD for NPL-H inverter using SVPWM.

3.3 RESULTS AND COMPARISON

3.3.1 Introduction

This chapter details the simulation of the NPL-H three-level dual three-phase inverter in a Simulink environment. As mentioned, a SPWM modulation scheme is used to generate the gate switching pulses. The L–L voltages and current generated at the three-phase outputs are measured along with the capacitor voltage ripple waveforms. Total harmonic contents present in the output waveforms are also calculated.

3.3.2 Simulation results

NPL-H three-level dual inverter topology mentioned above was simulated in Simulink environment based on the parameters mentioned in Table 3.13. Simulation setup is shown in Figure 3.15.

The output voltage and current waveforms, capacitor voltage waveforms, and switch current waveforms were observed along with the harmonic performance

Table 3.13 Simulation specifications

Parameter	*Value*
V_{DC}	450 V
Active power	20 kW (10 kW across each three phase)
f_{sig}	50 Hz
f_{sw}	10 kHz
C_{DC}	100 μF
Modulation index	1

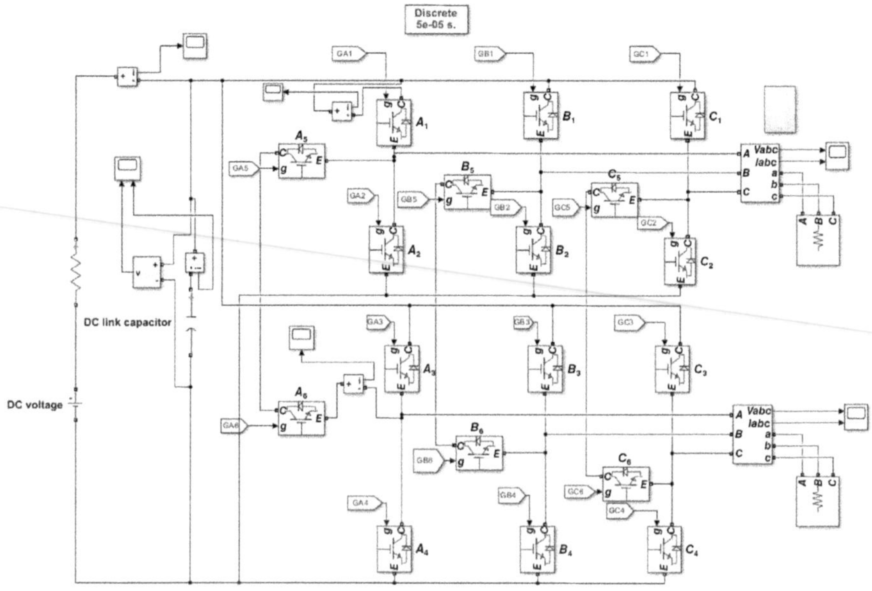

Figure 3.16 Simulation setup for NPL-H inverter.

for a pure resistive load. As mentioned in theory, a five-step waveform was observed. The output waveforms across a single three phase are shown in Figure 3.16. A slight deviation from actual V_{DC} value is observed in the peaks of the three-phase output voltage. This is due to the small resistance connected across the voltage source. A voltage peak of 435 V was observed. Similarly, current waveforms for one of the three phase outputs were measured and a peak current of 32 A was observed, as seen in Figures 3.17 and 3.18.

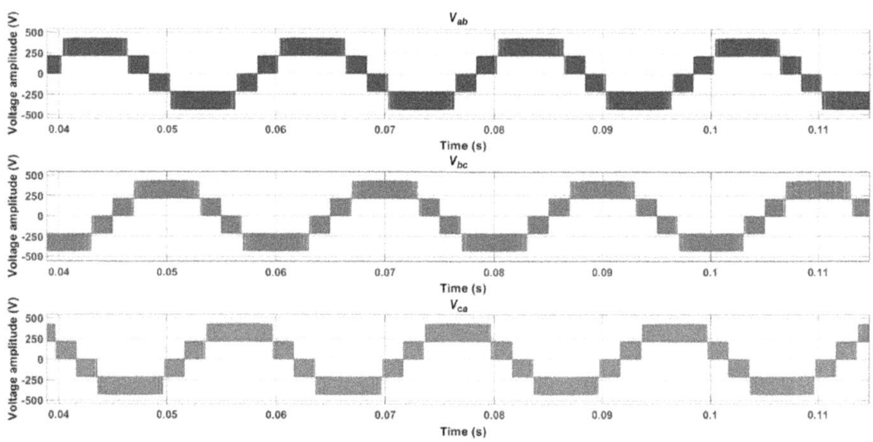

Figure 3.17 Output voltage waveform for three phase (upper).

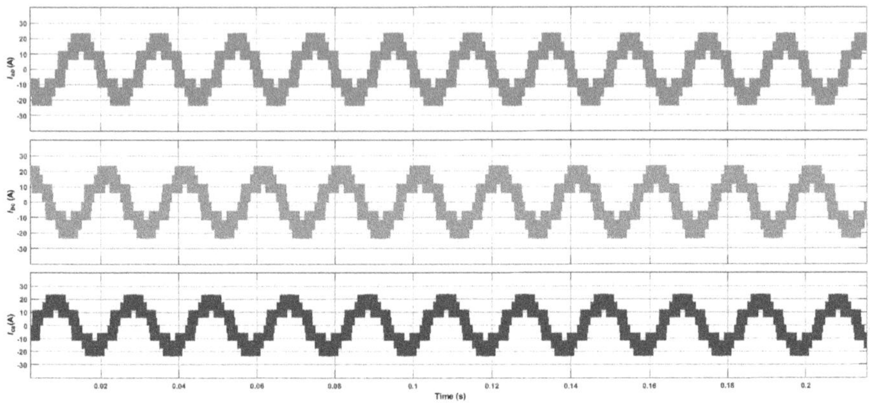

Figure 3.18 Output current waveform for three phase (upper).

The voltage waveform's THD was computed and displayed in Figure 3.19, and it came to be 35.24%. Significant harmonics were observed around the switching frequency range of 10 kHz. Pure resistive load also proved to be a reason for higher harmonics. Applying suitable filtering techniques could easily eliminate the higher-level harmonics. Fundamental voltage component was found to have rms voltage of 263 V. A different modulation scheme such as SVPWM or THIPWM is required to obtain a higher rms voltage for the fundamental frequency component.

The current waveform across switch A_1 is shown in Figure 3.20. It shows the current flow across the switches in every half-cycle like a half sinusoid. Figure 3.21 displays the voltage waveforms over the DC link capacitor. A voltage ripple of 10.6 V was observed, as shown in Figure 3.21, which is about 2.5%. From the analysis, it could be concluded that the inverter has good

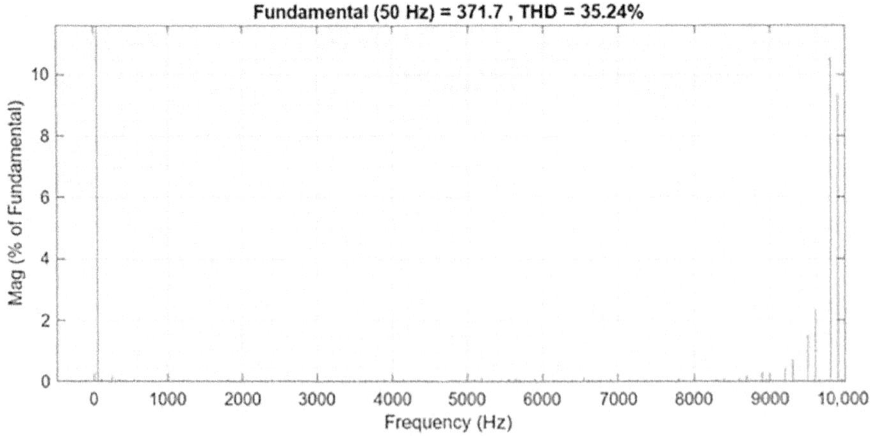

Figure 3.19 THD information for NPL-H inverter.

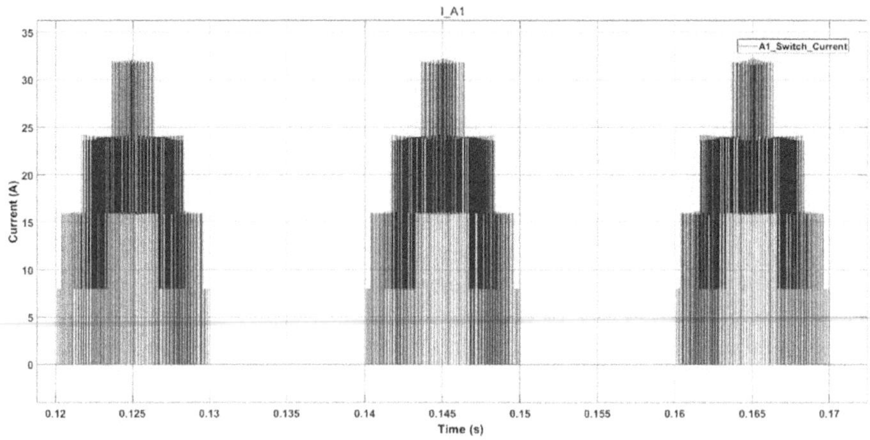

Figure 3.20 Current waveform across switch A1.

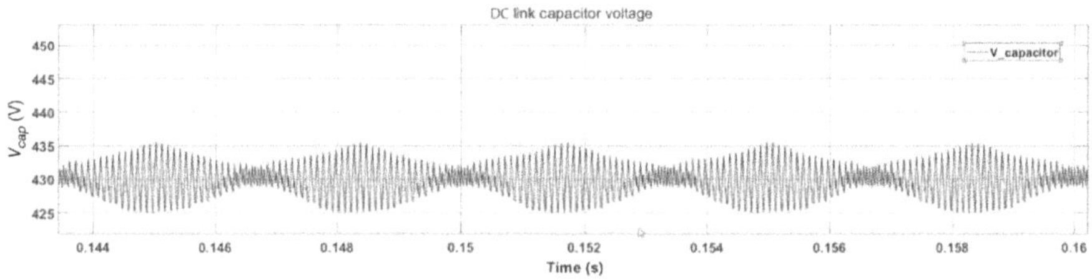

Figure 3.21 DC link voltage waveform.

performance with regard to voltage ripple and also has the obvious advantage of not needing the neutral point balancing.

3.3.3 Improvement and comparison

Earlier chapters introduced the NPL-H inverter topology using the SPWM modulation technique. One of the main issues of using SPWM technique is the underutilization of DC bus voltage. Therefore, the peak amplitude obtained for the output voltage is very less compared to the DC voltage input applied and hence it can be concluded that the SPWM technique is not the most effective in terms of using the DC bus. To overcome this, other modulation schemes are used to enhance the DC bus utilization. In this chapter, one such technique called THIPWM will be applied and the results are compared with the SPWM technique simulated earlier. Also, the NPL-H topology is compared in this chapter with other well-known multilevel inverter topologies, including T-type and NPC topologies.

3.3.3.1 Performance improvement through THIPWM

THIPWM is one technique employed to enhance the DC bus utilization in inverters. Based on the SPWM method used in the above, we could see that the DC bus utilization is less as visible from the peak magnitude of fundamental component which is only 371.7 V. By using THIPWM technique, DC bus utilization can be increased and the harmonic content could be reduced. The modulation signal for the THIPWM signal is shown in Figure 3.22.

After analyzing the output voltage waveform using THIPWM, it was observed that the fundamental voltage component has a peak of 405.3 V when compared to 371.7 V of that in SPWM technique. The THD content of inverter when THIPWM technique was employed is shown in Figure 3.23. An improvement in THD was also observed but at the cost of introducing lower order harmonics.

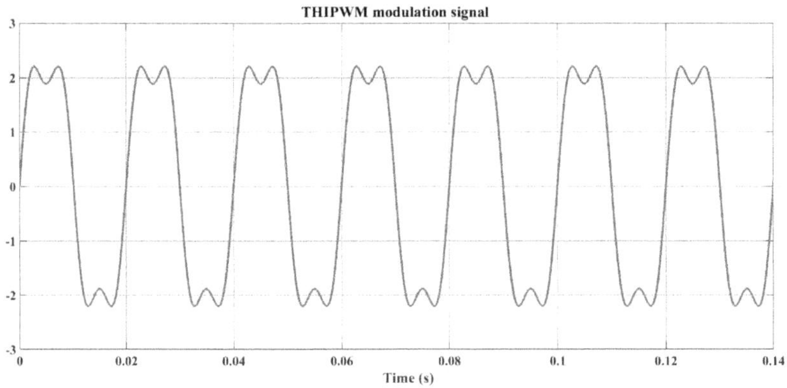

Figure 3.22 THIPWM modulation signal.

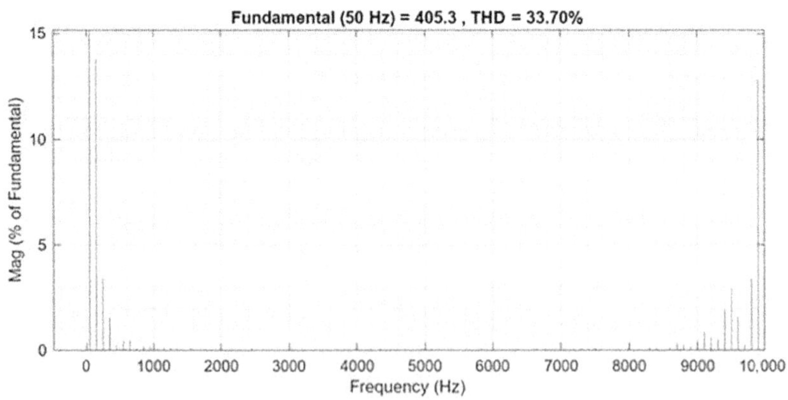

Figure 3.23 THD of NPL-H inverter using THIPWM technique.

3.3.3.2 *Comparison with existing topologies*

Characteristics of the existing well-known topologies such as T-type and NPC MLI's are simulated under the same conditions, as mentioned in Table 3.13 (except for the capacitor values), and the output waveforms and capacitor voltage ripples are measured.

For a dual NPC three-level inverter as illustrated in Figure 3.24, there are a total of 24 switches and 12 diodes compared to 18 switches in an NPL-H inverter. Hence, with a reduced number of switches, the three-level output voltage is obtained for NPL-H inverter (Figure 3.25).

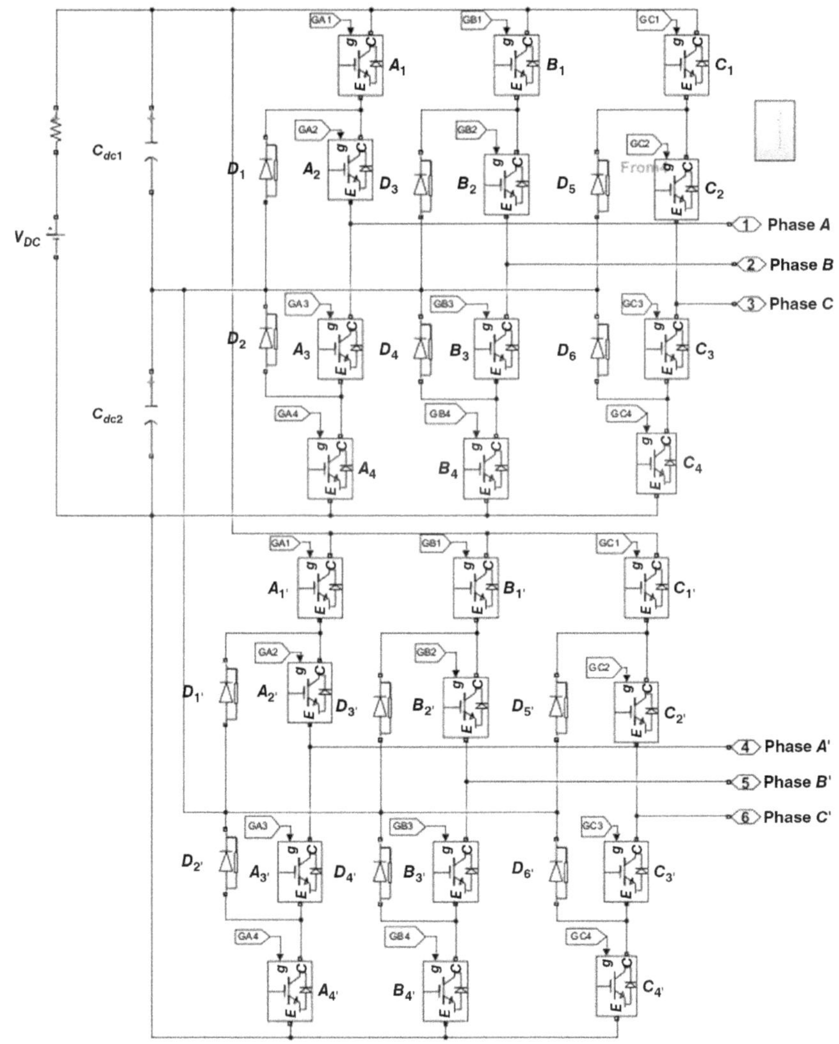

Figure 3.24 NPC dual three-phase three-level inverter.

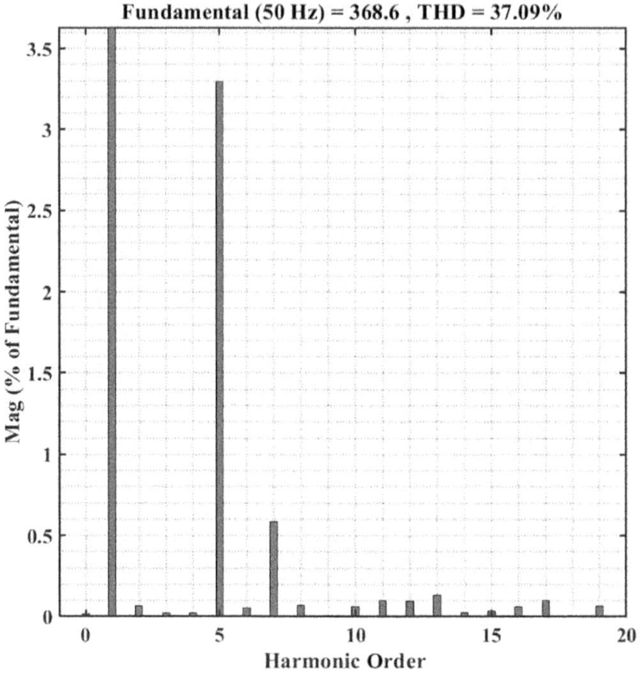

Figure 3.25 THD of NPC inverter.

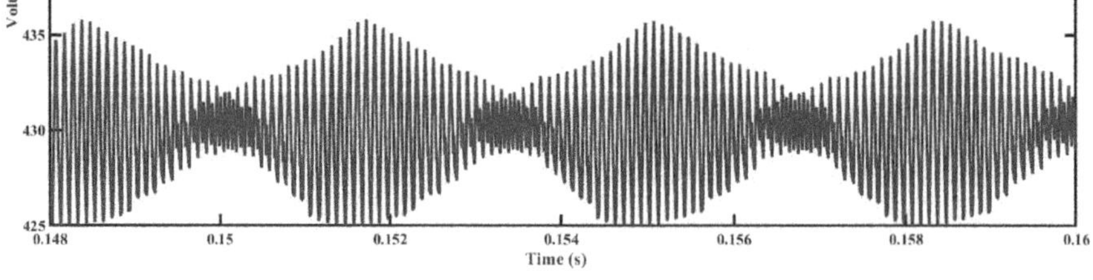

Figure 3.26 The NPC inverter's voltage ripple across the DC link.

For a voltage ripple of around 10 V across the DC link, as shown in Figure 3.26, two 200-F capacitors are required for dual NPC inverter compared to a single 100-F capacitor is NPL-H. Thus, the size of the NPL-H inverter is having lesser capacitance requirement and thus it improves the overall power density. Figure 3.25 shows the THD calculation for NPC inverter.

It could be seen that there is a significant harmonic content near to the fundamental and at the switching frequency. Overall, the THD of NPL-H is slightly better when compared to the NPC inverter [12].

For a dual T-type three-level inverter, as depicted in Figure 3.27, there are a total of 24 switches which is same as that of the NPC but without the diodes. Compared to the NPL-H topology which has 18 switches, dual T-type topology is slightly inferior in terms of the switching device count and which also results in additional switching or conduction losses.

After observing DC link ripple voltage across the capacitors, it was found that for a similar voltage ripple performance as NPL-H, two capacitors each of 200 F is needed. Thus, NPL-H offers significant improvement in terms of power density characteristics and harmonics when compared to T-type topology. Having multiple capacitors of higher capacity makes the inverter bulkier. Voltage ripple measurement for the T-type dual three-level inverter is depicted in Figure 3.28. A similar voltage ripple performance to that of NPL-H inverter could be seen. The harmonic distortion of the inverter is depicted in Figure 3.29.

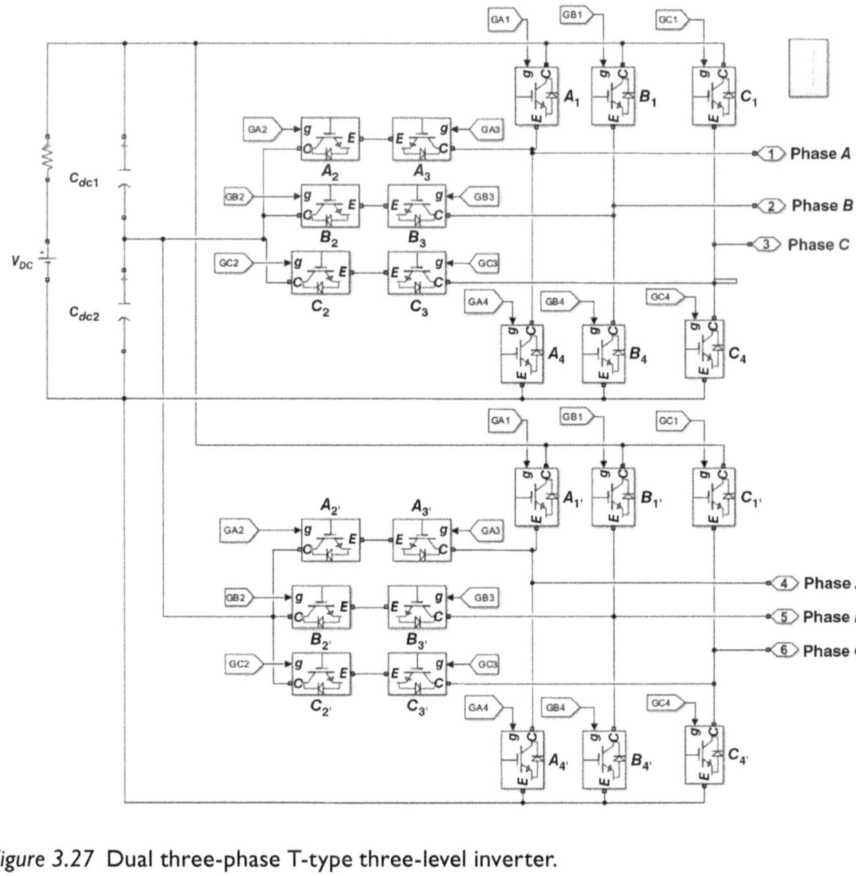

Figure 3.27 Dual three-phase T-type three-level inverter.

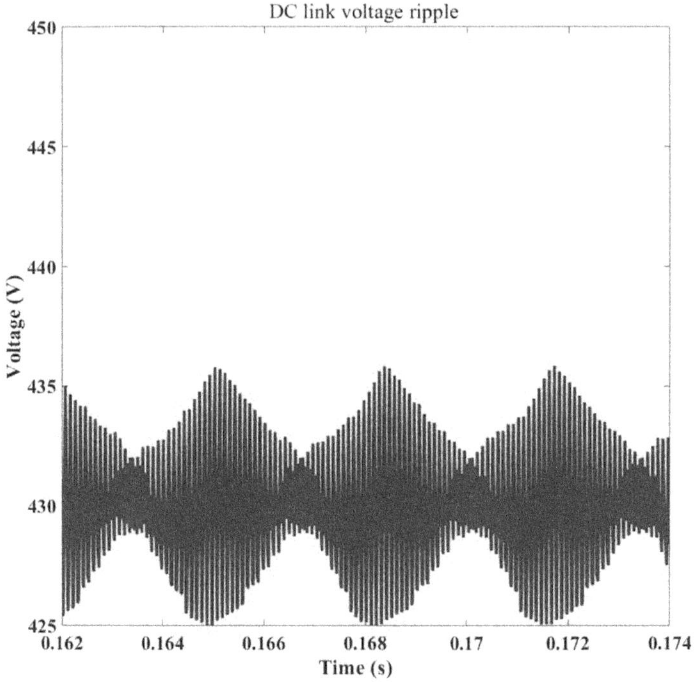

Figure 3.28 T-type inverter's DC link voltage.

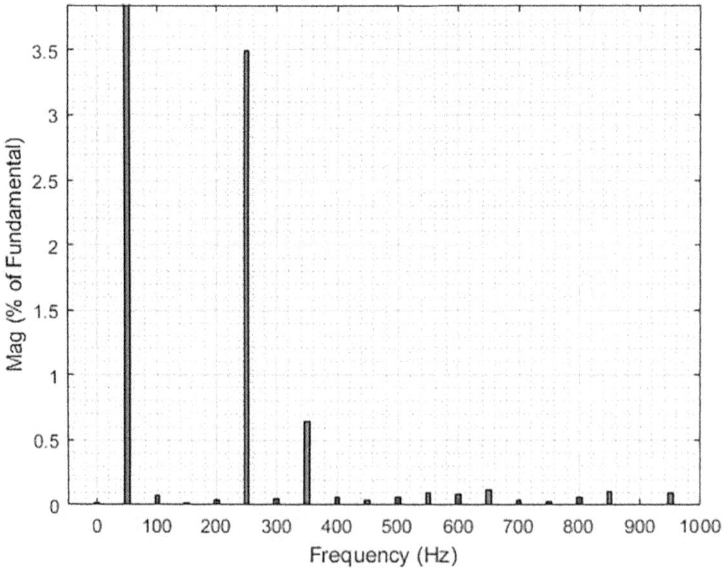

Figure 3.29 THD of T-type inverter.

From the harmonic content perspective, NPL-H has slightly less distortion than the T-type inverter. In summary, it was concluded that the NPL-H inverter offers superior voltage ripple characteristics with a single capacitance, thereby reducing the size of the inverter and does not have lower-level harmonics as compared with the NPC and T-type inverters.

3.4 CONCLUSION

In this project, NPL-H inverter was studied through simulation and the waveforms were observed through SPWM modulation. Then THIPWM technique was employed to improve the DC bus utilization. Later the inverter was compared with the known T-type and NPC MLI topologies and the advantages with respect to these inverters were documented. Advantages include having lesser number of switches, lesser capacitance values, and better harmonic performance compared to the standard topologies. Furthermore, a three-level SVPWM algorithm implementation was also done for the topology and the advantages were observed with respect to the SPWM implementation.

With respect to future work, suitable filtering or control techniques can be applied to reduce the THD within 5%. Since the SVPWM algorithm is developed, it could be used to control actual hardware. Other scope includes testing with a nonresistive load and observing the waveforms and distortion. Also, a closed-loop control using a motor load can also be studied using the SVPWM algorithm discussed in this project as the major application of the dual NPL-H three-level inverter is for motor drive applications.

REFERENCES

1. John Reimers, Lea Dorn-Gomba, Christopher Mak, and Ali Emadi, Automotive traction inverters: Current status and future trends, IEEE Transactions on Vehicular Technology, vol. 68, no. 4, pp. 3337–3350, 2019.
2. Mariusz Malinowski, K. Gopakumar, and Jose Rodriguez, Special section on multilevel inverters: Part II, IEEE Transactions on Industrial Electronics, vol. 57, no. 8, pp. 2550–2552, 2010.
3. Subhadeep Bhattacharya, Diego Mascarella, Gza Jos, Jean-Marc Cyr, and Jianhong Xu, A dual three-level T-NPC inverter for high-power traction applications, IEEE Journal of Emerging and Selected Topics in Power Electronics, vol. 4, no. 2, pp. 668–678, 2016.
4. Dhawal Dwivedi, Saurabh Singh, Shubham Kaushik, Eric Semail, Ngac K. Nguyen, and K. A. Chinmaya, T-type multilevel inverters: A comparative performance analysis, 2023 IEEE 3rd International Conference on Sustainable Energy and Future Electric Transportation (SEFET), Bhubaneswar, India, August 9–12, 2023, 6 pp.
5. V. Ramu, P. Satish Kumar, and G. N. Srinivas, Comparative analysis of multilevel inverters with various PWM techniques, Information and Communication Technology for Competitive Strategies (ICTCS 2021), Springer, Singapore, 2022.

6. Chithra Manivelan, A survey on multilevel inverter topologies and control schemes with harmonic elimination, 2020 International Conference on Electrotechnical Complexes and Systems (ICOECS), Ufa, Russia, October 27–30, 2023, 7 pp.
7. Xiaofeng Dong, Mikayla Benson, Kangbeen Lee, Lifang Yi, Woongkul Lee, and Jinyeong Moon, NPL.H: Neutral-point-less H-type 3-level inverter, IEEE Transactions on Power Electronics, vol. 39, no. 2, pp. 2522–2536, 2024.
8. Marija Vujacic, Obrad Dordevic, and Gabriele Grandi, Evaluation of DC-link voltage switching ripple in multiphase PWM voltage source inverters, IEEE Transactions on Power Electronics, vol. 35, no. 4, pp. 3478–3490, 2020.
9. Hoda Naderi, Neda Ghaderi, and Mohammad Abedini, Improving power quality and efficiency of multi-level inverter system through intelligent control algorithm, Soft Computing, vol. 28, pp. 12963–12976, 2024.
10. Y. Ayachi Amor, A. Kheldoun, B. Metidji, F. Hamoudi, and A. Merazka, Implementation of modified SVPWM for three-level inverter using STM32F4, 2018 International Conference on Electrical Sciences and Technologies in Maghreb (CISTEM), Algiers, Algeria, October 28–31, 2018.
11. T. A. Hussein and L. A. Mohammed, Detailed Simulink implementation for induction motor control based on space vector pulse width modulation SVPWM, Indonesian Journal of Electrical Engineering and Computer Science, vol. 22, no. 3, pp. 1251–1262, 2021.
12. Rutian Wang, Lan Ai, and Chuang Liu, A novel three-phase dual-output neutral-point-clamped three-level inverter, IEEE Transactions on Power Electronics, vol. 36, no. 7, pp. 7576–7586, 2021.

Chapter 4

Machine learning control techniques for DC–DC converters

S. Sunil Naik, Debashisha Jena, and Tukaram Moger

4.1 INTRODUCTION

Artificial intelligence (AI) is transforming the field of power electronics by enabling enhanced design, control, optimization, and maintenance. AI demonstrates the capability to enhance the competitiveness of products by optimizing global design, implementing intelligent control mechanisms, and estimating the health status of the system, and more. Consequently, research endeavors in power electronics can adopt a data-centric approach; it is advantageous for tackling complex and arduous scenarios.

Machine learning (ML) is engineered to autonomously unveil principles and patterns through experience from gathered data or iterative trial-and-error interactions. In power electronics, ML spans various domains, primarily categorized into design, control, and maintenance applications, as shown in Figure 4.1. The necessity of ML in DC–DC converters includes the following:

- In converter design, ML techniques encompass drive sizing and parameter identification tasks.

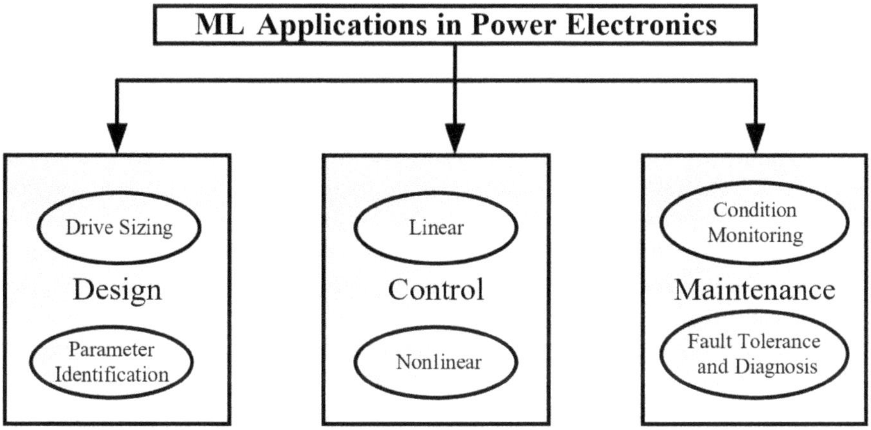

Figure 4.1 ML applications in power electronics.

DOI: 10.1201/9781003515807-4

- ML models can adopt either linearized or nonlinearized approaches tailored to specific application requirements for converter control.
- In maintenance, ML is vital in monitoring converter conditions, enhancing fault tolerance, and facilitating diagnosis.

Power converters have seen a rising utilization across industries and everyday settings. They play a vital role in regulating power transmission and transforming voltage and current. In industrial contexts, DC–DC converters and inverters will be crucial components in renewable energies, wireless power transfer methods, and DC-microgrids (MG) [1]. Similarly, power converters are ubiquitous, daily powering applications like electric vehicles (EVs) and solar photovoltaic (PV) systems.

Particularly, DC–DC converters play major roles in EVs. Since these converters are nonlinear time-variant systems, they represent a hard task for control design. DC–DC converters are variably structured in nature. The major controllers include linear, nonlinear, and AI controllers. The linear controllers are cascaded voltage and current loops to effectively manage power flow and provide stability. However, these controllers are typically designed to function within linear systems. For example, the proportional integral derivative (PID) controller repeatedly updates the error within the set point and effective output, adapting the control signals found on PID terms to maintain system behavior within desired bounds. However, it will struggle with complex control requirements and significant variations in change of system parameters. So nonlinear controllers are proved to be effective for the control of DC–DC converters. These controllers respond immediately to the transient conditions.

Nonlinear control methods encompass two prominent techniques—model predictive controller (MPC) and sliding mode controller (SMC). Nonlinear MPC is particularly notable for improving voltage regulation accuracy and control efficiency across a broader range of operating conditions. Moreover, distributed MPC becomes significant in scenarios characterized by interconnected DC–DC converters or distributed consistent power loads (CPLs) dispersed across diverse locations. In the domain of DC–DC converters, the application of SMC encounters various constraints. These include unwanted electromagnetic interference and mechanical strain on components, susceptibility to parameter alterations, constrained adaptability to nonlinear systems, intricate design requirements, and elevated switching frequencies near sliding surfaces.

Utilizing AI to manage power electronic converters presents a promising approach to tackle existing challenges. Key AI technologies in this field includes fuzzy logics, neural network (NN), and reinforcement learning (RL) [2]. Fuzzy logic control (FLC) methods are effective in handling system uncertainty but have limitations such as the absence of a method for updating internally, leading to limited flexibility to changing operational conditions. Additionally, their control law design relies on a specific level of expertise.

This chapter covers the various ML techniques existing in the literature. All these methods are implemented in controlling DC–DC converters to regulate the voltage, handle unpredictable changes in load power and parameter swings, even with untrained data, and improve performance, robustness, and adaptability.

The major contribution of this chapter includes the existing study of ML control techniques, which are mainly RL controllers. The non-isolated DC–DC buck converter is validated with linear, nonlinear, and ML control. This study emphasizes the differences between all types of control techniques. A comparison of all the existing RL controllers implemented in DC–DC converters is studied, and future scopes are given. Section 4.2 discusses the basic ML strategies, Section 4.3 discusses control strategies of DC–DC converters, Section 4.4 discusses RL controllers, and Section 4.5 is a case study. Sections 4.6, 4.7, and 4.8 will discuss comparison, future directions, and conclusions, respectively.

4.2 ML CONTROLLERS

ML aims to automate the discovery of patterns and principles by exposure to data or iterative interactions involving trial and error. Three primary types of ML methodologies are employed within power electronics—supervised, unsupervised, and RL. The categorization is shown in Figure 4.2.

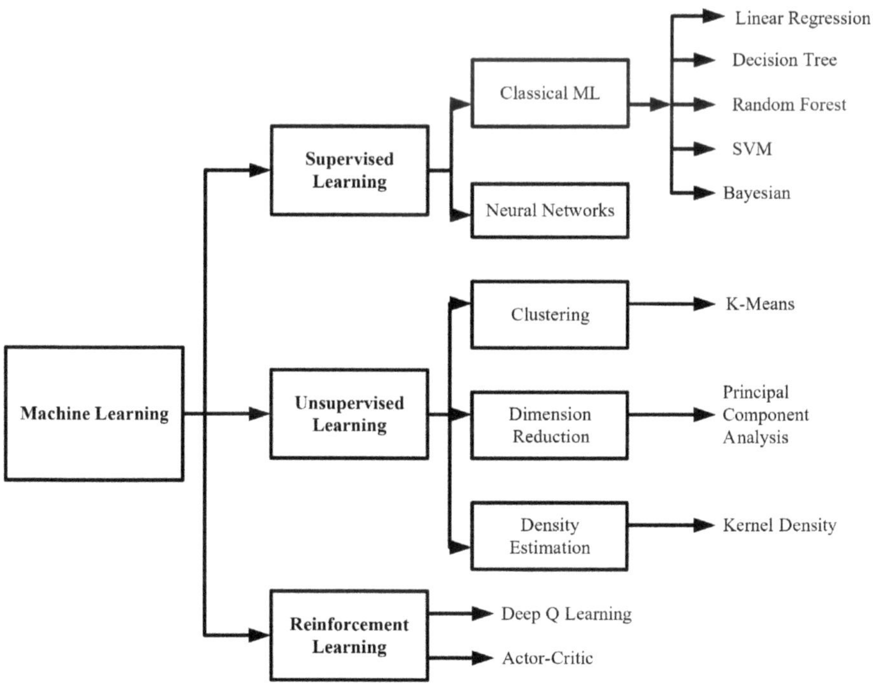

Figure 4.2 Classification of ML techniques.

4.2.1 Supervised learning

Supervised learning in power electronics uses input–output pairs to establish mappings, aiding in scenarios where system models are complex. Tasks include classification (e.g., fault diagnosis [3]) and regression (e.g., RUL prediction [4]). Once trained, the model can generalize to new data points, which is crucial for performance. Generalization accuracy is key, given the limited training data. Supervised learning facilitates automated pattern discovery in challenging domains like power electronics.

Supervised learning methods encompass connection-based (NN), probabilistic graphical, and memory-based (kernel) approaches. NN methods leverage learned knowledge through connection weights and structures. Research focuses on enhancing NN performance in power electronics, addressing noise robustness and dynamic performance. Techniques like fuzzy NN (FNN) and adaptive neuro-fuzzy inference system (ANFIS) improve robustness [5], while dynamic performance enhancements benefit applications such as intelligent controllers and RUL prediction.

Probabilistic graphical methods gather knowledge from data using diagrams that show input–output relationships, indicating conditional dependencies between decision variables. Based on Bayesian principles, these methods offer better interpretability than NNs. They excel at handling uncertainty and incomplete information. A popular example is the Bayesian network [6].

The training data is discarded after training for NN and graphical methods, while kernel methods retain and use the training data during testing. Gaussian processes and kernel schemes are computationally intensive as they use the entire training dataset in testing [7]. Sparse methods such as support vector machines (SVMs) and relevance vector machines (RVMs) are employed to alleviate the computational load. Use Bayesian methods for parameter estimation. Kernel methods require less training data than NN methods but have higher memory requirements and slower speed in testing, which is crucial for time-sensitive applications like control systems.

4.2.2 Unsupervised learning

Unsupervised learning is compared to supervised learning, which relies on input–output pairs; unsupervised learning has no predefined output data during training. In power electronics, unsupervised learning typically involves tasks like data clustering and compression.

Data clustering involves identifying patterns and grouping data into clusters based on similarities, enhancing data characteristics within clusters while distinguishing them from others. An illustration is the recognition of distinct health states from continual degradation data in the supervision of power electronics [8]. Data compression aims to reduce dataset complexity by extracting essential features, like using Principal Component Analysis (PCA), to maintain dataset integrity with fewer features [9].

These unsupervised learning algorithms typically perform preliminary data processing steps, often preceding subsequent analytics like fault diagnosis. While this stage is not mandatory, it significantly reduces computational load and enhances analytics precision.

4.2.3 RL

RL is a learning-based technique proficient in deducing optimum control policies except necessitating the system type or prior systems information. Consequently, these attributes endow RL controllers with a marked superiority over conventional control methods when governing the output voltage of DC–DC converters.

RL does not rely on training datasets like supervised or unsupervised learning; instead, it seeks to optimize actions to maximize rewards in dynamic tasks, learning through trial-and-error interactions with systems or simulations [10]. RL operates as a Markov decision process (MDP), developing a *Q*-table for optimal action selection based on condition variables to maximize future rewards [11]. It is advantageous for systems with limited knowledge or complex models. For example, RL is applied in maximum power point tracking (MPPT), deriving experience directly from system interactions rather than preexisting datasets.

RL represents the cutting-edge of ML techniques in power electronics, leveraging advancements in computing hardware for its implementation. Deep learning methods are currently utilized as powerful algorithms capable of extracting features effectively and accurately analyzing the nonlinear characteristics of power electronic systems. The study of RL techniques will focus more on ensuring expeditious stability, heightened efficiency, enhanced dynamic response, and bolstered reliability than conventional controllers.

4.3 CONTROL APPROACH FOR DC–DC CONVERTERS

This section discusses various control approaches for converters. The primary goal is to ensure that the output voltage of a DC–DC converter remains at its intended nominal values. Conventional control solutions often depend on either a small-signal model (SSM) or a large-signal model (LSM) for the topologies. Nevertheless, the SSM tends to be imprecise in the presence of substantial disturbances, while controllers relying on the large-signal system model may exhibit redundancy when dealing with smaller disturbances.

Linear controllers, like cascaded voltage and current loops, effectively handle power flow and stability. However, they are limited to linear systems. For instance, the PID control adjusts the control signal based on error but struggles with complex requirements and parameter variations.

Nonlinear control methods like MPC and SMC offer improved voltage regulation and efficiency compared to linear controllers. MPC is beneficial for diverse operating conditions, especially in interconnected systems. However, SMC in DC–DC converters faces challenges like electromagnetic interference, parameter sensitivity, and complex design requirements near sliding surfaces [12].

In DC–DC converters, deep learning controllers are advantageous because of their many useful features, such as their capacity to deal with nonlinearities, optimized control strategies, increased efficiency, improved fault tolerance, and enable data-driven and adaptive control methods. Because of these features, deep learning controllers are ideal for use in today's power electronic converters.

4.4 RL CONTROLLERS

As RL controllers play a major role in converters, here the focus is to study about these. Model-free and model-based RL algorithms are two distinct categories. The classification is shown in Figure 4.3. Algorithms such as actor-critic (AC), policy-based, and value-based are included in the model-free category.

The agent includes two components—the policy and the learning algorithm.

- The selection of actions in response to environmental observations is at the heart of policymaking, a strategic decision-making process. A DNN is often used to implement the policy.
- RL algorithm seeks to iteratively adjust the policy's parameters based on actions, observations, and rewards. The target of the algorithms is to determine the path of action that will yield the highest possible reward for completing the assignment.

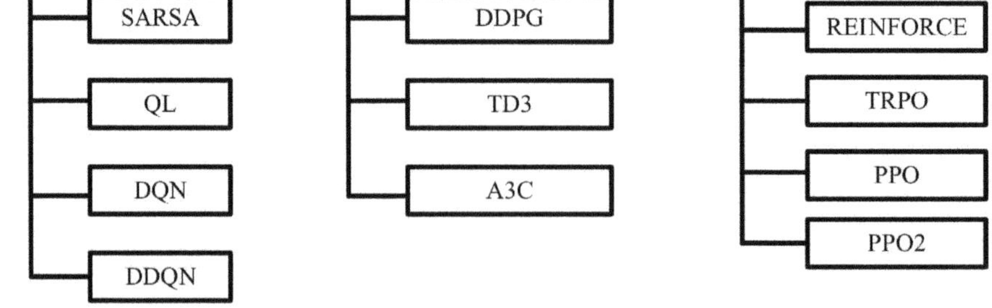

Figure 4.3 Classification of RL algorithms.

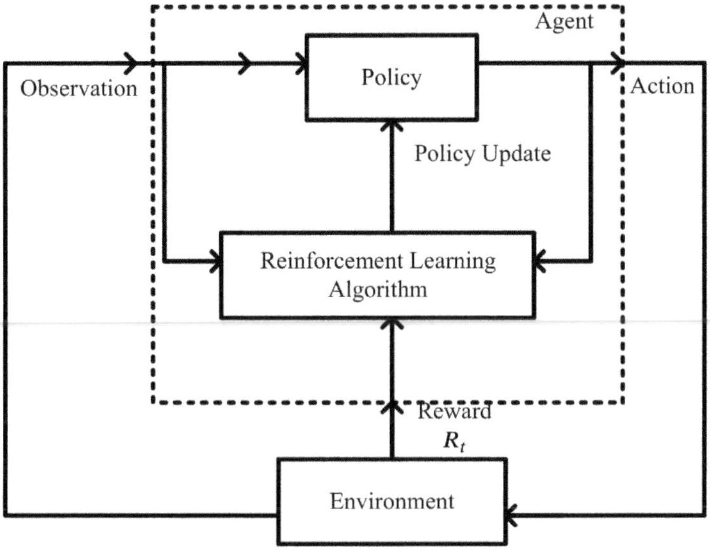

Figure 4.4 RL general representation.

Model-based RL methods depend on an environment model, acquired immediately or constructed progressively throughout acquiring knowledge. These algorithms have a substantial impact and enhance learning effectiveness by using the environment model. However, obtaining a precise representation of the world is often challenging, and creating a model of the environment while learning does not guarantee accuracy and convergence. In contrast, a broader and more extensive body of research on model-free RL algorithms exists. The general RL representation is depicted in Figure 4.4.

Value-based algorithms in model-free RL prioritize the modeling and valuation function estimation, which play a critical role in policy formation. SARSA and QL are traditional RL algorithms that depend on Q-value tables to generate value-based decisions. Both approaches utilize temporal difference (TD) algorithms to calculate Q-values, enhancing sampling efficiency. The main contrast between SARSA and QL is the way they handle exploration. SARSA is more focused on exploration, whereas QL is more non-exploratory. The above approaches are renowned for their exceptional convergence capabilities but can only be applied to issues with a limited number of states. To overcome this constraint, scientists devised the DQN technique, which integrates NNs with QL to effectively tackle challenges associated with continuous states. The DQN algorithm utilizes NNs to approximate Q-values, allowing it to effectively handle continuous state spaces.

RL policy-based algorithms are designed to derive the optimal policy directly, resulting in faster convergence. Examples are depicted in Figure 4.3.

Reinforce utilizes Monte Carlo methods to compute policy gradients, though it suffers from low sampling efficiency. TRPO, PPO, and PPO2 improve stability and convergence by adjusting hyper-parameters to limit update steps, ensuring better policies. These algorithms are effective for high-dimensional, continuous action spaces but have higher costs and lower sampling efficiency, with a risk of getting trapped into local optima.

AC systems have been developed by researchers to make use of the pros of both the policy and value-based methodologies. To concurrently learn the policies and the valued functions, the techniques in issue use two NNs—an actor and a critic network. Whereas actor network modifies policy according to the value function provided by the critic, the critic network trains the valued functions using TD techniques. Upgraded variants of AC algorithms, including deep-deterministic policy gradient (DDPG) and twin-delayed DDPG (TD3), are suggested to increase convergence. Constructed on the AC framework and DQN, DDPG effectively controls continuous action spaces. Various RL controllers are studied in the following section.

4.4.1 DNN control

The DQN employs NN approximants to depict the action-value function, comprising an input layer, two hidden layers, and an output layer. The DNN model for the DC–DC converter is shown in Figure 4.5; the early hidden layer features M-neurons, while the subsequent layer includes N–N neurons. The activation function applied in every hidden layer is the Rectified Linear Unit (ReLU) function, facilitating nonlinear transformations. Fully connected layers integrate the status and action spaces, enhancing the network's ability to learn complex relationships.

The size of the DNN input (n_l), hidden layers (n), and dimensionality of the DNN's output (m_l) determine the time complexity of the DNN method. The overall temporal complexity of all DNN layers is as follows: $O\left(\sum_{l=1}^{d} n_l \cdot m_l\right)$, where l is an index of a layer and d is the depth [13].

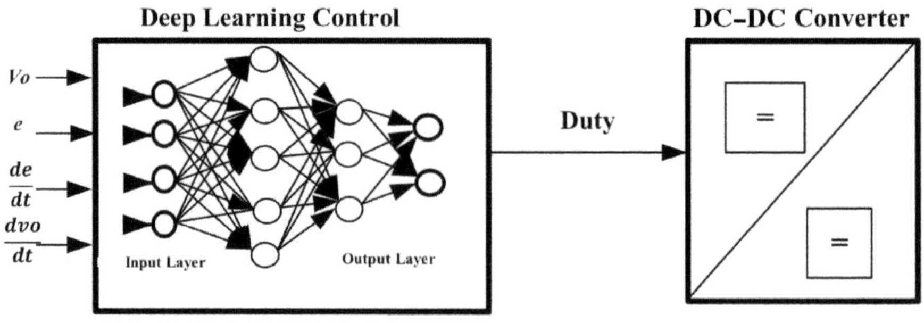

Figure 4.5 DNN model.

4.4.2 AC network

In RL applied to power electronic DC–DC converters, the AC network is crucial in optimizing control strategies, as depicted in Figure 4.6.

- *Actor Network:* The major role of the actor network is to select actions stationed on the system's current state. Its major function is to map the observed state (e.g., output voltage, error, and change in error) to a probability distribution over possible actions. This determines which control action to take to achieve the desired voltage regulation.
- *Critic Network:* It assesses the acts performed by the actor network. It extracts the current states and the actions chosen by actors as input, and outputs the expected valued function. This evaluation helps assess the quality of the actor's action [14].

4.4.3 Deep reinforcement learning (DRL) control

DRL controller in DC–DC converters mainly include two steps:

Step 1: Agent initially encounters difficulty attaining the required control performance despite undergoing offline training in the MATLAB RL environment. The DNN undergoes optimization using the DQN approach after several environmental interactions. This method enables the DNN to autonomously produce the pulse width modulation (PWM) duty based on the signals received from the environment.

Step 2: We modified the DNN to obtain the trained agent by utilizing a MATLAB function. In Step 1, this approach replicates the weight and bias values from the agent. Once the comprehensive strategy is successfully implemented, the module will generate the necessary response if the environment shows signs of this operating plan.

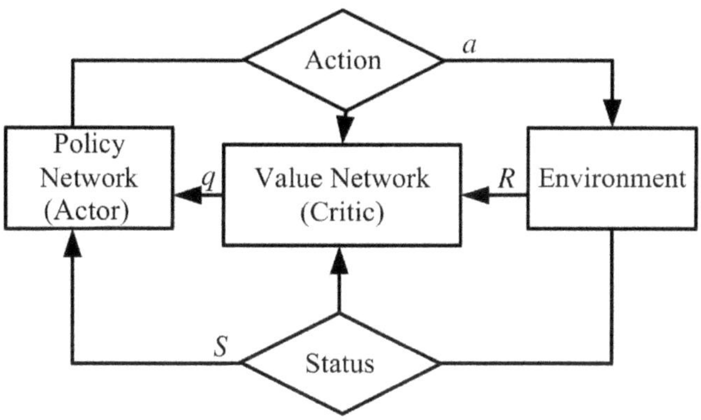

Figure 4.6 AC network.

The steps involved in controller design are as follows:

State Space (SS): v_o is the output voltage, and $e(t) = V_o(t) - V_{ref}$ error tracking, which is the fundamental indicator to get the state of systems shown:

$$S_t = \left\{ v_o(t),\, v_{o\_del}(t), \frac{dvo(t)}{dt}, e(t), e_{del}(t), \frac{de(t)}{dt} \right\}$$

Action Space (AS): The switching control serves as a reference point for the construction of a unique action space. The duty ratio is the approximate range of the ratio between the reference and actual input voltages. The three variables under consideration are the steady-state value ξ, the range of fluctuations ϕ, and the interval of change c. As a result, a unique range of actions is created, denoted as $A \in [\xi - \phi, \xi + \phi]$.

Reward System with Sub-Objectives: The rewards system incorporates the tracked error $e(t)$, representing the disparity between the current states and the desired control aim. The learning agent is instructed using two distinct goals, namely ϵ_1 and ϵ_2. The penalty coefficients are β_1, β_2, and β_3. In this case, we have the following controller reward function:

$$r = \begin{cases} \beta_1 - \beta_3.e(t), & \text{if } 0 \le |e(t)| < \epsilon_1 \\ \beta_2 - \beta_3.e(t), & \text{if } \epsilon_1 \le |e(t)| \le \epsilon_2 \\ -\beta_3.e(t), & \text{else} \end{cases} \tag{4.1}$$

In contrast, the adaptability of the DRL method, shown in Figure 4.7, allows for accelerated voltage recovery rates under varying working conditions, leading to adaptive optimal control outcomes [15].

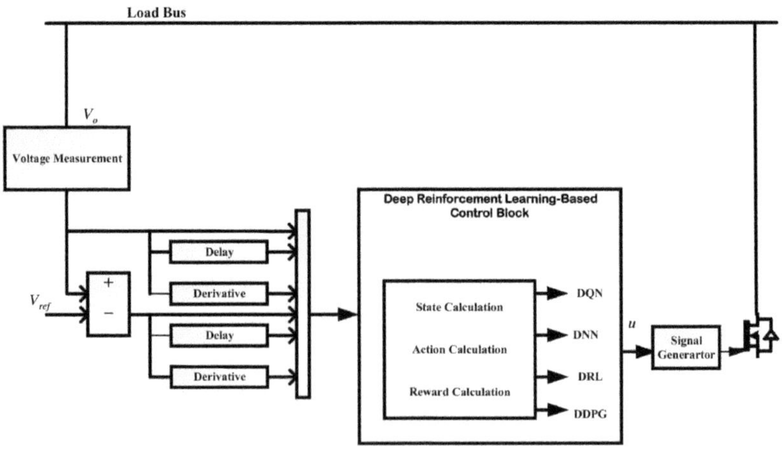

Figure 4.7 Voltage control by DRL.

4.4.4 DQN control

RL solves complex control challenges in nonlinear and stochastic systems. In RL, an AI agent learns to control a process by receiving its state and taking actions to maximize rewards from interacting environments.

The environment gives the agent *a* state s_t at each time step $t = 1,2,3, \ldots,$ and the agent responds by choosing an action A_t. Following this, the environment supplies the discount factor γ_{t+1} and the reward R_{t+1}.

The bus voltage regulation employs the DQN. The objective of the DQN is to enhance *Q*-values iteratively by reducing the disparity between the expected and targeted *Q* values. The schematic is shown in Figure 4.8. The critic comprises two different NNs—one that estimates the desired *Q*-value and the other that approximates the present *Q*-values $Q(s_t, a_t)$. The online network modifies the target network's parameters regularly after a predetermined number of episodes. The critic network examined by diminishing error derived from the Bellman equation, linking the present states and actions to an anticipated long-term reward system [13]:

$$L\left(\theta^{Q}\right)=\mathbb{E}_{(s_t,a_t,r_t,s_{t+1})}\left[\left(y_t-Q\left(s_t,a_t;\theta^{Q}\right)^2\right]\right] \tag{4.2}$$

This is the definition of the loss approach for critic, $L(\theta^Q)$. The symbol θ^Q denotes the parameter of the critic network. Variables *s* and *a* denote the state and action at a specific movement. The variable is denoted as s_t. r_t represents

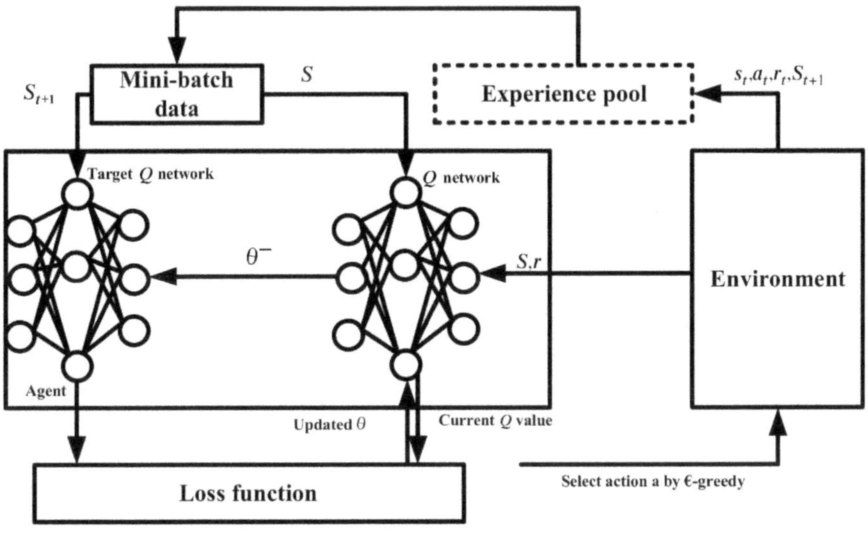

Figure 4.8 Deep Q-network algorithm.

the reward gained when action is taken in state s_t. The variable s_{t+1} represents the following state, and Q represents the determined target Q-value.

$$y_t = r(s_t, a_t) + \gamma \max_a Q'\left(s_{t+1}, a_{t+1}: \theta^{Q'}\right) \tag{4.3}$$

The discount factor γ balances benefits for now and those in the future, and $\theta^{Q'}$ shows the target network's parameters. Expedition is crucial in RL, where the agent occasionally chooses random actions to enhance environmental understanding. In this study, the DQN agent randomly selects actions with a probability ϵ, promoting exploration across the state space. Employing the *ϵ-greedy* method balances exploration and exploitation, deciding between maximum Q-value and random actions based on a formula [16].

$$a'_t = \begin{cases} \arg\max_a Q(s'_t, a'_t), & \text{if } p < \varepsilon \\ a'_{t-\text{ random},} & \text{in other case} \end{cases} \tag{4.4}$$

Let p be a randomly generated number between 0 and 1, and let ε be a randomly selected action from the action space. Furthermore, the effectiveness of the controller is greatly based upon the selection of the state, action, reward function, and exploration approach.

$$Q^{\pi}(s'_t, a'_t) = \mathbb{E}\left[r_t(s'_t, a'_t) + \gamma\left[Q^{\pi}(s'_{t+1}, a'_{t+1})\right]\right] \tag{4.5}$$

4.4.5 Proximal policy optimization control (PPO)

The data-driven MPC (DDMPC) method utilizes the PPO algorithm, which employs RL to attain optimal control. This approach does not necessitate an exact model of the plant. Instead, it depends on observational data to acquire an ideal strategy [17] proposed for the DC–DC buck boost converter.

The PPO algorithm in DDMPC employs two NNs:

- *Actor Network (AN):* It gives observed state s_t to a probability distribution over actions a. It determines which action to take but does not evaluate the action's quality.
- *Critic Network (CN):* Assesses the actor's decision-making process. It predicts the value function, which assesses the state's desirability and returns to the projected longer-term reward established on the state S_t and the action a_t.

The AC's main objective is to maximize the alternative goal-setting process:

$$L(\theta) = \hat{\mathbb{E}}_t\left[\min\left(r_t(\theta'')\hat{A}_t, \text{clip}\left(r_t(\theta''), 1-\epsilon, 1+\epsilon\right)\hat{A}_t\right)\right] \tag{4.6}$$

The anticipation model of the advantage function by $\hat{A}$, policy parameter θ'', and the probability ratio $r_t(\theta'')$ are given as

$$r_t(\theta'') = \frac{\pi_\theta(a_t \mid s_t)}{\pi_{\theta_{\text{old}}}(a_t \mid s_t)} \tag{4.7}$$

The AC network mainly maximizes the alternate objective function; it represents the proportion of the present policy to that of the previous policy. Using the present policy parameter π_θ and comparing them to the likelihood under previous policy parameters θ_{old}, this probability ratio compares the likelihood of doing an action a in state s at time t.

4.4.6 DDPG control

This method enhances the DQN. It demonstrates rapid convergence and high-reward performance when dealing with continuous state space and action spaces are s and a. The DDPG algorithm possesses three crucial characteristics:

- Utilizes DNNs as the architecture for the agent.
- Generates deterministic behaviors without employing sampling from the probability distribution.
- Applies policy gradients to update the policy network after each step.

This algorithm employs an AC architecture. This architecture utilizes an actor network that receives a state input, represented as s_t, and produces a deterministic action, represented as μ (s_t $\mid\theta\mu$). Conversely, a CN requires a state s_t and action as μ (s_t $\mid\theta\mu$) as input and finds associated action value $Q(s_{t+1}, \mu(s_t \mid \theta\mu \mid \theta^Q))$. The CN directs the alterations to the argument θ of the AN. To tackle the issue of unsteady convergence resulting from frequent updates of network arguments $\theta\mu$, this technique employs two identical networks for the actor and critic. These networks are known as the destination network and online network. This is depicted in Figure 4.9. Target networks employ soft updates, while online networks employ gradient updates.

$$y_i = r_i + \gamma Q'\left(s'_{i+1},\ \mu'\left(s'_{i+1} \mid \theta\ \mu'\right) \mid \theta Q\right) \tag{4.8}$$

$$L' = \frac{1}{K}\sum_{i=1}^{k}\left(y_i - Q'\left(s_i, a_i \mid \theta^Q\right)\right)^2 \tag{4.9}$$

$$\nabla_{\theta\mu} J^{\theta\mu} \approx \frac{1}{K}\sum_{i=1}^{k} \nabla_a Q\left(s, a \mid \theta^Q\right)\Big|_{a=\mu^\theta(s)} \nabla_{\theta\mu}\mu\left(s \mid \theta^\mu\right) \tag{4.10}$$

The critic network update makes use of the lossy function gradient. The method of updating is identical to training NN offline; it uses the Adam optimizer and a

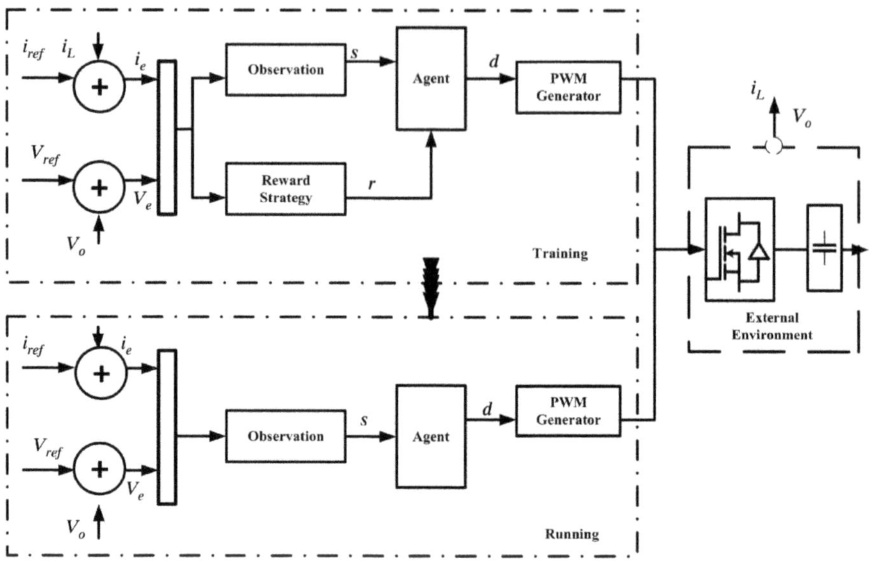

Figure 4.9 Deep-deterministic policy gradient.

loss function, represented by Equation (4.9), which considers the outputs of the target CN. The AN undergoes a gradient upgrade with different magnitudes, defined by the solution technique of the Bellman equation [4.10].

4.5 SIMULATIONS CASE STUDY WITH DC–DC BUCK CONVERTER

The buck converter model is shown with different control mechanisms in Figure 4.10. This section presents the results. It discusses the performance of various controllers applied to the buck converter model. First, a linear PI-lead compensator is implemented, and the corresponding output waveforms are analyzed, including Bode plots for both open- and closed-loop systems. Next, a nonlinear MPC is applied to the system. Finally, a ML-based controller is introduced, and its performance is compared to the other two controllers. The designed circuit parameters and controller parameters are shown in Tables 4.1 and 4.2.

4.5.1 PI-Lead compensator

In this feedback control system, the feedback mechanism constantly modifies the duty cycle to maintain the output voltage close to the target value. This procedure causes transients in the inductor current as well as the output voltage. The goal is to define a correlation between changes in output voltage, duty cycle, and input voltage, concentrating on the system's transient

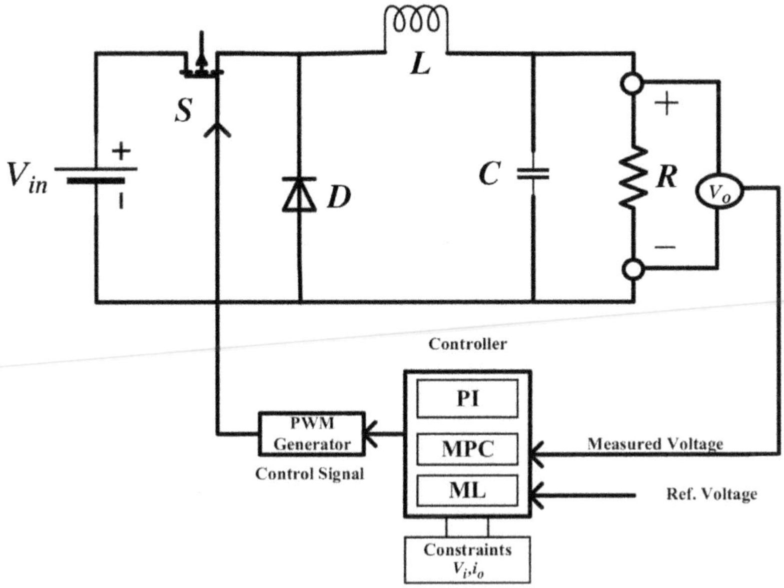

Figure 4.10 Block diagram for DC–DC buck converter model with controller.

Table 4.1 DC–DC buck converter (environment) parameters

Parameter	*Definition*	*Value*
V_{in}	Input voltage	200 V
V_{ref}	Nominal output voltage	100 V
L	Inductance value	2 mH
C	Capacitance	150 μF
R	Resistance	50 Ω
f_s	Switching frequency	10 kHz

behavior. In steady-state conditions, the average inductor voltage and capacitor current across a switching cycle stay stable, although this is not the case during transients [18].

The open-loop transfer function of the $\frac{\hat{v}}{\hat{d}}$:

$$\frac{\hat{v}}{\hat{d}} = \frac{100{,}000 * s + 1.333 * 10^7}{s^2 + 133.3 * s + 3.333 \times 10^6} \tag{4.11}$$

The compensated system is

$$t_v = \frac{-6.558 \times 10^{15} * s + 1.016 \times 10^{19}}{s^5 + 1.508 \times 10^5 * s^4 - 9.777 \times 10^8 * s^3 + 3.689 \times 10^{11} * s^2 - 3.337 \times 10^{15} * s} \tag{4.12}$$

Table 4.2 Controller simulation parameters

Controller	*Parameter*	*Value*
PI controller	Current loop gain k_{cp}, k_{ci}	0.02,100
	Voltage loop gain k_{cp}, k_{ci}	1,300
MPC controller	Prediction horizon (N_p)	20
	Control horizon (N_c)	15
	Output on weight error	10
	Sampling time	1 ms
DRL controller	Learning rate	0.001
	Discount factor	0.95
	Mini batch size	256
	Experience buffer length	2×10^5
	Exploration rate	0.1
	Training episode length	100
	Number of neurons (M)	64

Figure 4.11 will give the voltage mode control of the buck converter with sudden changes in the reference voltages. The system is tracking changes in the references. Here, the desired phase margin is 55° with a cross-over frequency of 62.8 rad/s. The system is stable, and both the uncompensated and compensation Bode analyses are analyzed.

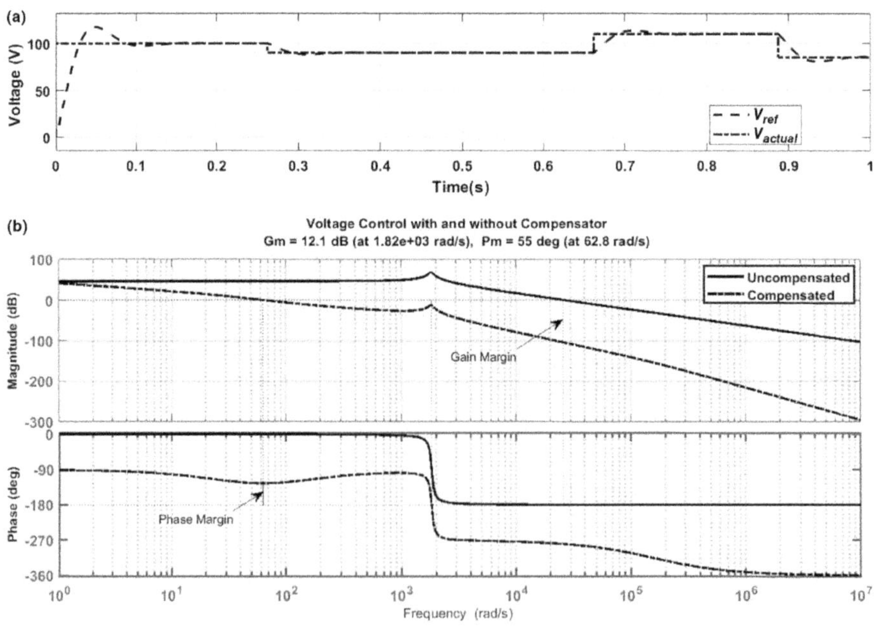

Figure 4.11 (a) Output voltage control with reference changes. (b) Bode diagram.

The open-loop transfer function of the $\frac{\hat{i}}{\hat{d}}$:

$$\frac{\hat{i}}{\hat{d}} = \frac{6.667\times10^{8}}{s^{2}+133.3*s+3.333\times10^{6}} \tag{4.13}$$

Compensated system is as follows:

$$t_i = \frac{7.513\times10^{14}*s^{2}+5.05\times10^{18}*s+6.6\times10^{20}}{s^{5}+3.07\times10^{5}*s^{4}+2.357\times10^{10}*s^{3}+4.16\times10^{12}*s^{2}+7.843\times10^{16}*s} \tag{4.14}$$

Figure 4.12 will give the current mode control of the buck converter with sudden changes in the reference current. The system is tracking changes in the references. Here, the desired phase margin is 55° with a cross-over frequency

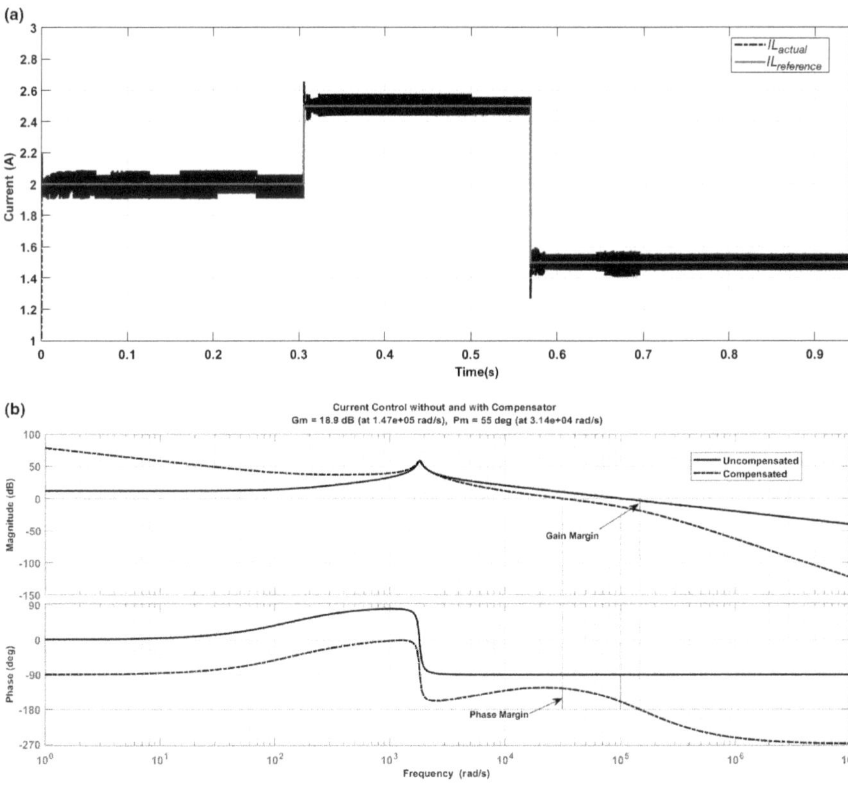

Figure 4.12 (a) Inductor current control with reference changes. (b) Bode diagram.

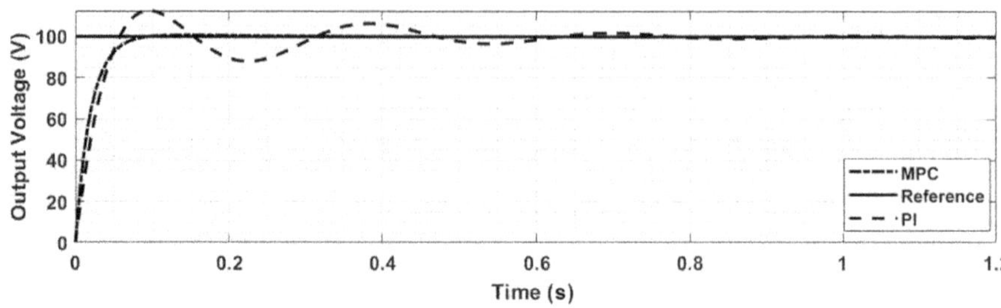

Figure 4.13 PI controller comparison with MPC.

of 31,400 rad/s. The gain margin is improved. The system is stable, and the uncompensated and compensation Bode analyses are analyzed.

4.5.2 MPC controller

The most appealing feature of MPC is its use of the system model to predict the future behavior of the system being controlled [19]. Figure 4.13 shows how the output voltage of a DC–DC buck converter is controlled using the MPC method and compared with PI controller, system constraints are incorporated into the controller design process, and the results are compared with those of a PID controller. Here the peak overshoot of the MPC is reduced compared to the PID.

4.5.3 DRL controller

DRL is an ML approach used to develop control strategies by interacting with an environment and receiving feedback through rewards or penalties [20]. For DC–DC buck converters supplying CPLs, DRL optimizes voltage regulation by adjusting parameters like duty cycles. The agent's actions are guided by real-time feedback, making DRL a powerful tool for achieving adaptive and efficient voltage regulation performance. The methodology is discussed in Section 4.3. A comparison is made between the DRL controller and the traditional PI-lead compensator, where the output DC bus voltage is regulated through the collaborative effort of an inner current loop and an outer voltage loop. The results of the comparison between the PI and DRL controllers are discussed. Although both controllers had their parameters tuned once and kept constant throughout the simulation, the PI controller faced challenges in maintaining optimal performance when inductance varied. In contrast, the DRL control method demonstrated consistent performance across various scenarios, highlighting its superior adaptability. The voltage response was anticipated to quickly return to the reference bus voltage (V_{ref}) with minimal overshoot. This is shown in

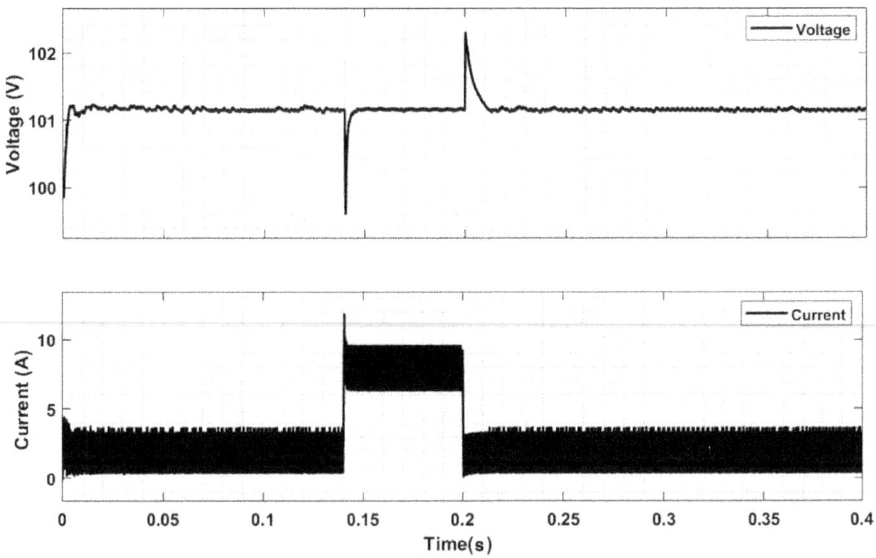

Figure 4.14 Converter output voltage and with DRL control.

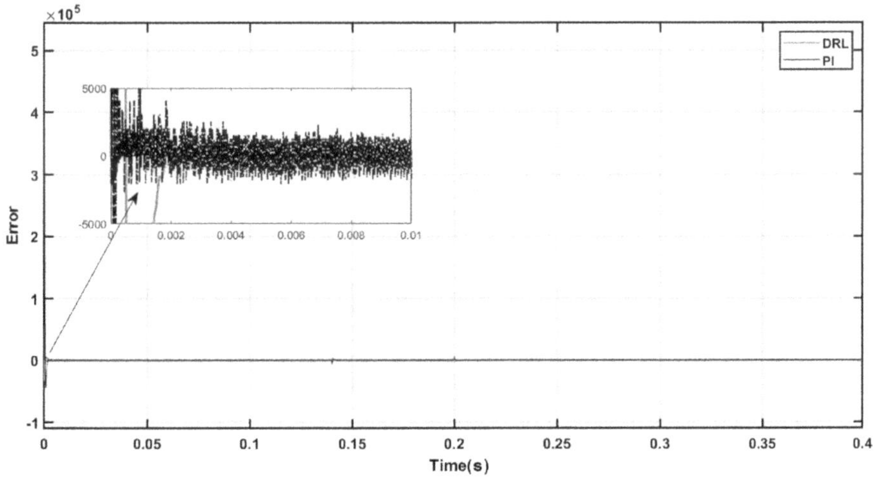

Figure 4.15 Comparison with DRL and PI control.

Figures 4.14 and 4.15. The robustness of the controller is also studied under external disturbances such as large load power variations.

The comparison of the linear, nonlinear, and ML(DRL) controllers is given in Table 4.3, which performs with respect to any type of DC–DC converter.

Table 4.3 Comparison of the PI, MPC, and DRL control

	PI	*MPC*	*DRL control*
Performance	Not optimal	Ideal with an accurate model	Near to optimal
Online computation cost	Less	Very higher	Less
Offline computation cost	System identification	Model identification	Policy and model identification
Reliance on model	Parameter tuning is needed	At all times	Only for training
Sensitivity to tuning	High	High	Low
Processor required	DSPs/FPGAs	Not suitable for application	DSPs/FPGAs

4.6 COMPARISON

This section compares various RL control algorithms utilized in DC–DC converters. All these methods offer effective solutions for achieving fast dynamics and accurate tracking while addressing large signal stability issues. Advanced control techniques can be applied to non-isolated and isolated converters to achieve high performance. The selection of these methods depends on their respective advantages and disadvantages in meeting various application requirements, as summarized in Table 4.4.

4.7 FUTURE RESEARCH DIRECTIONS

- Hybridization of traditional techniques with ML controllers to achieve better performance.
- To decrease the required data and training, transfer learning involves applying previously taught models to new tasks linked to them.
- In the realm of power electronics, real-time implementation pertains to the development of efficient algorithms and hardware to enable the application of ML in real-time scenarios.
- Future research may focus on improving training algorithms to reduce computational demands and training time, making the technology more viable for real-time applications.
- Further investigation is required to assess the efficacy of pure RL controllers in tackling stability issues in MGs. In addition, DC-MGs comprise many distributed generators (DGs) that are networked in parallel. This configuration presents difficulties in efficiently controlling voltage and guaranteeing appropriate current allocation among the parallel sources.

Table 4.4 ML control approaches from previous literature

Reference	*Method used*	*Pros*	*Cons*	*Applications*
[14]	DDPG-RL control	Enhanced performance in controlling many variables, improved capacity to regulate dynamics, increased resilience and adaptability, and adaptive optimization	Complex, sensitive to hyperparameters, stability issues	DC-MGs and EV
[13]	MDP+DQN	The settling period is much shortened when a disturbance occurs due to the rapid adaptability of DRL in a control method that has been proposed	Regardless of variations in the CPL, the DRL controller remains invariant concerning the number of network layers and the parameters of the reward function	DC-MG and RES
[17]	Proximal Policy Optimization (PPO) algorithm	DDMPC with hybrid the technique is robust in nature	Although the hybrid approach that combines the PPO algorithm with a PI controller or filtering mechanism demonstrates promising results regarding short settling time and stability, its robustness under various disturbances, parameter uncertainties, and noise conditions require further investigation	DC-MGs and data centers
[21]	MDNN	1. Intelligent and smart control 2. The goal is to detect and stop sensor-level false data injection (FDI) attacks	1. Historical data may not encompass all the anomalies that could arise within the systems 2. Simulated data is computationally intensive and guarantees a combination of typical and atypical data	DC-MGs

(*Continued*)

Table 4.4 ML control approaches from previous literature *(Continued)*

Reference	*Method used*	*Pros*	*Cons*	*Applications*
[22]	Twin-delayed DDPG (TD3)	Unlike DQN, which suits discrete tasks, this improves system stability and dynamics. TD3 is effective for continuous control	The reward function is not unique to dynamic conditions	DC-MGs
[23]	DDPG + iPI (DDPGiPI) based on a SM observer	It minimizes observer estimate error and effectively manages destructive impedance instabilities	The iPI + MPC scheme can steady voltage output despite the substantial modifications to CPLs. Nevertheless, their efficacy is restricted by the fact that the system output's stable error exceeds 2%	On board DC-MGs
[24]	DNN	This control method facilitates the converter's rapid attainment of new referral values, notably enhances performances and efficiency by substantially decreasing error time	Relying solely on DNNs with biased data can compromise result integrity and reliability	Vehicle–Vehicle (V2V) and Vehicle–Grid (V2G)
[25]	DDPG with AC NN model	Addresses issues of nonlinearity and negative impedance instability and a higher level of robustness	Training the AC NNs for DDPG entails significant data acquisition and computational resources, presenting hurdles in data gathering, processing, and model refinement	Marine power systems
[26]	Artificial neural network (ANN)	Due to their parallel structure, ANNs improve system model accuracy despite inaccurate parameters and offer a lower computational burden than MPCs	ANNs need meticulous calibration and verification to guarantee compatibility and dependable functioning	DC-MGs

(Continued)

Table 4.4 ML control approaches from previous literature *(Continued)*

Reference	*Method used*	*Pros*	*Cons*	*Applications*
[27]	PID+DDPG	Significantly enhances transient performance without relying on theoretical model derivations. It can be widely applied to other controllers and offers greater robustness due to its online learning capabilities	Increased complexity and not feasible for low-power applications	DC-MGs
[28]	DQN+Duty ratio mapping (DRM)	Proper voltage regulation, flexibility, and efficiency in unpredictable circuit parameters and diverse working environments	Issues with insufficient control precision and techniques for continuous interval learning can be explored further	Motion control and MGs systems
[29]	HOSMO+DDPG adaptive tuner	The standard double-loop PI controller takes longer to regulate the bus voltage and has more overshoot than this benchmark robust controller	Practical implementation can be challenging due to the need for real-time computation and the integration of multiple control strategies	DC-MGs

- Research into incorporating sophisticated RL algorithms, such as multi-agent RL and safe RL, into managing power electronic converters is highly recommended. Investigating the method of linking mathematical RL algorithms with physical power electronic converters is equally crucial.

4.8 CONCLUSION

This chapter explains the simulation case study of the buck converter performance with various types of controllers like linear, nonlinear, and ML type. It shows that ML type will give faster response than traditional controllers. This chapter also offers recent research on ML controllers, which are used to control the DC–DC converters applied in EV and DC-MG to overcome issues like regulation of output voltage, methods to stabilize DC-MGs at the source side with constant power loads (CPLs), design of the advanced model-driven control strategy, and to improve the efficiency of the converters. It compares classical nonlinear stabilization methods, highlighting their advantages and disadvantages, and explores the possible advantages of RL regulated as another approach. According to the research shown above, although traditional nonlinear techniques, when paired with disturbance observers, help resolve these concerns to achieve optimal performance, these techniques require an accurate model of the particular system in question. However, RL-based controllers like DDPG, PPO, and DQN can adapt to changing system conditions through online learning; they are not model-reliant.

REFERENCES

1. Li, X., Zhang, X., Lin, F., and Blaabjerg, F. (2021). Artificial-intelligence-based design for circuit parameters of power converters. IEEE Transactions on Industrial Electronics, 69(11), 11144–11155. https://doi.org/10.1109/TIE.2021.3088377
2. Zhao, S., Blaabjerg, F., and Wang, H. (2020). An overview of artificial intelligence applications for power electronics. IEEE Transactions on Power Electronics, 36(4), 4633–4658. https://doi.org/10.1109/TPEL.2020.3024914
3. Khomfoi, S. and Tolbert, L. M. (2007). Fault diagnostic system for a multilevel inverter using a neural network. IEEE Transactions on Power Electronics, 22(3), 1062–1069. https://doi.org/10.1109/TPEL.2007.897128
4. Alghassi, A., Perinpanayagam, S., and Samie, M. (2015). Stochastic RUL calculation enhanced with TDNN-based IGBT failure modeling. IEEE Transactions on Reliability, 65(2), 558–573. https://doi.org/10.1109/TR.2015.2499960
5. Grabowski, P. Z., Kazmierkowski, M. P., Bose, B. K., and Blaabjerg, F. (2000). A simple direct torque neuro-fuzzy control of PWM-inverter-fed induction motor drive. IEEE Transactions on Industrial Electronics, 47(4), 863–870. https://doi.org/10.1109/41.857966
6. Cai, B., Zhao, Y., Liu, H., and Xie, M. (2016). A data-driven fault diagnosis methodology in three-phase inverters for PMSM drive systems. IEEE Transactions on Power Electronics, 32(7), 5590–5600. https://doi.org/10.1109/TPEL.2016.2608842
7. Ali, S. H., Heydarzadeh, M., Dusmez, S., Li, X., Kamath, A. S., and Akin, B. (2017). Lifetime estimation of discrete IGBT devices based on Gaussian process. IEEE Transactions on Industry Applications, 54(1), 395–403. https://doi.org/10.1109/TIA.2017.2753722
8. Zhao, S., Makis, V., Chen, S., and Li, Y. (2018). Health assessment method for electronic components subject to condition monitoring and hard failure. IEEE Transactions on Instrumentation and Measurement, 68(1), 138–150. https://doi.org/10.1109/TIM.2018.2839938
9. Wang, T., Xu, H., Han, J., Elbouchikhi, E., and Benbouzid, M. E. H. (2015). Cascaded H-bridge multilevel inverter system fault diagnosis using a PCA and multiclass relevance vector machine approach. IEEE Transactions on Power Electronics, 30(12), 7006–7018. https://doi.org/10.1109/TPEL.2015.2393373

10. Glavic, M., Fonteneau, R., and Ernst, D. (2017). Reinforcement learning for electric power system decision and control: Past considerations and perspectives. IFAC-PapersOnLine, 50(1), 6918–6927. https://doi.org/10.1016/j.ifacol.2017.08.1217
11. Sutton, R. S. and Barto, A. G. (2018). Reinforcement Learning: An Introduction. MIT Press. https://doi.org/10.1109/TNN.1998.712192
12. Xu, Q., Vafamand, N., Chen, L., Dragičević, T., Xie, L., and Blaabjerg, F. (2020). Review on advanced control technologies for bidirectional DC/DC converters in DC microgrids. IEEE Journal of Emerging and Selected Topics in Power Electronics, 9(2), 1205–1221. https://doi.org/10.1109/JESTPE.2020.2978064
13. Cui, C., Yan, N., Huangfu, B., Yang, T., and Zhang, C. (2021). Voltage regulation of DC–DC buck converters feeding CPLs via deep reinforcement learning. IEEE Transactions on Circuits and Systems II: Express Briefs, 69(3), 1777–1781. https://doi.org/10.1109/TCSII.2021.3107535
14. Ye, J., Guo, H., Wang, B., and Zhang, X. (2024). Deep deterministic policy gradient algorithm based reinforcement learning controller for single-inductor multiple-output DC–DC converter. IEEE Transactions on Power Electronics, 39(4), 4078–4090. https://doi.org/10.1109/TPEL.2024.3350181
15. Mudiminchi, A. Y., Teeparthi, K., Mastanaiah, A., and Kolipaka, S. (2023). An intelligent deep reinforcement learning control for DC–DC power buck converter feeding a constant power load. 2023 IEEE 3rd International Conference on Sustainable Energy and Future Electric Transportation (SEFET), IEEE, pp. 1–4. https://doi.org/10.1109/SeFeT57834.2023.10245391
16. Tian, F., Cobaleda, D. B., and Martinez, W. (2022). Deep reinforcement learning for DC–DC converter parameters optimization. 2022 IEEE 31st International Symposium on Industrial Electronics (ISIE), IEEE, pp. 325–330. https://doi.org/10.1109/ISIE51582.2022.9831660
17. Prag, K., Woolway, M., and Celik, T. (2021). Data-driven model predictive control of DC-to-DC buck-boost converter. IEEE Access, 9, 101902–101915. https://doi.org/10.1109/ACCESS.2021.3098169
18. Ghosh, A., Banerjee, S., Sarkar, M. K., and Dutta, P. (2016). Design and implementation of type-II and type-III controller for DC–DC switched-mode boost converter by using K-factor approach and optimisation techniques. IET Power Electronics, 9(5), 938–950. https://doi.org/10.1049/iet-pel.2015.0144
19. Lakkanaboina, S. M., and Neeli, S. (2022). Model predictive and sliding mode controllers for the DC–DC boost converter. 2022 IEEE 2nd International Conference on Sustainable Energy and Future Electric Transportation (SeFeT), IEEE, pp. 1–6. https://doi.org/10.1109/SeFeT55524.2022.9909511
20. Yang, T., Cui, C., Zhang, C., and Yang, J. (2023). Robustness enhancement of DRL controller for DC–DC buck converters fusing ESO. Journal of Control and Decision, 12(1), 101–110. https://doi.org/10.1080/23307706.2023.2201587
21. Machina, V. S. P., Koduru, S. S., Madichetty, S., and Mishra, S. (2023). A novel standalone implementation of MDNN controller for DC–DC converter resilient to sensor attacks: A design approach. IEEE Journal of Emerging and Selected Topics in Power Electronics, 11(3), 2805–2815. https://doi.org/10.1109/JESTPE.2023.3242299
22. Rajamallaiah, A., Karri, S. P. K., and Sankar, Y. R. (2024). Deep reinforcement learning based control strategy for voltage regulation of DC–DC buck converter feeding CPLs in DC microgrid. IEEE Access, 12, 17419–17430. https://doi.org/10.1109/ACCESS.2024.3358412
23. Gheisarnejad, M., Farsizadeh, H., and Khooban, M. H. (2020). A novel nonlinear deep reinforcement learning controller for DC–DC power buck converters. IEEE Transactions on Industrial Electronics, 68(8), 6849–6858. https://doi.org/10.1109/TIE.2020.3005071

24. Madichetty, S., Banda, M. K., and Banda, S. K. N. (2023). Implementation of deep learning based bi-directional DC–DC converter for V2V and V2G applications in microgrid-an experimental investigation. Energies, 16(22), 7614. https://doi.org/10.3390/en16227614
25. Khooban, M. H. (2022). Smartenance DC–DC on-board power converters. IEEE Transactions on Circuits and Systems II: Express Briefs, 70(1), 191–195. https://doi.org/10.1109/TCSII.2022.3206230
26. Khan, H. S., Mohamed, I. S., Kauhaniemi, K., and Liu, L. (2021). Artificial neural network-based voltage control of DC/DC converter for DC microgrid applications. 2021 6th IEEE Workshop on the Electronic Grid (eGRID), IEEE, pp. 1–6. https://doi.org/10.1109/eGRID52793.2021.9662132
27. Hu, K., Zhang, X., and Ma, H. (2021). A novel proportion–integral–differential controller based on deep reinforcement learning for DC/DC power buck converters. 2021 IEEE 1st International Power Electronics and Application Symposium (PEAS), IEEE, pp. 1–4. https://doi.org/10.1109/PEAS53589.2021.9628495
28. Cui, C., Yang, T., Dai, Y., Zhang, C., and Xu, Q. (2022). Implementation of transferring reinforcement learning for DC–DC buck converter control via duty ratio mapping. IEEE Transactions on Industrial Electronics, 70(6), 6141–6150. https://doi.org/10.1109/TIE.2022.3192676
29. Huangfu, B., Cui, C., Zhang, C., and Xu, L. (2022). Learning-based optimal large-signal stabilization for DC/DC boost converters feeding CPLs via deep reinforcement learning. IEEE Journal of Emerging and Selected Topics in Power Electronics, 11(6), 5592–5601. https://doi.org/10.1109/JESTPE.2022.3189078

Chapter 5

Disturbance rejection controller of high-performance PMBLDC motor drives for electric vehicle applications

Prashant Kumar

5.1 INTRODUCTION

With the global push to target "Net Zero" to address climate change and greenhouse gas emissions reduction, the transportation electrification trend is emerging [1]. As the automobile sector is expanding tremendously, issues such as traffic congestion, environmental harm, and excessive fuel consumption are becoming prominent social challenges. Electric transportation is considered necessary for a sustainable world since it allows for less energy consumption and provides better control [2]. With the goal of reducing CO_2 emissions and environmental concerns, especially in urban areas, governments are promoting initiatives to support the use of electric vehicles (EVs) [3]. The rapidly increasing demand for clean, abundant energy has increased the desire to participate in vehicular electrification. By 2030, it is projected that EVs will replace one-third of all road vehicles. In the real world, EVs have the scope to provide a better approach to ecological transportation, particularly if they are being driven using renewable power sources, thus eliminating the need for further petroleum. EVs would be able to meet the demand for electrical power due to the increased use of electronic technologies to improve driving comfort, performance characteristics, emission levels, and security. The main considerations with electric cars are high efficiency, robustness, and compact size in power electronic devices and motor drives [4]. The fail-safe functionality of the powertrain of an EV comprises electric drive motors, voltage-source inverters (VSIs), battery packs (BPs), and sensor systems.

In the context of EVs, the term low power can be relative and subjective, depending on the specific characteristics of the EVs being considered. Typically, EVs are classified based on their power output or their performance specifications, such as their acceleration, top speed, and range. If the EVs being studied in the thesis are high-torque, low-speed vehicles, such as electric scooters or electric bicycles, then the power output of the motor may not be very high, and iron loss may not play a significant role in determining motor performance. In such cases, the focus of the analysis may be on other factors such as efficiency, battery life, and ride quality. However, if the EVs being studied are low-torque, high-speed vehicles, such as electric cars, then

DOI: 10.1201/9781003515807-5

the power output of the motor could be much higher, and iron loss could become a more significant factor in determining motor performance. In such cases, the thesis may need to consider the effect of iron loss on the motor efficiency, temperature rise, and overall performance. Therefore, it is important for the thesis to provide a clear definition of what is meant by low-power EVs and to specify the torque and speed characteristics of the EVs being studied. This would help to ensure that the analyses and conclusions are relevant to the specific type of EVs being studied. Additionally, in EVs where the torque and speed are high enough to make iron loss a significant factor in determining motor performance, it would help to illustrate the importance of considering iron loss in motor design and control. This example could be used to demonstrate the effect of iron loss on motor performance and the benefits of using advanced control techniques to mitigate its effects.

The architecture of the electric motor has a significant impact on the vehicle's dynamics and the type of power electronic converter used to govern the operating characteristics [4, 5]. To yield the desired performance indices of an electric powertrain at reliable and affordable services, there is an increasing demand for efficient motor drives. The main features of an electrical motor drive in transportation electrification are higher torque density and power density, a wide-scale operating speed range, and intermittent overload performance [6]. Low acoustic noise and minimal torque ripple are other essential aspects of the motor drive system. The motor drive system is a key element since it is the primary power source in the electric propulsion mechanism. As a result, finding a high-performance motor drive system has been a priority for several years [7]. Several electric motor types have been developed in terms of their effectiveness for transportation applications [5]. Numerous research studies have been carried out to improve the fundamental design and control systems of electric propulsion systems. In the beginning, DC motors were used in EVs; however, the commutator and brushes necessitate additional services. Due to the integrated commutators and brushes, DC drives are limited in dependability and inappropriate for maintenance-free performance. Owing to the advancements in motor control technology and PM material, induction motors (IMs), switched reluctance motors (SRMs), and permanent magnet (PM) motors are now being used in EVs [8]. The usually needed properties could be acquired by IM and SRM; however, permanent magnet brushless DC motors (PMBLDCMs) deliver superior efficiency and torque density. In SRM, the acoustic noise generated is widely highlighted as a major drawback. During low rotational speeds, acoustic noise is mostly caused by vibrations created by torque ripples. Their design and control are difficult and subtle. As employed in EV, IM has numerous limitations such as lower efficiency and higher volume as compared to permanent magnet synchronous motors (PMSMs). The operating voltage of the drive motor of an EV is desired to be low for safety aspects and battery power control. Since thick stator winding conductors are required at low operating voltages, the dimension of the IM extends with the power rating. Such high-volume IM is impractical for traction applications. With the advent of high-energy PM materials has led to the usage of PMSMs in EV propulsion systems. The traditional PMM is the best choice when it comes to cost, efficiency, control, and reliability. There is a growing interest toward using PMM as a drive motor in EVs. Owing to its high efficiency and power density, PM motors are deployed in heavy-load and light-load EVs such as trains, buses, cars, e-rikshaws, e-scooters, etc. The PMM is widely utilized in industry for a variety of purposes, and there are a number of effective control schemes to choose from. EV industries use PM-based traction motors like the PMSM and PMBLDCM. Particularly in low-power vehicles, the motor topology of PMSM is costlier and has a lower power density than that of PMBLDCM. The PMBLDCM propulsion technology is extremely appealing for usage in less expensive EVs because of the low-cost drive arrangement.

PMBLDCMs are generally preferred for low-power applications such as small-scale EVs and high-accuracy industrial deployments because of their enhanced torque–speed waveforms, improved torque-to-weight ratio, higher efficiency, superior control sensitivity level, and fewer oscillations [9]. A schematic diagram for PMBLDCM-driven EV topology is shown in Figures 5.1 and 5.2. The torque ripple in PMBLDC motors is a drawback that must be acknowledged, though it impairs performance and leads to serious problems such as undesired noise and vibrations in machinery. To overcome these challenges, three integrated Hall-effect position sensors are added to the PMBLDC motor. However, additional electrical and mechanical configurations are required for this integration. The possible occurrence of a position sensor malfunction raises questions about the system's resilience and operational capabilities under such situations.

Under PMBLDCM's usual operating circumstances, the inverter commutes every 60 electrical degrees and needs a six-step conduction mode regulation. Torque ripple and insufficient efficiency are the results of commutation faults. Highly accurate uses are limited by the torque ripple's vibration, acoustic noise, and speed changes. Furthermore, PMBLDCM motor torque ripple causes

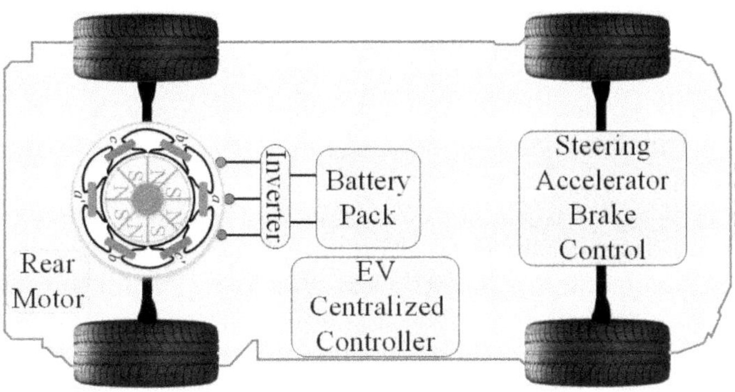

Figure 5.1 A schematic diagram for PMBLDCM-driven EV topology [10].

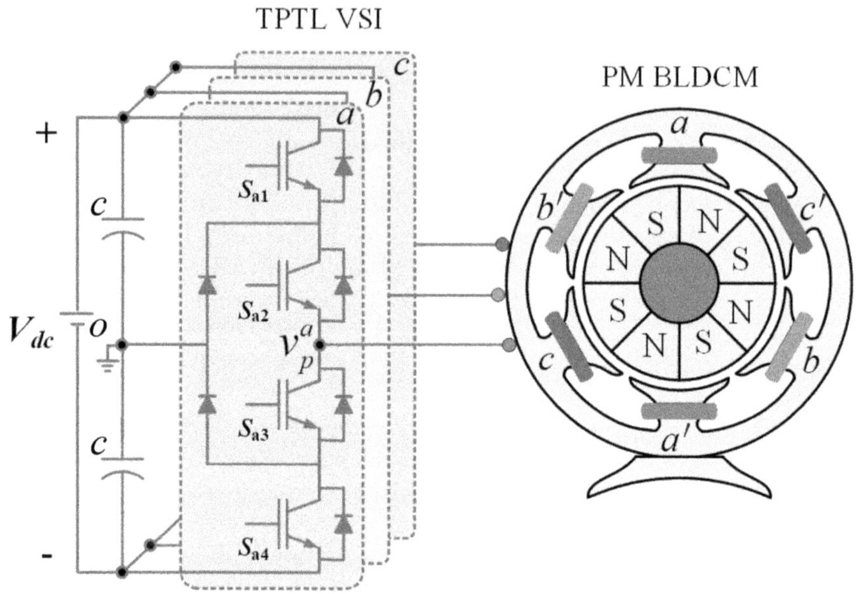

Figure 5.2 A schematic diagram for PMBLDCM-driven EV topology.

current ripples to rise, which in turn increases copper losses and decreases motor performance. Therefore, a key focus of the present research is reducing copper loss to improve motor torque efficiency with decreased torque ripple. The maximum torque per ampere (MTPA) management approach has been studied in this chapter to improve motor performance and minimize copper loss. Although the quadrature (q) axis delivers a flux-controlling behavior, the direct (d) axis stator current component is thought to be the torque-controlling component that produces a cross-coupling impact. Therefore, MTPA can be attained by either producing the stator current exhibiting a petal-shaped waveform that takes into account both dq-axis components or by eliminating the d-axis stator current component [10].

In essence, the MTPA technique described in Ref. [11] that uses q-axis current estimates to increase the PMBLDCM torque and lower the d-axis current to zero. By calculating the rotor angle of the PMBLDCM model to enhance the output torque with the predetermined value of the stator current, MTPA is accomplished in Ref. [12]. Various offline and online parameter estimate approaches are described in Refs. [13] and [14] to increase the accuracy of MTPA angle computations. Nevertheless, due to the significant differences in PMBLDCM electrical properties, it is not possible to accurately estimate the MTPA angle during heavy load function. In order to solve torque ripple, reliable PMBLDCM modeling that takes into account a variety of design and material characteristics is more essential. The strength of PM degrades due to factors such as temperature and magnetic saturation. The capacity of PMBLDCM decreases due to iron loss, which includes both hysteresis and eddy current losses. A number of iron loss-based PMBLDCM models have been investigated in research for the purpose of iron loss impact forecasting and evaluation. Taking iron loss into account, this chapter presents a revised MTPA control technique.

Uncontrolled instability and interruptions plague PMBLDCM, which takes iron loss into account. A rise in torque ripple and a decline in speed effectiveness are common examples of improper closed-loop control performance caused by it. These disruptions are not handled by the well-liked control schemes such as indirect current control [15], direct torque control (DTC) [12], and model predictive control [16] that would not overcome the issues of disturbances. Some of the disturbance-based control strategies, such as disturbance observer (DO) [17], extended state observer (ESO) [18], and active disturbance rejection control (ADRC) [19, 20], have been used for the last 20 years. For PMBLDCM drives, precise torque control along with speed ruggedness is significantly more crucial. When it comes to controlling complexity, linear ESO (LESO) is better than nonlinear ESO. Reducing the need for precise dynamic system modeling, LESO-based state feedback regulation compensates for entire disturbances with little tracking error. Even though it doesn't add complexity to the algorithm, a phase-locking loop observer (PLLO)-based ADRC enhances the disturbance attenuation capability. In light of the fact that iron loss represents a disruption to the initial model, this study proposes an ADRC approach. In industrial use, sensor-based methods have been used for years to determine the mechanical rotor position. These systems typically include a separate encoder with higher resolution or an integrated Hall-effect sensor. Nevertheless, present studies are primarily concerned with the implementation of sensorless rotor position/speed estimate techniques as a result of rising expenses and mechanical maintenance requirements. Current sensorless techniques described in research use the flux linkage approach, phase-locked loop (PLL) technique, third harmonic technique, and the return electromagnetic force (EMF) technique to determine speed. Nevertheless, in order to calculate the rotor position/speed of PMBLDC, which permits sensorless function, this chapter studies the existing ADRC-based disturbance-estimating approach equivalent to back EMF-utilizing ESO. In order to obtain proper tracking, including disturbances caused by load torque fluctuation and iron loss impact, a predicted speed is delivered as feedback to the speed ADRC.

An improved modulation controller for a three-phase, three-level (TPTL) neutral point clamped (NPC) VSI is presented in this chapter. Multiple real-world advantages are associated with the TPTL-NPC VSI, including more leeway in choosing switching vectors, less voltage stress on semiconductor switches $\frac{dv}{dt}$, better total harmonic distortion (THD), and less torque ripple.

The following are the key points covered in this chapter:

- As described in the above section, a study of the MTPA control leads to the establishment of a PMBLDCM model that takes iron loss into account as a disturbance.

- Through observation of the disturbances corresponding to load torque and iron losses for the PMBLDCM drive, an ADRC-based speed controller is presented. Additionally, to enable sensorless function, the speed is approximated using normalized back EMF disturbance and provided.
- By determining which disturbance corresponds to the back EMF of PMBLDCM, a new ADRC-based inner loop current control is developed to reduce the current and torque ripple.
- Provides specifics on the modeling and investigation used to validate the suggested speed ADRC and the present ADRC with LESO and PLLO layout.

5.2 MODELING OF PMBLDC MOTOR CONSIDERING IRON LOSS

The variables a, b, c represent the three-phase signals with 120 electrical degrees phase difference, $\alpha\beta$ represent the synthesized parameters and d–q frame is the direct and quadrature axis parameter that rotates simultaneously. The transformations from the abc-frame to the $\alpha\beta$-frame and to the dq-frame using general parameter X can be obtained using Equation (5.1) [10]. Here, $X \in v, i, e, \lambda$.

$$X^{\alpha\beta} = X^{\alpha} + jX^{\beta} = \sqrt{\frac{2}{3}} \sum_{\substack{k=1 \\ l \in \{a,b,c\}}}^{3} X^{l} e^{j\frac{2\pi}{3}(k-1)} \tag{5.1}$$

$$X^{dq} = X^{d} + jX^{q} = e^{-j\theta_r} X^{\alpha\beta}$$

θ_r denotes the rotor angular orientation of PMBLDC motor in rad. v, i, e, and λ represent voltage, currents, stator-phase back EMF, and flux linkage, respectively (Figure 5.3). The equivalent circuit representation of BLDC motor in dq-frame has been illustrated in Figure 5.4 considering irons losses. The BLDC motor in abc-frame is developed as Equation (5.2) and synthesized in $\alpha\beta$ as Equation (5.3) using Equation (5.1) [21].

$$v_s^{abc} = R_s i_s^{abc} + L_s \frac{di_m^{abc}}{dt} + \omega_r \lambda_s^{abc} \tag{5.2}$$

$$v_s^{\alpha\beta} = R_s i_s^{\alpha\beta} + L_s \frac{di_m^{\alpha\beta}}{dt} + \omega_r \lambda_s^{\alpha\beta} \tag{5.3}$$

By applying the transformation factor $e^{-j\theta_r}$ to Equation (5.3), the PM-BLDC model is converted into the dq reference frame, as expressed in Equation (5.4):

$$v_s^{dq} = R_s i_s^{dq} + L_s \frac{di_m^{dq}}{dt} + j\omega_r L_s i_m^{dq} + \omega_r \lambda_s^{dq} \tag{5.4}$$

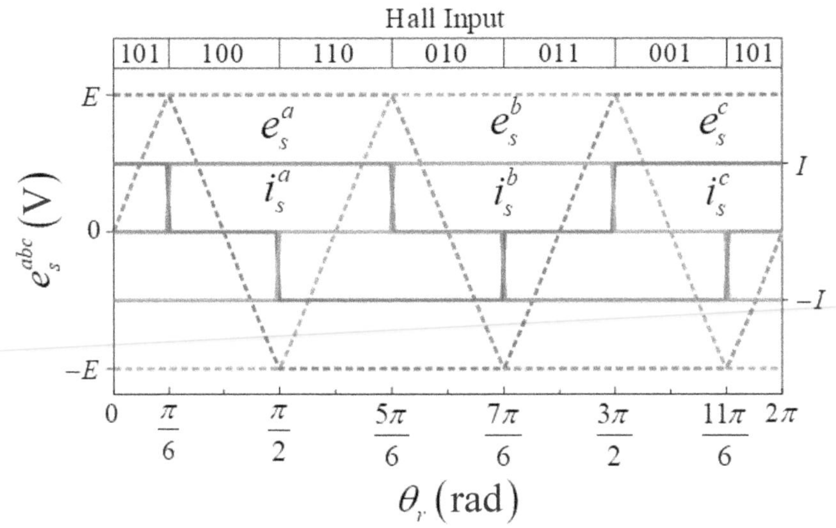

Figure 5.3 Stator back EMFs and stator-phase currents at different rotor electrical positions.

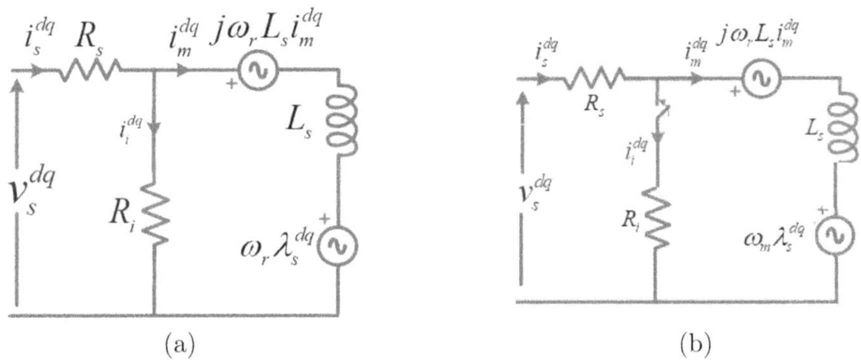

Figure 5.4 Equivalent circuit of PMBLDCM in (a) *abc*-frame and (b) *dq*-frame considering R_i effect.

In steady state, the stator current (i_s^{dq}) of PMBLDCM can be represented in terms of the magnetizing (i_m^{dq}) and iron loss (i_i^{dq}) component as in Equation (5.5).

$$i_s^{dq} = i_m^{dq} + i_i^{dq} = C^{-1} i_m^{dq} + \frac{\omega_r}{R_i} \lambda_s^{dq} \tag{5.5}$$

Here, $\lambda_s^{dq} = e_s^{dq} / \omega_m$ represents the stator flux linkage in the *dq* reference frame, defined as the ratio of the stator back EMF (e_s^{dq}) to the motor's angular speed (ω_m).

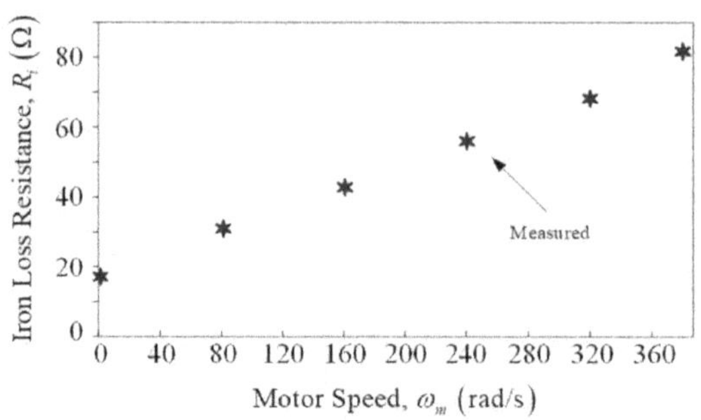

Figure 5.5 R_i at different ω_m.

The iron loss resistance (R_i) is determined following the method outlined in Ref. [22], and a linear fit to the experimental data is illustrated in Figure 5.5. The influence of R_i, representing iron losses, is captured by the complex coefficient $C = k_c e^{-j\theta_c}$. Here, $k_c = 1/\sqrt{1+(\omega_r L_s/R_i)^2}$ and $\theta_c = \tan^{-1}(\omega_r L_s/R_i)$ represent the attenuation magnitude and phase shift, respectively, introduced by the iron loss. In this context, ω_r denotes the rotor electrical speed, and L_s is the stator inductance. Figure 5.4 displays a comparable circuit model of PMBLDCM in the dq-plane while accounting for iron losses. According to Urasaki et al. [22], the iron loss resistance (R_i) is computed and shown as the measured data's linear approach in Figure 5.5.

Iron loss is a type of power loss that occurs in magnetic materials when they are subjected to a time-varying magnetic field, such as the rotating field in an AC motor. Iron loss is typically divided into two categories—hysteresis loss and eddy current loss. Hysteresis loss is caused by the reversal of magnetization in the iron core of the motor, and it increases with the frequency of the magnetic field. Eddy current loss is caused by the flow of currents in the iron core induced by the time-varying magnetic field, and it also increases with the frequency of the magnetic field. At low speeds, the frequency of the magnetic field is low, and therefore, the iron loss is relatively small. However, at high speeds, the frequency of the magnetic field increases, and the iron loss becomes significant. This is why iron loss has a more significant effect on motor performance at high speeds. The dependence of resistance on speed is quite low unless the speed is very high. This means that at low and moderate speeds, the effect of iron loss on the motor performance is small and can be neglected. Furthermore, the results shown support the assertion that the effect of iron loss on the motor performance is significant. The % improvement with the consideration of iron loss can be observed in presented results. If iron loss were not a significant factor, we would not expect a larger difference between the two control schemes. Therefore, for the rated speed considered in this work, the effect of iron loss on motor performance should be considered, and considering its effect is necessary for accurately analyzing the motor behavior.

Assuming $f_e^{dq} = -e_s^{dq}/L_s$ as the unknown perturbations, $f_i^{dq} = -\left(C^{-1}/\tau_s + j\hat{\omega}_r\right)i_s^{dq}$ is the known perturbance obtained from the PMBLDC motor parameters. $\hat{\omega}_r$ denotes the estimated ω_r. The current calculations of the PMBLDC motor are established as Equation (5.6) [11]:

$$\frac{di_s^{dq}}{dt} = b_v v_s^{dq} + f_i^{dq} + f_e^{dq} \tag{5.6}$$

Here, $b_v = C^{-1}/L_s$ represents the voltage (v_s^{dq}) control gain. The developed motor torque (T_{em}) considering iron loss can be expressed as Equation (5.7) [21]:

$$T_{em} = \Re\left(\bar{\lambda}_s^{dq} i_m^{dq}\right) = \Re\left(C\bar{\lambda}_s^{dq} i_s^{dq}\right) - T_i \tag{5.7}$$

In Equation (5.7), $C = 1$ and $T_i = 0$ signify that the iron loss is ignored, thus $T_{em} = T_{es} = \Re\left(\bar{\lambda}_s^{dq} i_s^{dq}\right) = \lambda_s^d i_s^d + \lambda_s^q i_s^q$ is achieved. The T_i is the additional T considering iron loss that varies from sectors $S_1,\ldots,S_6$ can be expressed as Equation (5.8):

$$T_i\big|_{S_1,\ldots,S_6} = \frac{\omega_m}{R_i}\left|\lambda_s^{dq}\right|^2_{S_1,\ldots,S_6} \tag{5.8}$$

The electromagnetic torque T_{em} is linearly dependent on the q-axis current (i_s^q). Thus, a constant torque can be generated by regulating i_s^q while maintaining the d-axis current (i_s^d) at zero. However, when this approach is applied to a PMBLDCM with a non-sinusoidal back EMF profile (λ_s^{dq}), it leads to undesirable torque pulsations. These arise from harmonic components in the back EMF, causing λ_s^d and λ_s^q to vary with the rotor's electrical position, as illustrated in Figure 5.3.

Considering iron loss and setting $i_s^d = 0$, the electromagnetic torque from Equation (5.8) is modified as follows:

$$T_{em} = k_t i_s^q - T_i \tag{5.9}$$

where $k_t = k_c\left|\lambda_s^{dq}\right|\sin(\theta_r + \theta_c)$ is the effective torque constant accounting for iron loss.

Assuming mechanical losses are negligible, the shaft torque of the PMBLDCM is derived from T_{em} and expressed using the reference q-axis current $i_s^{q^*}$. The mechanical dynamics are then given by

$$\frac{d\omega_m}{dt} = \frac{1}{J}T_{em} - \frac{1}{J}T_L - \frac{B}{J}\omega_m = b_i i_s^{q^*} + f_\omega^m \tag{5.10}$$

where $b_i = \frac{k_t}{J}$ is the current control gain and the total disturbance term is defined as follows:

$$f_{\omega}^{m} = -\frac{1}{J}\left[k_t(i_s^{q*} - i_s^{q}) + T_i + T_L + B\omega_m\right]$$

Here, T_L is the load torque, J is the moment of inertia, and B is the damping coefficient of the motor.

5.3 SENSORLESS OPERATION AND CURRENT CONTROL DESIGN

ADRC is a relatively new and advanced control technique that can provide better disturbance rejection and robustness compared to traditional control strategies. ADRC involves estimating and compensating for both internal and external disturbances in the system, leading to improved performance and stability. Therefore, the use of ADRC in this chapter could be a significant contribution, as it may have improved the disturbance rejection capability and robustness of the motor control system, leading to better performance in terms of efficiency, torque ripple reduction, and stability. Additionally, the ADRC technique is relatively easy to implement and does not require detailed knowledge of the system dynamics, making it suitable for a wide range of applications. In summary, the motivation to go for ADRC could be to improve the performance of the motor control system, address any limitations or shortcomings of the DO-based predictive control technique, and take the benefits of ADRC, such as improved disturbance rejection and robustness. Trajectories of $e_s^{\alpha\beta}$ and $\left\{i_m^{\alpha\beta*}, i_s^{\alpha\beta*}\right\}$ for six sectors is shown in Figure 5.6.

The dynamic of the current loop is indeed much faster than the speed loop; it can be justified that the d-axis current component (i_s^d) can be assumed to be zero at any given instant. In a PMBLDCM control system, the i_s^d component is responsible for controlling the flux or magnetic field produced by the motor. The i_s^q component, on the other hand, is responsible for producing torque. If the dynamic of the current loop is much faster than the speed loop, it means that the current loop can quickly respond to changes in the load or other disturbances. This would allow the current loop to quickly adjust the i_s^q current component to produce the desired torque while also ensuring that the i_s^d component is maintained at zero. Assuming that the flux component of the stator current is already regulated to zero, the current controller can focus on regulating only the torque-producing i_s^q component, which is directly related to the motor speed. This would allow for better control of the motor speed and torque, without the need to also control the i_s^d component. However, it is important to note that the assumption of zero i_s^d is only valid when the flux

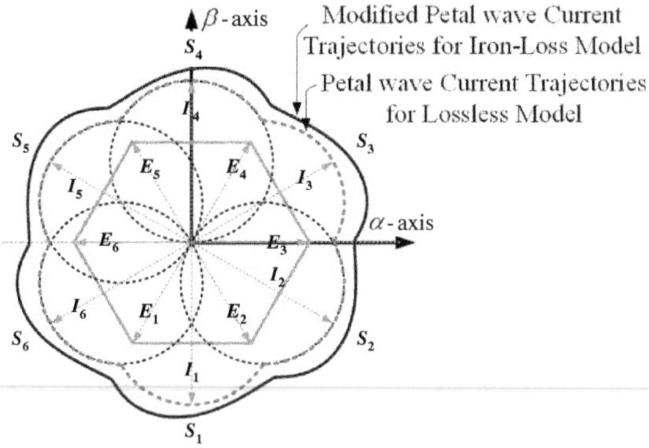

Figure 5.6 Trajectories of $e_s^{\alpha\beta}$ and $\left\{i_m^{\alpha\beta*}, i_s^{\alpha\beta*}\right\}$ for six sectors [21].

component is already regulated to zero. If the flux component is not regulated properly, it could cause issues such as motor saturation and increased losses. Therefore, proper control of both the i_s^q and the i_s^d components is necessary for optimal motor performance.

The estimation of the input gain is a critical component in the performance of ADRC, and obtaining an accurate initial guess is important for optimal performance. The critical gains for speed ADRC and current ADRC have been obtained using a trial and error approach, followed by fine-tuning through simulation and experimental results. This approach involved varying the gains until satisfactory performance is achieved, and then fine-tuning the gains based on the results. There are a few factors to be considered for selecting PMBLDCM with ADRC: First, PMBLDCM drives, making them less suitable for low power applications where cost and simplicity are important considerations. However, it is important to note that the choice of motor and control scheme will depend on the specific requirements and constraints of the application.

This part offers rotor location estimation and a high-performance current control method that replaces traditional proportional-integral (PI)-based speed and current controllers using two first-order ADRCs. Figure 5.7 shows an operational block diagram of these controllers together with the rotor orientation estimate. The internal current ADRC receives its i_s^{q*} from the outer speed ADRC, which functions as a speed regulator. The motor's torque ripple is decreased by controlling i_s^{dq} and tracking back EMF disturbances using the inner current ADRC. To accomplish such achievements, the following presumptions are taken into account:

- Disturbances are defined as nonlinearities that correlate to saturation and deformities in the PMBLDCM stator and rotor.
- The first and second derivatives of disturbances have uniform boundaries [23].

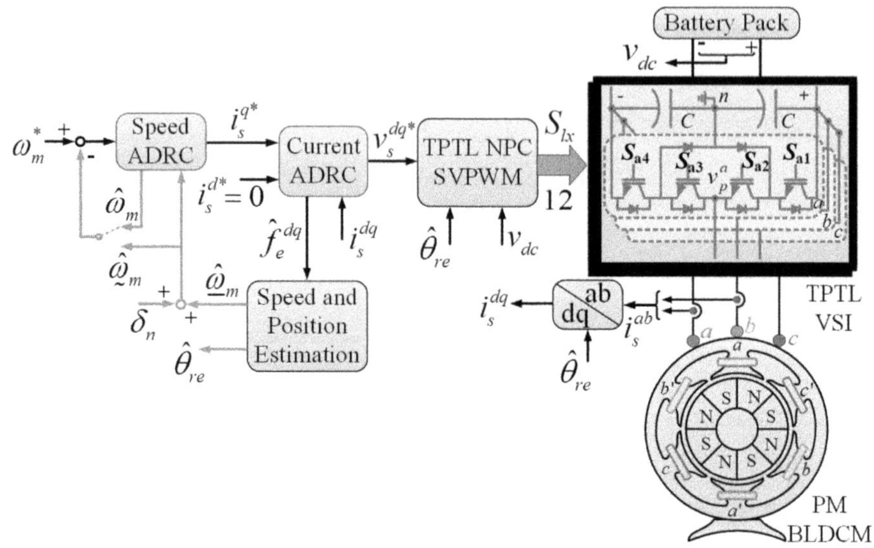

Figure 5.7 PMBLDC motor drives control using ADRC.

- Since the speed loop is slower than the current loop's evolving, the current IDs are regarded as "zero." A risk threshold that is two to three times the PMBLDCM's rated current limits the current i_s^d.
- Due to the direct relationship between load torque changes and current, a significant change in load torque could cause the current limit to be violated. So, it is safe to presume that T_L, the load torque, varies along a range that's connected to i_s^q [24].

5.3.1 Linear ADRC-based current controller design

The current ADRC layout takes into account the first-order disturbed system [11]. The control input v_s^x, where bv is an approximation of input gain, is required to be used to regulate the current i_s^x ($x \in d, q$). The following calculations provide the basis for designing the current ADRC utilizing the LESO- and PLLO-based external change spectator, with $\hat{f}_e^x$ the predicted state of the PMBLDCM model.

LESO design:

$$\begin{cases} \dfrac{d\hat{i}_s^x}{dt} = b_v v_s^x + f_i^x + \hat{f}_e^x + \beta_1^x \varepsilon_i^x \\ \dfrac{d\hat{f}_e^x}{dt} = \beta_2^x \varepsilon_i^x, \quad x \in \{d,q\} \end{cases} \tag{5.11}$$

PLLO design:

$$\begin{cases} \dfrac{d\hat{i}_s^x}{dt} = b_v v_s^x + f_i^x + \hat{f}_e^x \\ \dfrac{d\hat{f}_e^x}{dt} = \beta_1^x \dfrac{d\varepsilon_i^x}{dt} + \beta_2^x \varepsilon_i^x \end{cases}, \quad x \in \{d,q\} \tag{5.12}$$

In order to be assessed using the error margins and system stability factors, the observer gains β_1^x and β_2^x must be satisfied. An ADRC theory-based feed-forward compensation term for PMBLDCM's input might be the predicted external disturbance [25]. Thus, the following design can be used for the PMBLDCM's final control law:

$$v_s^x = v_{s,0}^x - b_v^{-1}\left(f_i^x + \hat{f}_e^x\right) \tag{5.13}$$

where $v_{s,0}^x$ is the plant input generated by the initial control law. Substituting Equation (5.13), the perturbed system in Equation (5.6) is reduced to Equation (5.14):

$$\frac{di_s^x}{dt} = b_v v_{s,0}^x \tag{5.14}$$

5.3.1.1 Stability and tracking performance analysis

By choosing a basic proportional current controller as the feedback controller, the required closed-loop behavior is attained.

The predicted disturbance $\hat{f}_e^x$ and the directed disturbance $\hat{f}_e^x$ are related through the transfer function in Equation (5.14) using the characteristic polynomial of $(s^2 + \beta_1^x s + \beta_2^x)$. Additional comparisons are made between the resulting polynomial and the second-order undamped system $((s + \omega_o^x)^2)$, where ω_o^x is the undamped natural frequency. At $-\omega_o^x$, all eigenvalues are assigned as the foundation for this modeling [11].

$$\frac{\hat{f}_e^x(s)}{f_e^x(s)} = \begin{cases} \dfrac{\beta_2^x}{s^2 + \beta_1^x s + \beta_2^x}, & \text{using LESO} \\ \dfrac{\beta_1^x s + \beta_2^x}{s^2 + \beta_1^x s + \beta_2^x}, & \text{using PLLO} \end{cases}, \quad x \in \{d,q\} \tag{5.15}$$

An observer bandwidth of $\beta_1^x = 2\omega_o^x$ is used to assess the stability of the current ADRC. To guarantee a quicker dynamic tracking of disturbance swings, as shown in Figure 5.8a, the bandwidth of the LESO should be maintained sufficiently larger. Under this situation, the sampling delay and sensitivity to sensors can be acceptable. Based on a comparison with Figure 5.8a and b, it can be observed that the PLLO system can provide the required performance

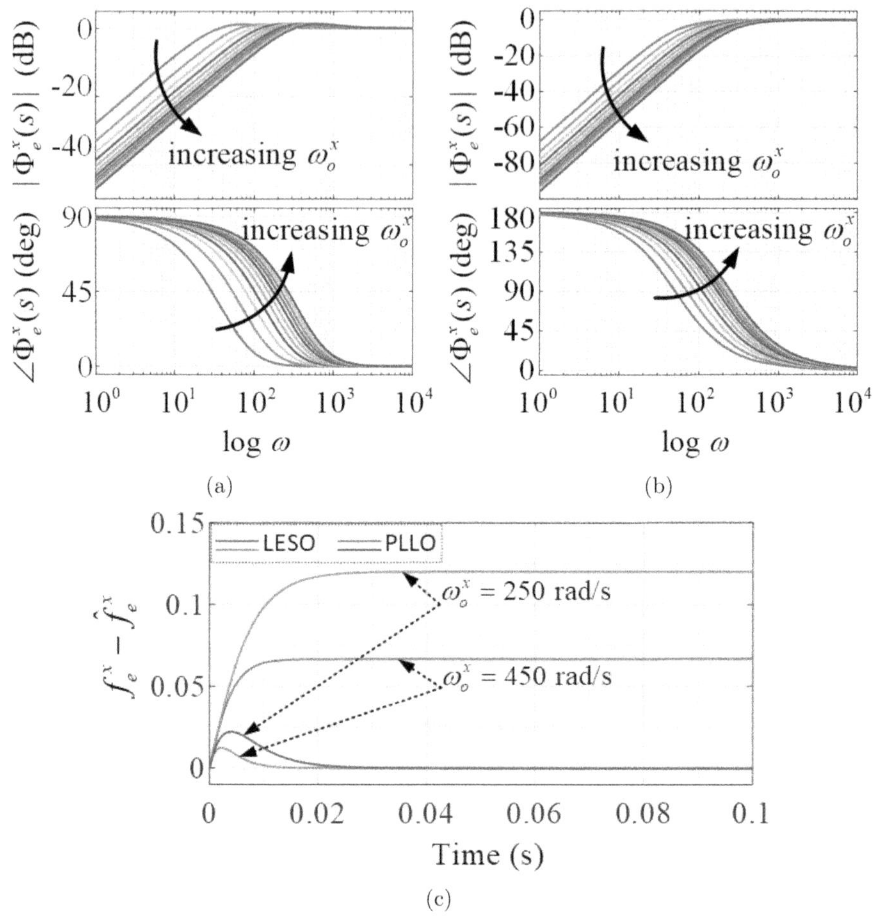

Figure 5.8 Bode plot: (a) LESO and (b) PLLO. (c) Time plot [11].

at nearly half the bandwidth of the LESO system. To determine the current ADRC's steady-state reaction, the error transfer function of the perturbation (Φ_e^x (s)) can be derived using Equation (5.16).

$$v_{s,0}^x = k_p^x \left(i_s^{x*} - i_s^x\right), \quad x \in \{d,q\} \tag{5.16}$$

k_p^x represents the proportional gain obtained as a trade-off between the steady-state and transient performance. The transfer function presented in Equation (5.15) is described as the estimated disturbance $\hat{f}_e^x$ to the reference signal f_e^x. The standard characteristics equation is expressed as $(s^2 + \beta_1^x s + \beta_2^x)$. The resulting polynomial is further compared with a second-order undamped system, represented by $\left(s+\omega_o^x\right)^2$, where ω_o^x denotes the undamped natural frequency. This parameterization approach ensures that all observer eigenvalues are placed at $-\omega_o^x$.

Based on Equation (5.15), the time/frequency-domain output of the external perturbation estimate error is achieved under a ramp-change stimulation (with a slope of 75) of the required $\hat{f}_e^x$. As illustrated in Figure 5.4a and b, a frequency–domain analysis is given with amplitude and phase of Φ_e^x (s) at different ω_o^x varying from 50 to 450 rad/s with a rise of 20 rad/s. Additionally, Figure 5.4c shows a time-domain characteristic providing details on bandwidth choice, bringing stable-state error into account. Results shown in Figure 5.4 indicate that the PLLO's bandwidth significantly affects the reliability of the disturbance prediction compared to LESO. The monitoring speed rises, and the disturbance estimate error falls with increased bandwidth.

To increase the capability of the current ADRC, the parameter k_p^x is selected to be 20%–33% of ω_o^x once the precise ω_o^x has been selected [26]. This estimation is only relevant when the settling time parameter of the transient response is used to compute k_p^x. The stability of the current controller is analyzed using ESO and Phase-Locked Loop Observer (PLLO) with bandwidths set at $\beta_1^x = 2\omega_o^x$. The gain $\beta_2^x = (\omega_o^x)^2$ defines the natural frequency for both observers. A higher LESO bandwidth is recommended to enhance the dynamic response and enable rapid tracking of disturbances, as illustrated in Figure 5.8a. However, this comes with increased sensitivity to sensor noise and sampling delays, which must remain within tolerable limits. As shown in Figure 5.8b, similar performance can be achieved with the LESO operating at nearly half the bandwidth. To assess the steady-state behavior of the current-loop ADRC, the disturbance error transfer function, $\Phi_e^x(s)$, is expressed by Equation (5.17) [11]:

$$\Phi_e^x(s) = \frac{f_e^x(s) - \hat{f}_e^x(s)}{f_e^x(s)} = \begin{cases} \dfrac{s\left(s+\beta_1^x\right)}{s^2+\beta_1^x s+\beta_2^x}, & \text{using LESO} \\ \dfrac{s^2}{s^2+\beta_1^x s+\beta_2^x}, & \text{using PLLO} \end{cases} \tag{5.17}$$

5.3.2 Speed estimation and control of PM-BLDC motor drives

5.3.2.1 Sensorless operation

As illustrated in Figure 5.7, this section provides a calculation of the motor speed and rotor position based on the calculated disturbances $\hat{f}_e^{dq}$ equivalent to back EMF in dq-plane. It is necessary to adjust the predicted q-axis unidentified external disturbance $\hat{f}_e^q$ for subsequent predictions since it includes the rotor position error details. The standardized $\hat{f}_e^d$ is controlled to zero using a tracking controller using a PLL. Irrespective of the basic working frequency, the predicted back EMF normalization assures a consistent linear dynamic response from the monitoring controller. By eliminating the phase delay associated with the complicated cutoff frequency selection procedure, this suggested method reduces the necessity for LPF in comparison to the traditional back EMF-based rotor position estimation. Additionally, the suggested estimation approach implements flat/linear ADRC using a simultaneously revolving frame rather than a fixed reference frame.

5.3.2.2 Speed ADRC design

The first-order disturbed system shown in Equation (5.8) receives input in the form of the calculated speed $\hat{\omega}_m$ and excess measurement noise δ_n. For a speed controller, a disturbance rejection controller for estimation based on LESO and PLLO can be explored using Equations (5.18) and (5.19) [11]:

LESO design

$$\begin{cases} \dfrac{d\hat{\omega}_m}{dt} = b_i i_s^{q*} + \hat{f}_\omega^m + \beta_1^m \varepsilon_\omega^m \\ \dfrac{d\hat{f}_\omega^m}{dt} = \beta_2^m \varepsilon_\omega^m \end{cases} \tag{5.18}$$

PLLO design

$$\begin{cases} \dfrac{d\hat{\omega}_m}{dt} = b_i i_s^{q*} + \hat{f}_\omega^m \\ \dfrac{d\hat{f}_\omega^m}{dt} = \beta_1^m \dfrac{d\varepsilon_\omega^m}{dt} + \beta_2^m \varepsilon_\omega^m \end{cases} \tag{5.19}$$

Here, $\hat{\omega}_m, \hat{f}_\omega^m$ denotes the estimated ω_m, f_ω^m.

5.3.2.3 Stability and tracking accuracy evaluation

The basic linear control feedback law can be used to locate the reference speed (5.16). With the chosen control gains, the speed controller's closed-loop transfer function is formulated as $\hat{\omega}_m$ or $\hat{\omega}_m$, described as Equation (5.20):

$$\frac{\omega_m(s)}{\omega_m^*(s)} = \frac{b_i k_p^m}{s + b_i k_p^m} \tag{5.20}$$

In order to guarantee the ADRC's tracking efficiency, a prompt and accurate assessment of external disturbance is required. As shown in Equation (5.19), the frequency-domain transfer function between the real amount $\hat{f}_w^m(s)$ and the predicted amount $\hat{f}_w^m(s)$ of the external disturbance can be found. By choosing $\beta_1^x = 2\omega_o^m$ and $\beta_2^x = \left(\omega_o^m\right)^2$, the characteristic polynomial of LESO/PLLO can be expressed as $s^2 + \beta_1^m s + \beta_2^m = \left(s + \omega_o^m\right)^2$, where ω_o^m is the undamped natural frequency of speed observer [11].

$$\Phi_\omega^m(s) = \frac{\hat{f}_\omega^m(s)}{f_\omega^m(s)} = \begin{cases} \dfrac{\beta_2^m}{s^2 + \beta_1^m s + \beta_2^m}, & \text{using LESO} \\ \dfrac{\beta_1^m s + \beta_2^m}{s^2 + \beta_1^m s + \beta_2^m}, & \text{using PLLO} \end{cases} \tag{5.21}$$

In LESO and PLLO, ω_o^x is set to 500 and 200 rad/s, respectively, to ensure the appropriate speed control loop stability. Figure 5.8a and b shows a Bode plot diagram, while Figure 5.8c shows a Nyquist plot using LESO and PLLO, respectively, for the purpose of analyzing noise elimination and disturbance elimination for speed ADRC.

In a comparison of the two systems' speed ADRC performance, PLLO offers superior disturbance cancellation at low frequencies, even though both systems have comparable noise-reduction capabilities. When the two observers' disturbance cancellation characteristics are same, the LESO helps to reduce high-frequency noise even more effectively.

5.4 RESULTS AND DISCUSSIONS

Both the current ADRC and the proposed sensorless speed ADRC are evaluated by means of simulation analysis. Table 5.1 lists the settings for the simulation. The sample time is $10\,\mu s$ and the dead time is $1\,\mu s$. The gain parameters are $k_p^d = 95, k_p^q = 115$ (for both LESO and PLLO), k_p^m (LESO) = 98.5, k_p^m (PLLO) = 52.6.

5.4.1 Simulation results

The performance of the MTPA-ADRC and MTPADTC are compared through the use of MATLAB/Simulink simulations of the PMBLDCM drive. Figure 5.9 displays the simulation results for T_{es}, load torque (T_L), ω_{m}, ω_{m}^*, i_s^q, i_s^d, and active and reactive power (P_e, Q_e) under various issues, including positive step change in speed and load, negative step change in load and speed. The motor is set to run at $\omega_m^* = 500\,\text{rad/min}$ without any load at $t = 0$ s.

Table 5.1 Parameters specifications

3Φ, 8 Pole, $V_{dc} = 240$ V, $P_{rated} = 3.5$ kW, $\omega_{rated} = 1{,}500$ rpm
$T_{rated} = 20\,\text{N}\cdot m$, $\lambda_n = 0.175\,\text{V}\cdot\text{s/rad}$, $R_s = 2.4\,\Omega$
$L_s = 8.5\,\text{mH}$, $J = 0.089\,\text{kg}\cdot\text{m}^2$, $B = 10^{-3}\,\text{N}\cdot\text{m}\cdot\text{s/rad}$

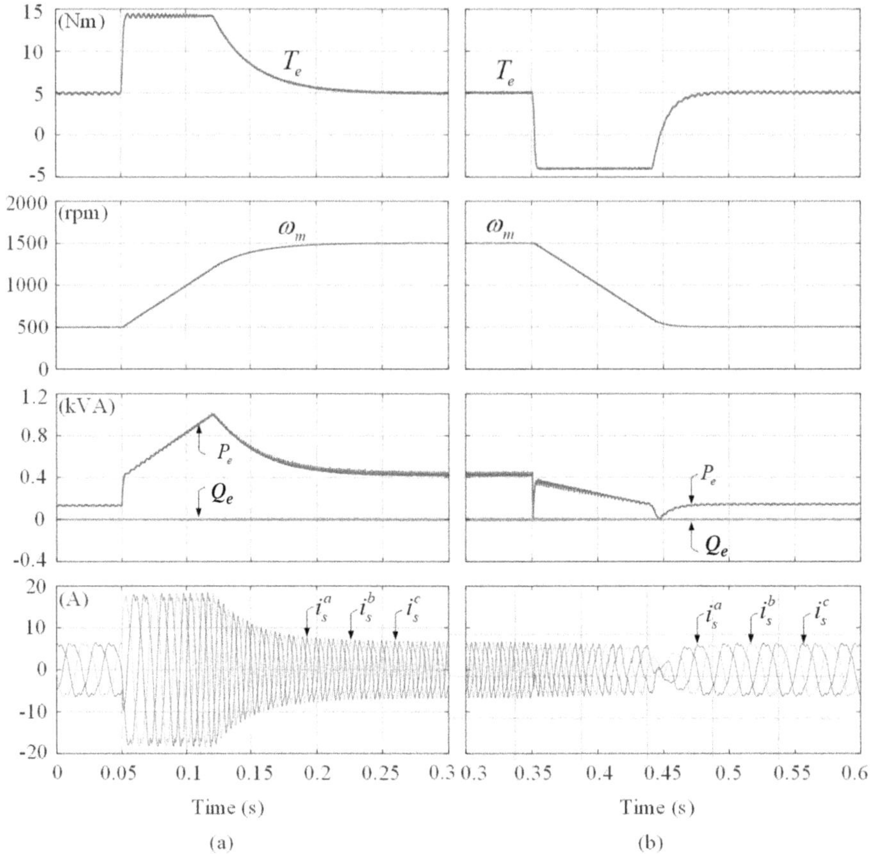

Figure 5.9 Simulation results of T_e, ω_m, P_e, Q_e, and i_s^{abc} during (a) acceleration and (b) deceleration.

At $t = 0.2$ s, a step change in load torque of $T_L = 5$ N·m is provided. The suggested sensorless speed ADRC reacts quickly and provides precise speed estimations as the load changes. The suggested performance is evaluated against MTPA-DTC, which offers quicker dynamics but more torque ripple. At t = 0.3 s, a step shift in ω_m from 500 rpm to 1,500 rpm with a load of $T_L = 5$ N·m is provided. It takes 0.17 s for the PMBLDCM to stabilize.

The torque and power requirements of PMBLDCM attain their maximum during the speed change. The simulation results of T_e, P_e, Q_e, and i_s^{abc} with timescale expansion are given in Figure 5.10. These waveforms are presented at steady state with $\omega_m = 1{,}500\,\text{rpm}$ and $T_e = 5\,\text{N} \cdot m$. It is observed that the torque ripple is low and is about 2%. The speed variation is very small and limited to 0.015%. In comparison to MTPA-DTC, it is also noted that the suggested MTPA-ADRC exhibits less torque ripple during speed changes. When the torque demand is met, the i_s^q nullifies the i_s^d. To study the reaction

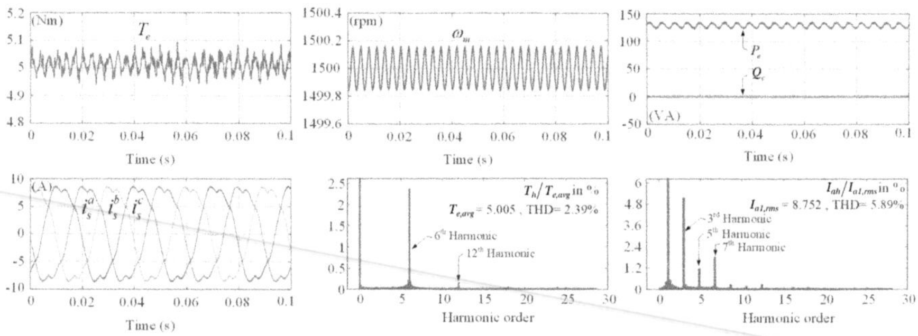

Figure 5.10 Simulation results for $\hat{\omega}_m$, T_{es}, P_e, Q_e, i_s^{abc}, and harmonic analysis of T_{es} & i_s^{abc}.

of the suggested sensorless ADRC system, PMBLDCM is unloaded and run at a steady-state speed of 1,500 r/min.

These waveforms in steady state with $\omega_m = 1{,}500\,\text{rpm}$ and $T_e = 5\text{N}\cdot m$ are shown in Figure 5.10 with timescale expansion. It is observed that the torque ripple is low and is about 2%. The speed variation is very small and limited to 0.015%. Both controllers show a little rise in ω_m following unloading. Nevertheless, the MTPA-ADRC settles ω_m more quickly than the MTPA-DTC. As illustrated in Figure 5.10, a negative step change in ω_m is commanded from 1,500 r/min to 0 r/min at $t = 0.7$ s. It takes 0.17 s for the PMBLDCM to reach zero speed.

Both systems exhibit regenerative behavior during deceleration. The existing ADRC and the proposed speed-based ADRC demonstrate faster dynamics, as the speed closely tracks the reference with minimal steady-state error and an improved transient response. Along with the harmonic spectra of T_{es} and $i_s^{a,b,c}$, Figure 5.10 displays the performance results of ω_m, ω_m^*, T_{es}, T_L, P_e, and Q_e. Reductions in losses are achieved by matching the dynamics of torque development with the active power requirement and zero reactive power. The 6th and 12th harmonics in T_{es} are 2.3% and 0.1%, respectively. MTPA-ADRC makes up 1.53% of THD, whereas MTPA-DTC makes up 2.39%. This shows that compared to MTPA-DTC, the suggested MTPA-ADRC has less torque ripple. Furthermore, the steady-state error of MTPA-ADRC is just 0.015%. A THD of 5.83% indicates that the current waveform resembles like petal-wave structure.

5.5 CONCLUSION

A suitable loss model for the PMBLDC motor was established to ensure control stability of the EV throughout the entire speed range under varying roadway conditions. A sensorless method employing a DO is implemented for the drive control with high accuracy, considering the iron loss component. An improved disturbance rejection control technique is proven for torque ripple-free performance of EVs with minimized losses. The influence of iron loss on the ideal current vector has been examined, and the adjusted optimal current reference vector guarantees reduced torque ripple and diminished losses. The effective driving behavior utilizing the proposed model is illustrated through simulation results. The results demonstrate the adequacy of the method for a 2.5-kW PMBLDCM proposed for EV application.

REFERENCES

1. D.-J. Kim, B. Kim, C. Yoon, N.-D. Nguyen, and Y. I. Lee, "Disturbance observer-based model predictive voltage control for electric-vehicle charging station in distribution networks," in *IEEE Transactions on Smart Grid*, vol. 14, no. 1, pp. 545–558, 2023.
2. J. E. Oncken and B. Chen, "Real-time model predictive powertrain control for a connected plug-in hybrid electric vehicle," in *IEEE Transactions on Vehicular Technology*, vol. 69, no. 8, pp. 8420–8432, 2020.
3. P. Rajan and S. Jeevananthan, "Toward energy sustainability: Domestic power station, tariff acquiescent EMS, and procedure to rejuvenate petrol scooter into electric scooter for accelerated participation of rural consumers in energy demand management," in *IEEE Journal of Emerging and Selected Topics in Power Electronics*, vol. 11, no. 4, pp. 4442–4452, 2023.
4. A. Emadi, Y. J. Lee, and K. Rajashekara, "Power electronics and motor drives in electric, hybrid electric, and plug-in hybrid electric vehicles," in *IEEE Transactions on Industrial Electronics*, vol. 55, no. 6, pp. 2237–2245, 2008.
5. Z. Q. Zhu, and D. Howe, "Electrical machines and drives for electric, hybrid, and fuel cell vehicles," in *Proceedings of the IEEE*, vol. 95, no. 4, pp. 746–765, 2007.
6. S. P. Nikam, V. Rallabandi, and B. G. Fernandes, "A high-torque-density permanent-magnet free motor for in-wheel electric vehicle application," in *IEEE Transactions on Industry Applications*, vol. 48, no. 6, pp. 2287–2295, 2012.
7. M. Falahi, H.-M. Chou, M. Ehsani, L. Xie, and K. L. Butler-Purry, "Potential power quality benefits of electric vehicles," in *IEEE Transactions on Sustainable Energy*, vol. 4, no. 4, pp. 1016–1023, 2013.
8. K. T. Chau, C. C. Chan, and C. Liu, "Overview of permanent-magnet brushless drives for electric and hybrid electric vehicles," in *IEEE Transactions on Industrial Electronics*, vol. 55, no. 6, pp. 2246–2257. 2008.
9. V. Viswanathan and J. Seenithangom, "Commutation torque ripple reduction in the BLDC motor using modified SEPIC and three-level NPC inverter," in *IEEE Transactions on Power Electronics*, vol. 33, no. 1, pp. 535–546. 2018.
10. P. Kumar, D. V. Bhaskar, U. R. Muduli, A. R. Beig, and R. K. Behera, "Iron-loss modeling with sensorless predictive control of PMBLDC motor drive for electric vehicle application," in *IEEE Transactions on Transportation Electrification*, vol. 7, no. 3, pp. 1506–1515, 2021.
11. P. Kumar, A. R. Beig, D. V. Bhaskar, K. A. Jaafari, U. R. Muduli, and R. K. Behera, "An enhanced linear active disturbance rejection controller for high performance PMBLDCM drive considering iron loss," in *IEEE Transactions on Power Electronics*, vol. 36, no. 12, pp. 14087–14097, 2021.
12. A. Khazaee, H. Abootorabi, G. Arab Markadeh, and H. Mosaddegh, "MTPA strategy for direct torque control of brushless DC motor drive," in *IEEE Transactions on Industrial Electronics*, vol. 68, no. 8, pp. 6692–6700, 2020. doi: 10.1109/TIE.2020.3009576.
13. K. Li and Y. Wang, "Maximum torque per ampere (MTPA) control for IPMSM drives based on a variable-equivalent-parameter MTPA control law," in *IEEE Transactions on Power Electronics*, vol. 34, no. 7, pp. 7092–7102, 2019.

14. A. Shinohara, Y. Inoue, S. Morimoto, and M. Sanada, "Direct calculation method of reference flux linkage for maximum torque per ampere control in DTC-based IPMSM drives," in *IEEE Transactions on Power Electronics*, vol. 32, no. 3, pp. 2114–2122, 2017.
15. G. Buja, M. Bertoluzzo, and R. K. Keshri, "Torque ripple-free operation of PM BLDC drives with petal-wave current supply," in *IEEE Transactions on Industrial Electronics*, vol. 62, no. 7, pp. 4034–4043, 2015.
16. K. Xia, Y. Ye, J. Ni, Y. Wang, and P. Xu, "Model predictive control method of torque ripple reduction for BLDC motor," in *IEEE Transactions on Magnetics*, vol. 56, no. 1, pp. 1–6, 2020.
17. A. Apte, V. A. Joshi, H. Mehta, and R. Walambe, "Disturbance-observer-based sensorless control of PMSM using integral state feedback controller," in *IEEE Transactions on Power Electronics*, vol. 35, no. 6, pp. 6082–6090, 2020.
18. Z. Xu, T. Zhang, Y. Bao, H. Zhang, and C. Gerada, "A nonlinear extended state observer for rotor position and speed estimation for sensorless IPMSM drives," in *IEEE Transactions on Power Electronics*, vol. 35, no. 1, pp. 733–743, 2020.
19. L. Qu, W. Qiao, and L. Qu, "An enhanced linear active disturbance rejection rotor position sensorless control for permanent magnet synchronous motors," in *IEEE Transactions on Power Electronics*, vol. 35, no. 6, pp. 6175–6184, 2020.
20. H. V. Nguyen, T. Vo-Duy, and M. C. Ta, "Comparative study of disturbance observer-based control and active disturbance rejection control in brushless DC motor drives," *2019 IEEE Vehicle Power and Propulsion Conference (VPPC)*, pp. 1–6, 2019.
21. P. Kumar, D. V. Bhaskar, U. R. Muduli, A. R. Beig, and R. K. Behera, "Disturbance observer based sensorless predictive control for high performance PMBLDCM drive considering iron loss," in *IEEE Transactions on Industrial Electronics*, vol. 69, no. 6, pp. 5442–5452, 2022.
22. N. Urasaki, T. Senjyu, and K. Uezato, "Investigation of influences of various losses on electromagnetic torque for surface-mounted permanent magnet synchronous motors," in *IEEE Transactions on Power Electronics*, vol. 18, no. 1, pp. 131–139, 2003.
23. Y. Wang, H. Yu, and L. Y. Liu, "Speed-current single-loop control with overcurrent protection for PMSM based on time-varying nonlinear disturbance observer," in *IEEE Transactions on Industrial Electronics*, vol. 69, no. 1, pp. 179–189, 2021. doi: 10.1109/TIE.2021.3051594.
24. L. Li, G. Pei, J. Liu, P. Du, L. Pei, and C. Zhong, "2-DOF robust $H\infty$ control for permanent magnet synchronous motor with disturbance observer," in *IEEE Transactions on Power Electronics*, vol. 36, no. 3, pp. 3462–3472, 2021.
25. S. Ahmad and A. Ali, "Unified disturbance estimation based control and equivalence with IMC, PID: Case study on a DC–DC boost converter," in *IEEE Transactions on Industrial Electronics*, vol. 68, no. 6, pp. 5122–5132, 2020. doi: 10.1109/TIE.2020.2987269.
26. Z. Gao, "Scaling and bandwidth-parameterization based controller tuning," *Proceedings of the 2003 American Control Conference*, IEEE, pp. 4989–4996, 2003.

Chapter 6

Mitigating power quality challenges

MqSZS grid-tied level-3 battery charger control with SVPWM for EV charging infrastructure

Anish Tiwari and Anandita Chowdhury

NOMENCLATURE

BESS	Battery energy storage systems
CPD	Custom power devices
DC	Direct current
EMI	Electromagnetic interference
EV	Electric vehicles
FACTS	Flexible alternating current transmission systems
FS-MPC	Finite set model predictive control
G2V	Grid-to-vehicle
M-qSZSI	Modified quasi-switched Z-source inverters
PLL	Phase lock loop
SAPF	Shunt active power filter
STATCOM	Static synchronous compensators
SVPWM	Space vector pulse width modulation
SZSI	Switched Z-source inverters
THD	Total harmonic distortion
V2G	Vehicle-to-grid
VSI	Voltage source inverters

6.1 INTRODUCTION

Modern power networks face several possibilities and problems because of the growing popularity of electric vehicles (EVs) and the grid's integration of renewable energy sources. Infrastructure for EV charging must be dependable and effective as the globe shifts to renewable energy sources. In addition, utilizing renewable energy sources like wind and solar is essential to cutting carbon emissions and improving energy sustainability. To decrease intermittency and assure grid stability, improved energy storage technologies are necessary for the integration of renewable energy sources. When renewable energy sources are scarce or there is a surge in demand, battery energy storage systems (BESS) are essential for storing excess energy from

 DOI: 10.1201/9781003515807-6

renewable sources and delivering it when required. Additionally, BESS offers auxiliary services to assist grid operations, improving the overall efficiency and dependability of the system [1–4]. Within the framework of EV charging infrastructure, BESS performs several functions. First, it makes it possible to integrate renewable energy sources into EV charging stations in an effective manner, which minimizes greenhouse gas emissions and lessens reliance on fossil fuels. Second, BESS is essential to grid support, especially when it comes to controlling power quality problems that arise during the charging process and reducing voltage swings brought on by EV charging loads. Finally, BESS makes it easier for vehicle-to-grid (V2G) activities to occur. This enables EVs and the grid to exchange power in both directions, enabling grid services like frequency management and demand response [5]. To address power quality issues and ensure grid stability, devices like static synchronous compensators (STATCOMs) with integrated energy storage offer enhanced control capabilities and flexibility in responding to dynamic grid conditions [3, 4]. The integration of BESS with flexible alternating current transmission systems (FACTS) devices has been proposed [2]. By providing reactive power support and voltage regulation, these integrated systems improve grid stability and overall power quality [3, 4]. However, challenges remain in optimizing the performance of EV charging infrastructure and ensuring seamless integration with the grid. Issues such as harmonic distortion, voltage fluctuations, and unbalanced grid conditions pose significant obstacles to the efficient operation of EV charging stations. Custom power devices (CPDs) are deployed to mitigate these challenges by maintaining harmonic levels within specified limits and improving power quality [6]. Furthermore, advanced control algorithms and optimization techniques are required to manage the complex interactions between EV charging loads, renewable energy sources, and grid operations [7].

In this context, power electronics and innovative converter topologies become crucial. Switched Z-source inverters (SZSIs) and their modified variants [8] offer unique advantages for EV charging infrastructure, including bidirectional power flow, improved voltage regulation, and enhanced grid support capabilities [9]. These inverters, coupled with advanced control strategies such as finite set model predictive control (FS-MPC), enable efficient and reliable operation of EV charging stations while ensuring grid stability and power quality [10]. BESS has several uses when it comes to EV charging infrastructure. First off, it makes it possible to integrate renewable energy sources into EV charging stations more effectively, which lowers reliance on fossil fuels and lowers greenhouse gas emissions. Second, BESS are essential to grid support. They are especially important in controlling power quality problems that arise during the charging process and reducing voltage variations brought on by EV charging loads. Finally, BESS enables bidirectional power transmission between EVs and the grid, facilitating V2G operations and enabling grid services like frequency management and demand response [11].

The integration of dynamic load devices in electrical distribution systems has intensified power quality concerns, notably increasing reactive power and harmonic distortion. To address these challenges, this chapter proposes an MPC strategy combined with space vector pulse width modulation (SVPWM) for a shunt active power filter (SAPF) based on a seven-level cascaded H-bridge converter [12]. This innovative approach leverages the adaptability of MPC and the computational efficiency of SVPWM to effectively mitigate harmful harmonic distortion. Compared to conventional control algorithms, this solution overcomes limitations such as high computational demands and variable switching frequencies. SVPWM is particularly significant in managing nonideal grid conditions and protecting EV battery performance, especially under rising power demand. By optimizing switching sequences in power converters, SVPWM reduces detrimental harmonic distortion, which is essential in environments where excessive harmonics may decrease operational efficiency and potentially damage electrical components, including EV batteries. Integrated with MPC, SVPWM enhances the control system's adaptability to fluctuating load conditions and grid quality variations. This ensures a stable power supply to EV batteries and mitigates risks such as over-voltage and voltage dips, which could otherwise compromise battery health. Furthermore, the improved system efficiency promoted by SVPWM supports faster and safer charging and extended battery lifespan, thus enabling reliable EV performance in increasingly complex and demanding power networks [13].

Pires et al. [14] emphasize the critical importance of fault-tolerant capabilities in power converters, particularly in applications where reliability is crucial due to risks to human safety, environmental stability, and economic security. They note that in recent years interest has surged in enhancing the reliability and fault tolerance of power converters and their control methods across various sectors, including industrial systems, EVs, and renewable energy. These applications often incorporate advanced technologies and functionalities that, while beneficial, add complexity and potential risks, underscoring the need for robust fault-tolerant strategies. Traditionally, three-phase voltage source inverters (VSIs) have been central to power conversion due to their efficiency, cost-effectiveness, fast dynamic response, and flexible control. To increase fault tolerance in VSIs, the authors report that researchers have introduced various modifications, such as additional redundant components, multi-parallel configurations, and control adjustments to enable continued operation during faults, though often with trade-offs like reduced power output and added system complexity. To address inherent limitations in traditional VSIs, including the need for additional gate drive circuitry and dead-time intervals, the authors propose a fault-tolerant approach based on a q-ZSI, designed for high-reliability applications requiring continuous operation under power device failures. By augmenting the classic q-ZSI topology with additional non-active components, their design maintains operation during open-circuit faults, minimizing output voltage loss and ensuring balanced three-phase AC currents with low total harmonic distortion (THD) through inverter reconfiguration and the q-ZSI's inherent boost capability.

This streamlined structure eliminates the need for additional passive components or dead-time circuits, further enhancing reliability.

This chapter is organized to provide a thorough overview of the issues and solutions related to the integration of clean energy sources into EV charging infrastructure. A detailed explanation of the IEEE 13-bus test feeder system is given in Section 6.2 along with a discussion of how load profiles affect bus voltages at connected nodes. This helps to identify the issue of an unbalanced and distorted grid, which can cause power oscillations and negative sequence currents. Section 6.2.1 delves into the mathematical background behind active power oscillations, deriving expressions for reference currents and providing a theoretical foundation for understanding how these oscillations occur and how they can be mitigated. Section 6.2.2 describes the control algorithms used in the system, explaining their complexity, and suggesting the FS-MPC theory as an effective solution, highlighting the advantages of FS-MPC in managing power quality and ensuring stable grid operations. Section 6.2.3 introduces the SVPWM theory and discusses the section selection process, elaborating on the practical implementation of SVPWM in enhancing the performance of inverter systems used in the EV charging infrastructure. Section 6.3 presents the results and analysis of the implemented system, including a detailed discussion of performance outcomes that showcase the effectiveness of the proposed solutions in real-world scenarios. The final section 6.4 provides concluding remarks, summarizing the key findings and implications of the study, underscoring the significance of integrating advanced control strategies and power electronics in developing sustainable and resilient EV charging systems. Following the main content, the chapter includes an Acknowledgements section, recognizing contributions and support from various individuals and institutions, and a comprehensive list of references detailing the sources and literature cited throughout the chapter.

6.2 THEORETICAL BACKGROUND

The EV charging cycle of a single unit, as illustrated in Figure 6.1, reveals that a peak-to-peak current of 16 A drawn from the feeder node induces the injection of multiple order harmonics. These harmonics significantly degrade power quality at adjacent buses. Additionally, the presence of nonlinear loads exacerbates these effects, leading to further system distortion. During peak hours, the combined impact of these loading conditions results in a range of power quality issues, including voltage sags, swells, fluctuations, and power oscillations on the direct current (DC) side. A key concern is that the connected converter lacks control over the feeder node voltage, limiting its ability to mitigate these disturbances effectively. The converter can either ensure the injected current remains sinusoidal or dampen power oscillations reflected in the DC bus, thereby maintaining a constant DC link. This capability is crucial for enhancing efficiency during unbalanced grid operations. Figure 6.2 illustrates the typical harmonic amplitude and phase angle characteristics of a nonlinear load

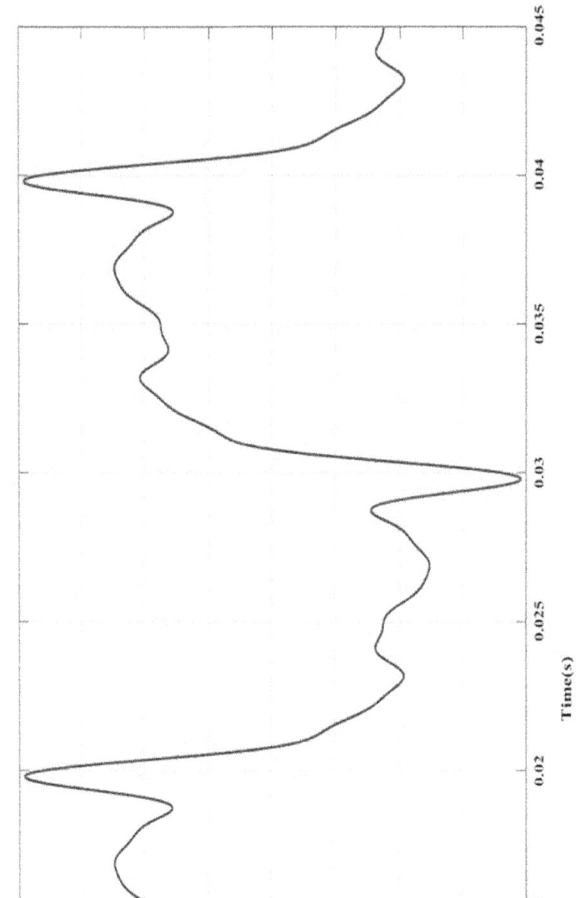

Figure 6.1 Charging and discharging profile for a typical EV battery charger [15].

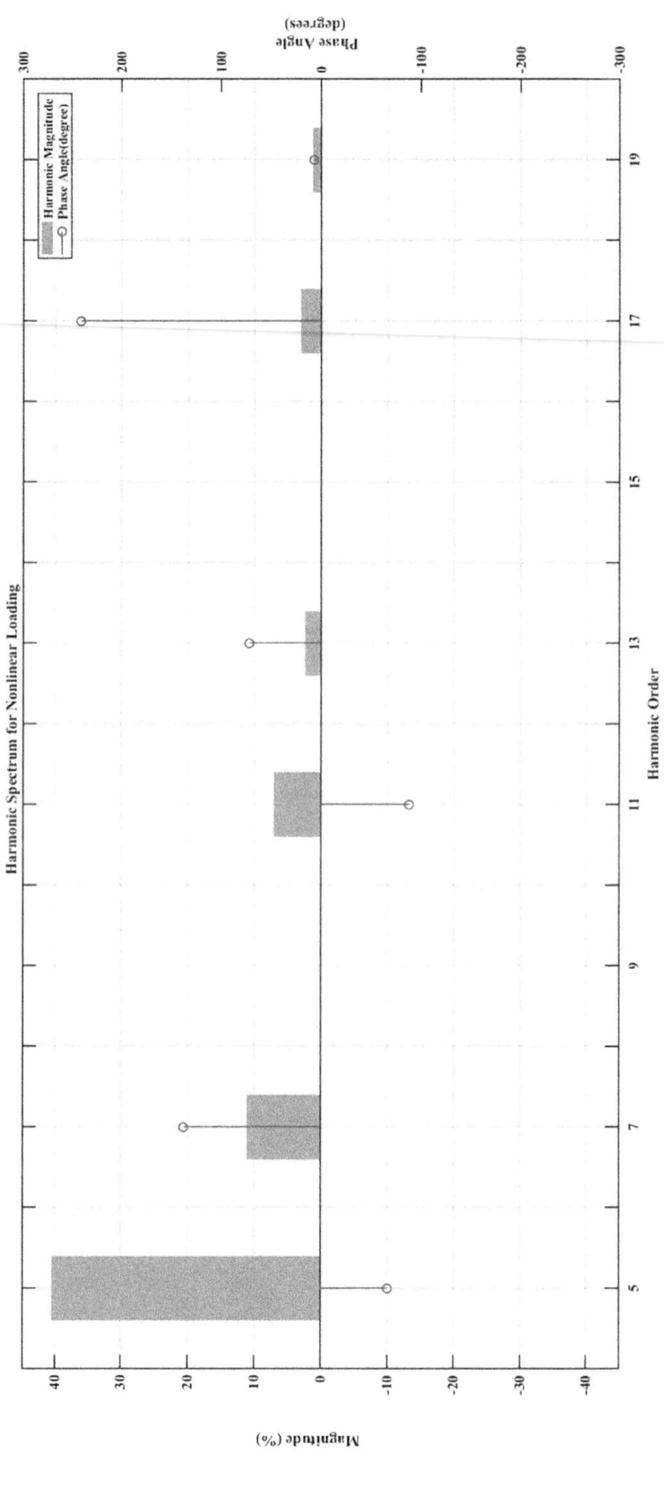

Figure 6.2 Amplitude and phase harmonic spectrum for typical nonlinear load.

profile. This profile is used to model harmonic loading for harmonic load flow analysis, which examines the impact of harmonic loads on neighboring buses.

The impact of EV charging stations on the IEEE 13 Bus Test Feeder is analyzed, with a focus on harmonic distortion and voltage regulation in adherence to the IEEE 519™-2014 standard. It presents the results of this analysis, highlighting harmonic magnitudes and $THD_{(v)}$ across different scenarios. In the baseline scenario without EV charging stations, harmonic distortion levels and voltage magnitudes remain within acceptable limits. However, the introduction of a single EV charger at node 646 results in significant deviations. Specifically, the overall THD and many harmonic orders exceed permissible limits, with a voltage magnitude of 1.378 per unit indicating an overvoltage condition. When EV chargers are connected at both nodes 611 and 646, compounded effects become evident: Node 611 experiences excessive harmonics with a voltage magnitude of 0.9326 undervoltage per unit, while node 646 maintains similar overvoltage conditions. In the scenario with EV chargers at nodes 611, 646, and 633, almost all nodes exhibit THD limit violations. Node 611 remains undervoltage at 0.9326 per unit, while nodes 646 and 633 show voltage magnitudes of 1.378 and 1.571 per unit, respectively. These substantial harmonic distortions and voltage irregularities across multiple scenarios underscore the need for effective management of EV charging integration. The analysis highlights specific nodes vulnerable to these issues, guiding the design and implementation of mitigation strategies to ensure stability, reliability, and compliance with power quality standards in distribution networks with high EV charger penetration. Figure 6.3 illustrates these findings graphically, providing a clear visual representation of the harmonic magnitudes and $THD_{(v)}$ for different nodes and EV charging scenarios [16].

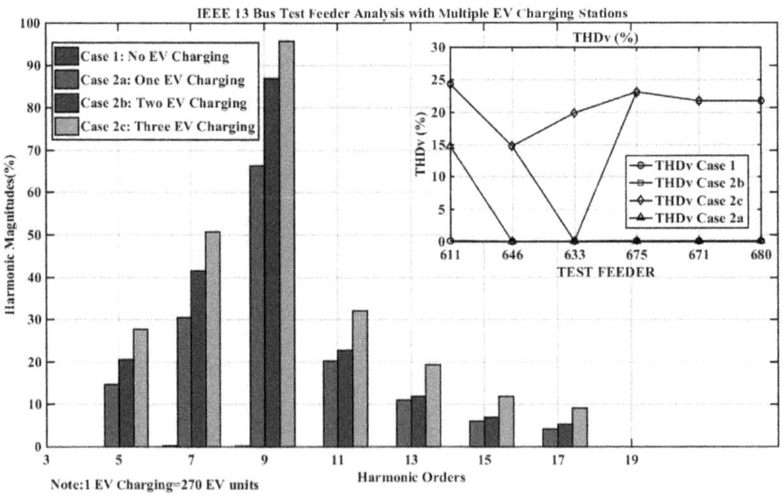

Figure 6.3 IEEE-13 Bus Test Feeder with EV charging stationed at different nodes for power quality.

One of the articles [17], focus on adoption of DC grid in residential buildings so ensure minimal impact on the nodes by shifting the charging load to local generating solar panels.

The data shown is crucial for understanding the extent to which EV charging stations can impact power quality and the necessary measures to mitigate these effects in compliance with IEEE 519™-2014 standards.

6.2.1 Power oscillation theory

The current harmonic content of the modified quasi-switched Z-source inverters (M-qSZSI) architecture increases significantly when there are imbalanced grid voltages. Furthermore, both the active and reactive power exhibit a double line-frequency ripple, which may have a negative effect on the power quality [18, 19]. The sum of the positive and negative sequence components in the synchronous reference frame may be used to represent the unbalanced grid voltages and currents:

$$e_{dq}^{+} = e_d^{+} + je_q^{+} \tag{6.1}$$

$$e_{dq}^{-} = e_d^{-} + je_q^{-} \tag{6.2}$$

$$i_{dq}^{+} = i_d^{+} + ji_q^{+} \tag{6.3}$$

$$i_{dq}^{-} = i_d^{-} + ji_q^{-} \tag{6.4}$$

Next, the grid-side power may be stated as follows:

$$S = P + jQ = \left(e_{dq}^{+}.e^{jwt} + e_{dq}^{-}.e^{-jwt}\right) * \left(i_{dq}^{+}.e^{jwt} + i_{dq}^{-}.e^{-jwt}\right) \tag{6.5}$$

By extending Equation (6.5), we may represent the active and reactive powers in Equations (6.6) and (6.7) as follows:

$$P = p_0 + p_{c2}\cos(2wt) + p_{s2}\sin(2wt) \tag{6.6}$$

$$Q = q_0 + q_{c2}\cos(2wt) + q_{s2}\sin(2wt) \tag{6.7}$$

$$\begin{Bmatrix} p_0 \\ p_{c2} \\ p_{s2} \\ q_0 \\ q_{c2} \\ q_{s2} \end{Bmatrix} = \frac{3}{2} \begin{Bmatrix} e_d^{+} & e_q^{+} & e_d^{-} & e_q^{-} \\ e_d^{-} & e_q^{-} & e_d^{+} & e_q^{+} \\ e_q^{-} & -e_d^{-} & -e_q^{+} & e_d^{+} \\ e_q^{+} & -e_d^{+} & e_q^{-} & -e_d^{-} \\ e_q^{-} & -e_d^{-} & e_q^{+} & -e_d^{+} \\ -e_d^{-} & -e_q^{-} & e_d^{+} & e_q^{+} \end{Bmatrix} * \begin{Bmatrix} i_d^{+} \\ i_q^{+} \\ i_d^{-} \\ i_q^{-} \end{Bmatrix} \tag{6.8}$$

The following conditions can be used (Equation (6.8)) to reduce oscillations in both reactive and active powers (Equations (6.9)–(6.12)):

$$p_0 = P_{ref} \tag{6.9}$$

$$p_{c2} = 0 \tag{6.10}$$

$$p_{s2} = 0 \tag{6.11}$$

$$q_0 = Q_{ref} \tag{6.12}$$

This allows the reference current to be calculated as follows (Equations (6.13) and (6.14)):

$$i_{dref}^{*} = \frac{v_q^{\perp}}{v_d * v_q^{\perp} - v_d^{\perp} * v_q} * P_{ref} + \frac{2v_q}{v_d^2 + v_q^2 + \left(v_d^{\perp}\right)^2 + \left(v_q^{\perp}\right)^2} * Q_{ref} \tag{6.13}$$

$$i_{qref}^{*} = \frac{-v_d^{\perp}}{v_d * v_q^{\perp} - v_d^{\perp} * v_q} * P_{ref} - \frac{2v_d}{v_d^2 + v_q^2 + \left(v_d^{\perp}\right)^2 + \left(v_q^{\perp}\right)^2} * Q_{ref} \tag{6.14}$$

6.2.2 Finite set model predictive control algorithm

The overall control configuration of the MSQZS level 3 off-board charger is presented here. These control strategies are designed to efficiently manage voltage imbalanced operating conditions, employing PI controller and a finite-set predictive algorithm. To determine the grid angle, grid voltages and currents are measured and processed through phase lock loop (PLL). This enables the transformation of three-phase quantities into two-phase quantities using Park's transformation. Errors can then be managed either through a PI controller (Figure 6.4) or by generating an optimal vector using FS-MPC control (Figure 6.5), which is subsequently transformed back into three-phase quantities for SVPWM gate pulse generation. This method is advantageous as it eliminates the need for PI controllers or modulators. As indicated in Ref. [10], the finite set MPC technique significantly reduces power ripple in single-phase three-leg quasi-converters under imbalanced conditions, offering a distinct advantage over conventional PI control. Additionally, by incorporating two variables in the cost function, this approach achieves an extended prediction horizon, enhancing harmonic distortion performance while reducing computational load by approximately 50%.

The finite set MPC technique lowers current distortion, improving stability and performance. The working principle of this method is shown in Figure 6.6. As a general system variable, the current is sampled during a period of T_s, where k is the number of control actions, indicated by S_n, for which n ranges from 1 to m. The current measured value $i_{p(t(k))}$ and the control

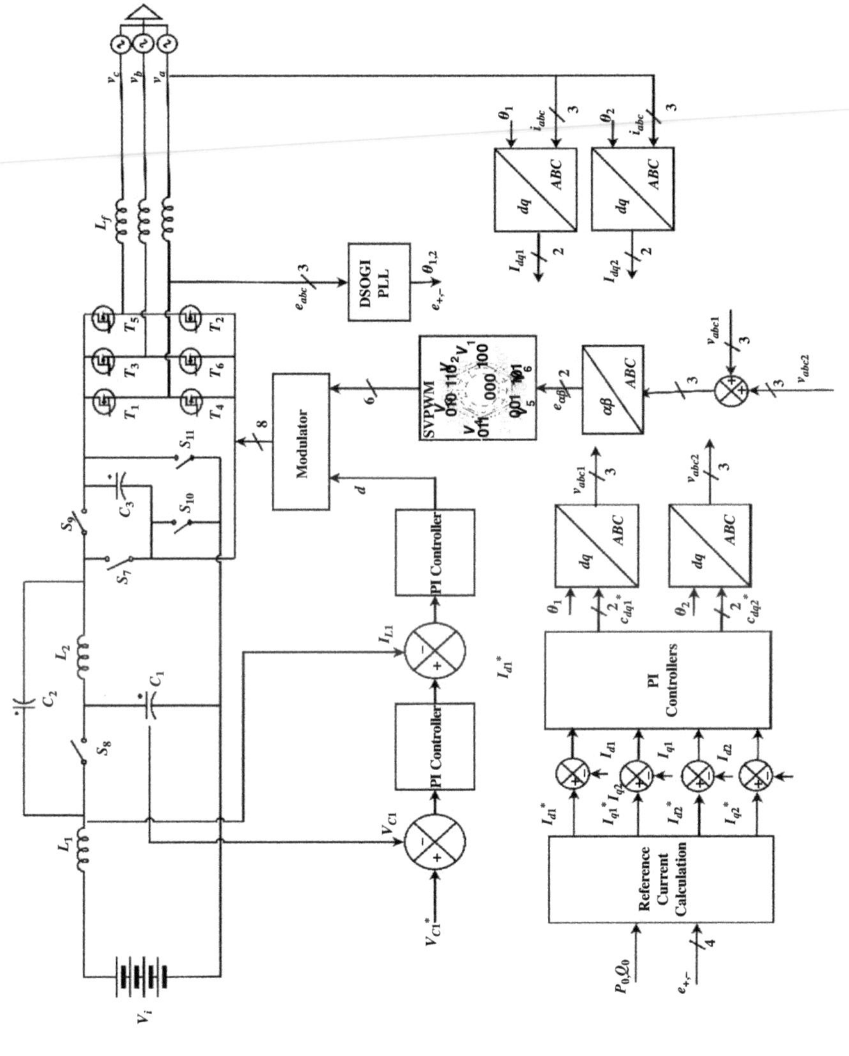

Figure 6.4 Control algorithm of PI control-based SVPWM for active power oscillation damping.

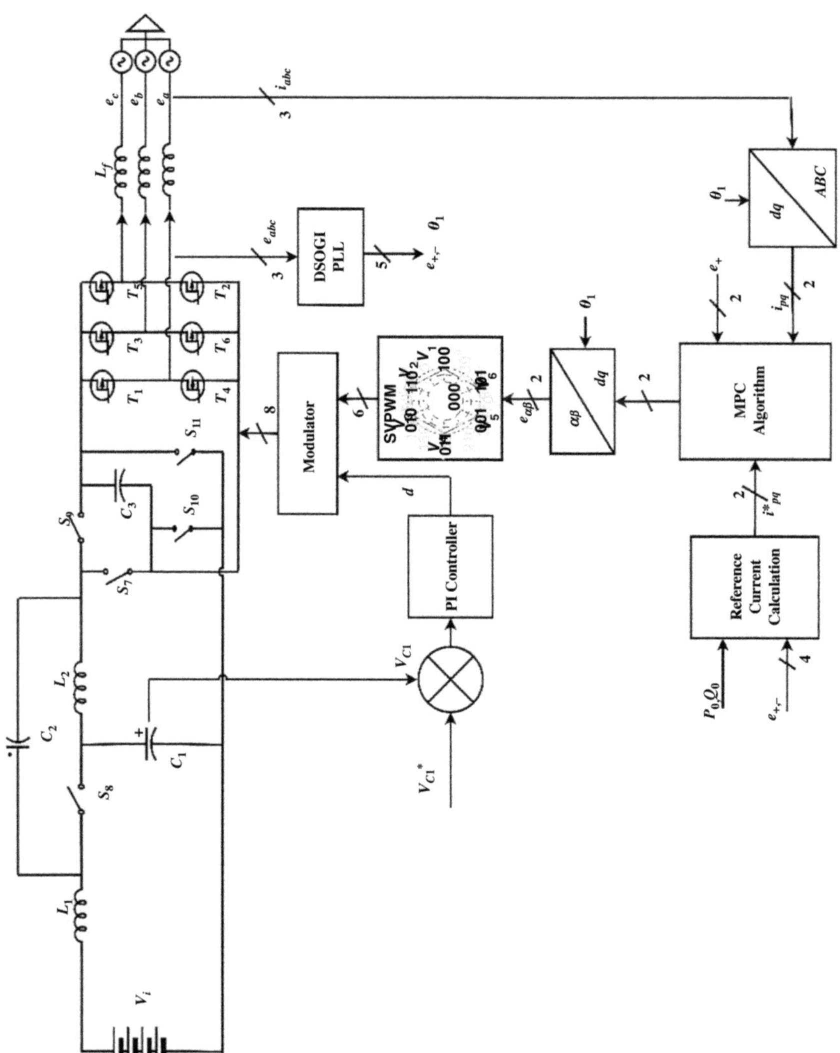

Figure 6.5 Control algorithm of finite set MPC-based SVPWM for active power oscillation damping.

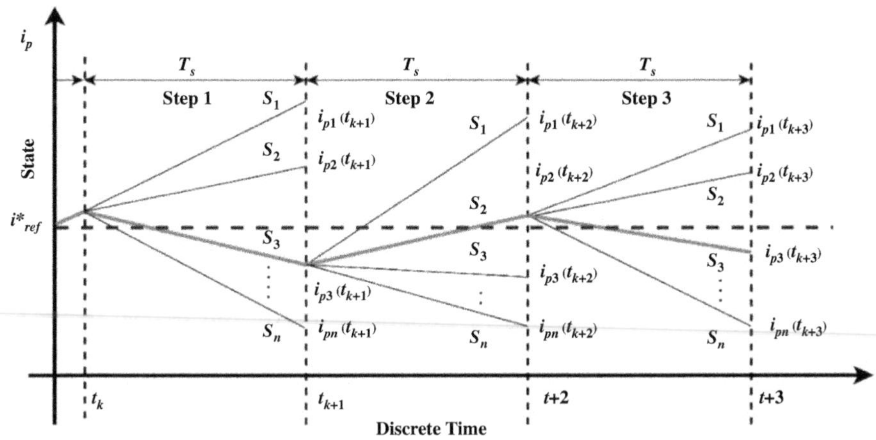

Figure 6.6 Operating principle for finite set model predictive control.

actions determine the predicted current function $i_{pn(t(k+1))}$. By reducing the absolute error between the predictive function and the reference, the algorithm chooses the function that is closest to the reference value. As can be seen in Equations (6.15) and (6.16), the first predicted values of the grid current vectors are produced:

$$i_{d(m+1)} = \left(1 - \frac{R_f \cdot T_s}{L_f}\right) i_{d(m)} + \frac{T_s}{L_f} \cdot \left[V_{inv,d(m)} - v_{d(m)}\right] \tag{6.15}$$

$$i_{q(m+1)} = \left(1 - \frac{R_f \cdot T_s}{L_f}\right) i_{q(m)} + \frac{T_s}{L_f} \cdot \left[V_{inv,q(m)} - v_{q(m)}\right] \tag{6.16}$$

The quadratic cost function, Equation (6.17), is defined as follows:

$$C = C_d + C_q = \left|i^*_{d(m+1)} - i_{d(m+1)}\right|^2 + \left|i^*_{q(m+1)} - i_{q(m+1)}\right|^2 \tag{6.17}$$

As seen in Equations (6.18) and (6.19), minimizing the cost function entails zeroing out its partial derivatives:

$$\frac{\partial C}{\partial V_{inv,d}} = 0 \tag{6.18}$$

$$\frac{\partial C}{\partial V_{inv,q}} = 0 \tag{6.19}$$

The minimum grid voltage vector in the DQ-domain can be derived as follows:

$$V_{d,min} = -\frac{L_f}{T_s} * \left[\left(1 - \frac{R_f \cdot T_s}{L_f} \right) i_{d(m)} + \frac{T_s}{L_f} \cdot v_{d(m)} - i^*{}_{d(m+1)} \right] \tag{6.20}$$

$$V_{q,min} = -\frac{L_f}{T_s} * \left[\left(1 - \frac{R_f \cdot T_s}{L_f} \right) i_{q(m)} + \frac{T_s}{L_f} \cdot v_{q(m)} - i^*{}_{q(m+1)} \right] \tag{6.21}$$

The voltage vectors described in Equations (6.20) and (6.21) are transformed using Park's transformation to generate three-phase reference voltages for the SVPWM algorithm. These reference voltages are then compared to the switching signals, producing the appropriate switching pulses for the MSQZSI configuration. The complete flow of the proposed predictive control algorithm is outlined in Figure 6.7.

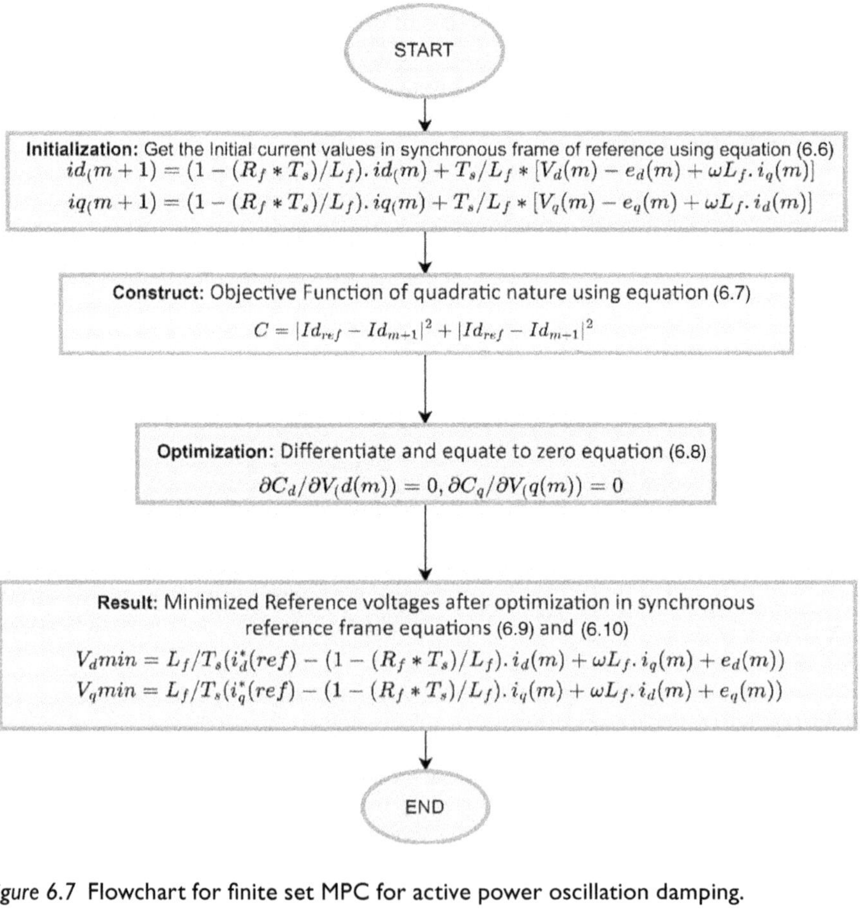

Figure 6.7 Flowchart for finite set MPC for active power oscillation damping.

6.2.3 Space vector modulation

Combining the three closest space vectors to create the reference voltage vector V_{ref} is the aim of SVPWM. These vectors are V_0 (000), V_1 (100), and V_2 (110) in Sector I. All three switches are off in V_0, so there is no line-to-line voltage. Phase A's top switch is on in V_1, and phases B and C's bottom switches are also on, creating a voltage vector that points in the direction of phase A. In V_2, a voltage vector is created between phases A and B by turning on the top switches for phases A and B and the bottom switch for phase C. Figure 6.8a shows typical switching sequence in Sector I: $V_0 \rightarrow V_1 \rightarrow V_2 \rightarrow V_7 \rightarrow V_7 \rightarrow V_2 \rightarrow V_1 \rightarrow V_0$. The zero vector V_0 is applied initially for $T_0/2$

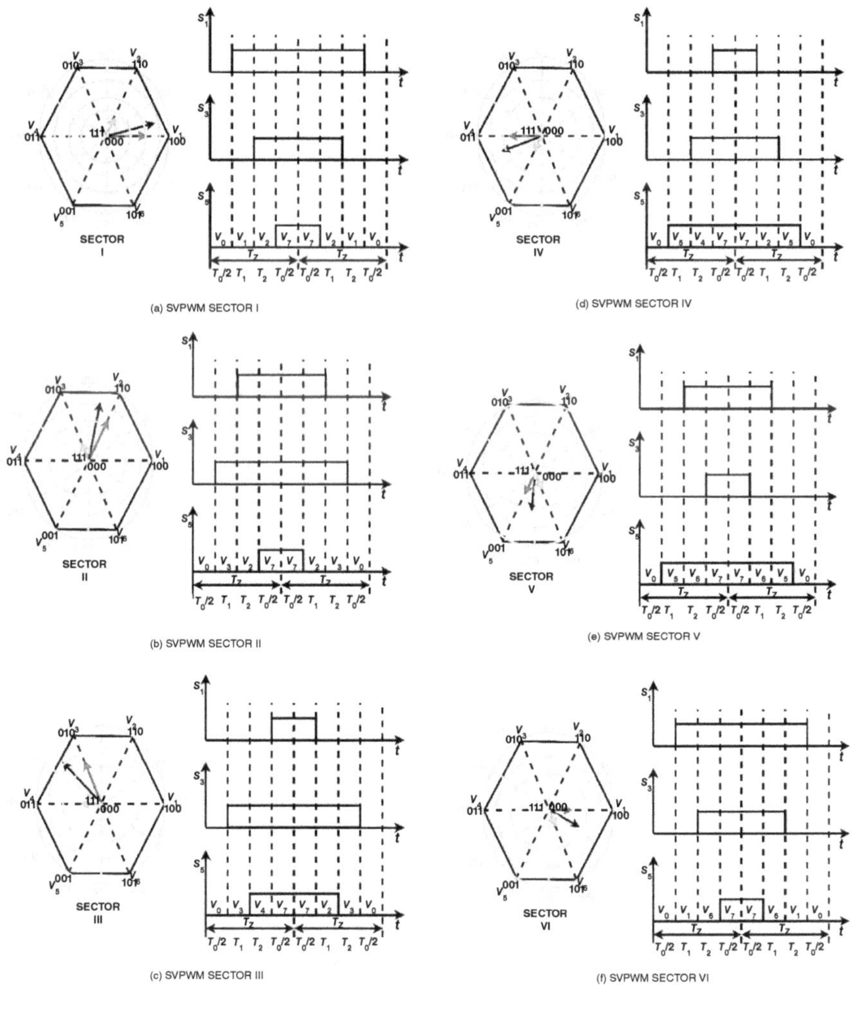

Figure 6.8 SVPWM timing sequence diagram.

duration to balance the switching cycle. Active vector V_1 is applied for T_1 duration, aligning the synthesized vector closer to V_{ref}. Active vector V_2 is applied for T_2 duration, further adjusting the vector towards V_{ref}. The zero vector V_7, where all top switches are on, is applied for $T_0/2$ to symmetrically balance the cycle. The active vectors V_2 and V_1 are reapplied in reverse order to maintain symmetry and minimize switching losses. Finally, zero vector V_0 is reapplied to complete the switching period.

In Figure 6.8b, to equalize the switching cycle, V_0 (all switch off) is applied for $T_0/2$ initially. Next, V_3 is applied to T_1, creating a voltage vector going in the direction of phase B with the top switch for phase B and the bottom switches for phases A and C turned on. Next, V_2 is applied for T_2, creating the voltage vector between phases A and B, with the top switches of phases A and B and the bottom switch of phase C turned on. To symmetrically balance the cycle, V_7 (with all top switches on) is applied for $T_0/2$ at the end. Sector II switching orders are typically as follows: $V_0 \rightarrow V_3 \rightarrow V_2 \rightarrow V_7 \rightarrow V_7 \rightarrow V_2 \rightarrow V_3 \rightarrow V_0$. The initial V_0 is applied for $T_0/2$ to start with no voltage, followed by V_3 for T_1, aligning the synthesized vector closer to V_{ref}. Then, V_2 is applied for T_2, further adjusting the vector toward V_{ref}. V_7 is applied for $T_0/2$ to symmetrically balance the cycle and is reapplied for $T_0/2$ to continue the balancing. V_2 and V_3 are then reapplied in reverse order to maintain symmetry and minimize switching losses. Finally, V_0 is applied again for $T_0/2$ to complete the switching period with no voltage. This sequence ensures V_{ref} is synthesized accurately while minimizing harmonic distortion and optimizing switching losses. The voltage reference vector the V_{ref} of Sector III for SVPWM algorithm is positioned between 120° and 180° as shown in Figure 6.8c. This encompasses the active voltage vectors V_3 and V_4, with V_7 as the adjacent vector. The timing sequence begins with V_0, followed by V_3, then V_4, and finally V_7. This sequence then reverses, starting with V_7 and proceeding through V_4, V_3, and back to V_0. This method of sequencing the voltage vectors ensures precise control over the switching states of the inverter. This, in turn, enables efficient and low-distortion generation of output voltages suitable for AC motor control and other power electronic applications.

The objective shown in Figure 6.8d is to apply the closest space vectors, V_5 (001), V_4 (011), V_0 (000), and V_7 (111), to synthesize the reference voltage vector V_{ref}. To balance the cycle, V_0, which starts the switching sequence with all switches off, is applied for $T_0/2$. Subsequently, V_5 is applied for T_1, creating a voltage vector going in the direction of phase C, with the top switch of phase C and the bottom switch of phases A and B both on. Next, V_4 is applied for T_2, forming a voltage vector between phases B and C with the top switches of phases B and C and the bottom switch of phase A turned on. In order to symmetrically balance $T_0/2$, V_7 (with all top switches on) is applied at the end. This typical Sector IV switching sequence: $V_0 \rightarrow V_5 \rightarrow V_4 \rightarrow V_7 \rightarrow V_7 \rightarrow V_4 \rightarrow V_5 \rightarrow V_0$ - ensures the synthesized voltage vector accurately tracks the reference V_{ref} while minimizing harmonic distortion and switching losses. The precise, balanced application of these space vectors enables efficient, reliable operation of grid-tied converters, essential for applications like EV charging and renewable energy integration.

Sector V seeks to synthesize the reference voltage vector V_{ref} using the nearest space vectors (Figure 6.8e) V_5 (001), V_6 (101), V_0 (000), and V_7 (111). With all switches off, V_0 is applied for $T_0/2$ initially to balance the switching cycle. Subsequently, V_5 is applied to T_1 to produce a voltage vector that moves toward phase C. To do this, turn on the bottom switches for phases A and B and the top switch for phase C. Next, V_6 is applied for T_2, turning on the top switches of phases A and C as well as the bottom switch of phase B, to create a voltage vector between phases A and C. When all top switches are activated, V_7 is applied for $T_0/2$ to complete the cycle's symmetric equilibrium. Sector V often switches in the following order: $V_0 \rightarrow V_5 \rightarrow V_6 \rightarrow V_7 \rightarrow V_7 \rightarrow V_6 \rightarrow V_5 \rightarrow V_0$ to align the synthesized vector closer to V_{ref}; this sequence starts with V_0 for $T_0/2$, which begins with no voltage. Next, V_5 for T_1 was applied. Then, V_6 is applied for T_2 to tune the vector toward V_{ref} further. To symmetrically balance the cycle, V_7 is applied for $T_0/2$ and then reapplied for $T_0/2$ to keep the balancing going. Then, to preserve symmetry and reduce switching losses, V_6 and V_5 are applied again in the opposite sequence. To finish the switching phase without any voltage, V_0 is finally applied again for $T_0/2$.

Figure 6.8f for Sector VI is strategically designed to optimize inverter control by precisely positioning the reference voltage vector V_{ref}. This 60° sector spans from 300° to 360°, utilizing the active voltage vectors V_6 and V_7, with V_1 serving as the adjacent vector. The timing sequence begins with V_0, then transitions to V_1, V_6, and V_7, before reversing the order back to V_7, V_6, V_1, and V_0. This cyclic sequence ensures seamless transitions between the active vectors, generating the desired output voltage waveform with exacting timing control. This meticulous approach is crucial for high-efficiency, low-harmonic-distortion applications like AC motor drives and other power electronic systems. The SVPWM approach achieves higher inverter performance by deliberately selecting and coordinating various voltage vectors within six sectors. This system can create the required output voltage with remarkable accuracy, reducing distortion and optimizing energy efficiency by precisely aligning the V_{ref}. The waveform quality is further improved by the symmetrical, back-and-forth sequencing of the active and zero vectors, which smooths out any possible discontinuities. SVPWM is an effective tool for advanced motor control and other high-performance power conversion applications because it allows for comprehensive, real-time control over the inverter switching patterns. For many three-phase systems, sectors with a 60° span provide the perfect working range, enabling the inverter to dynamically modify V_{ref}, while maintaining excellent control throughout. To evaluate the performance of the control algorithms, MPC-SVPWM and PI-SVPWM, the simulation employed the design parameters for q-MSZS converter system shown in Table 6.1.

Table 6.1 Design parameters for MSZS topology used in the simulation

S. No.	*Parameters*	*Values*
1	Inductors L_1 and L_2	2.5 mH, 10 A
2	Capacitors C_1, C_2, and C_3	470 uF, 400 V
3	Filter inductors	8 mH, 10 A

6.3 RESULTS AND DISCUSSION

The simulation results in Figure 6.9 show the single-phase unbalance of the PI-SVPWM control algorithm implemented on a grid-tied level 3 EV charger connected to the MSZS grid. As expected, the voltage imbalance caused by disturbances at adjacent nodes occurs at 2 seconds, resulting in visible active and reactive power oscillations in the waveform. Despite using four PI control loops on the AC side, the control is not precise and effective, and the overall system complexity has increased significantly.

As shown in Figure 6.10, the PI SVPWM control algorithm injects a very high level of current harmonics, which violates the standard limits for

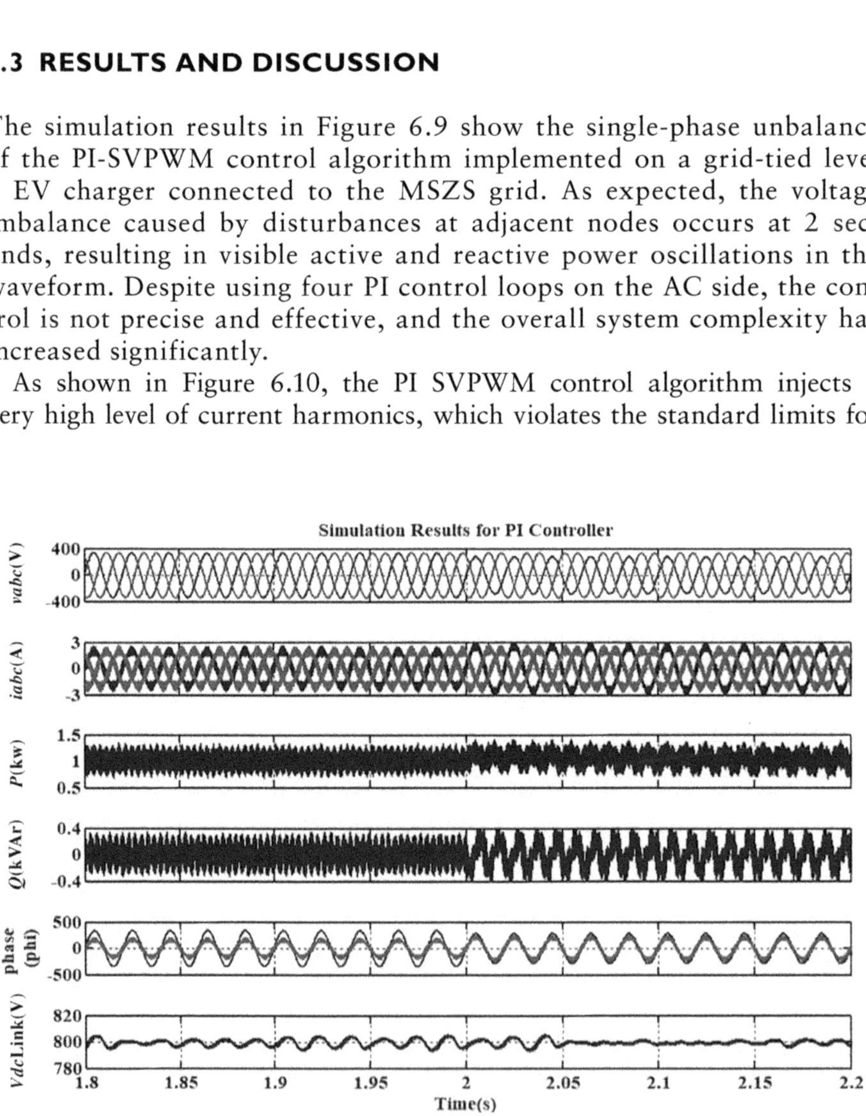

Figure 6.9 Simulation results for PI SVPWM control-based algorithm during single-phase unbalance (20% A sag) grid operation.

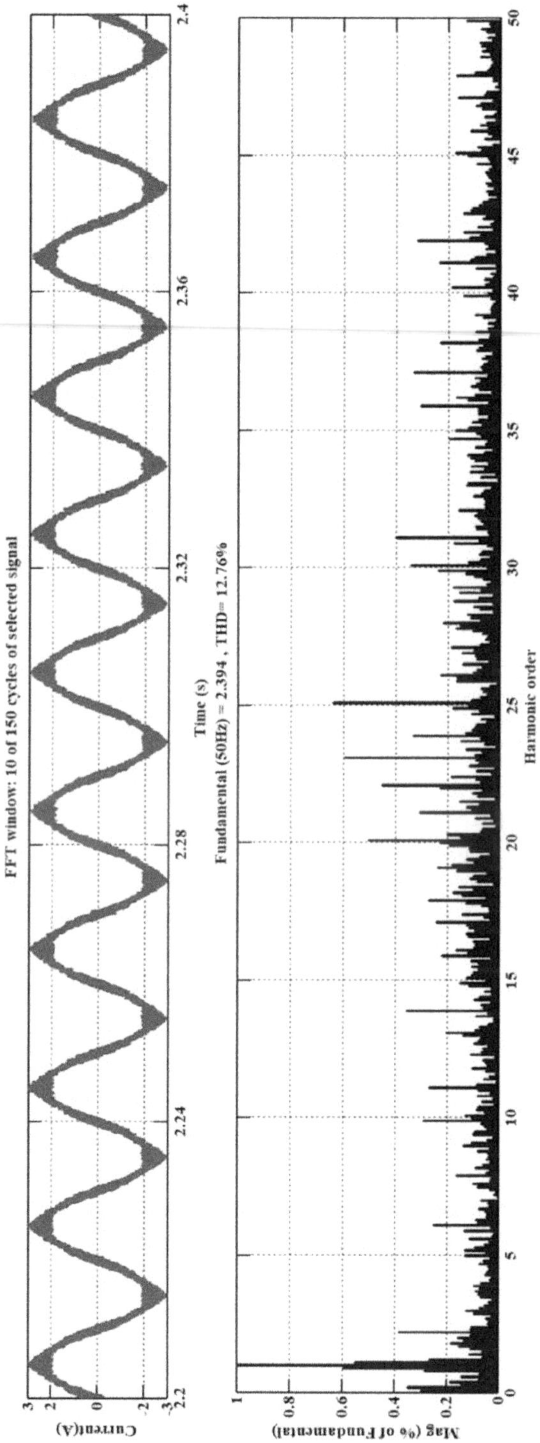

Figure 6.10 FFT analysis of current waveform for PI control-based algorithm during single-phase unbalance (20% A sag).

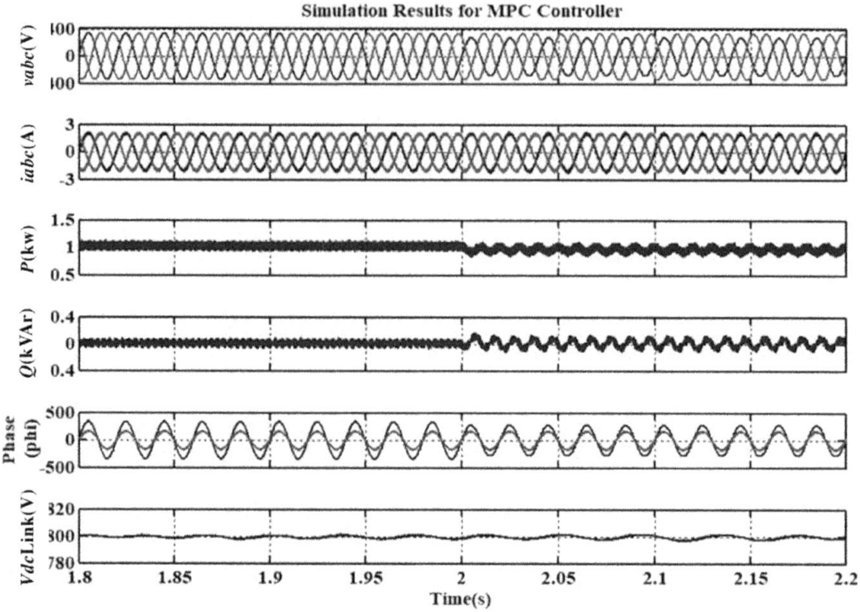

Figure 6.11 Simulation results for FS-MPC SVPWM-based algorithm during single-phase unbalance (20% A sag) grid operations.

harmonic content in grid currents. This could potentially lead to serious power quality issues if left unaddressed.

The FSMPC SVPWM technique greatly reduces oscillations in the active and reactive phases of the MSZS charger, as shown in Figure 6.11.

The grid current harmonic content shown in Figure 6.12 falls within the IEEE 519™-2014 standards for bus voltage harmonic limits in the AC microgrid distribution system. Additionally, the charger exhibits appropriate behavior even under unbalanced operating conditions when using the FS-MPC SVPWM algorithm.

Multiple phase imbalance circumstances can arise from the placement of EV charging stations in proximity, as seen in Figure 6.13. As the PI controller is harder to tune, this added complexity makes the PI SVPWM control even more challenging. Because of the impact of active power oscillations on the DC bus, the findings show that the PI SVPWM control technique causes more disruption to the DC link than the FCMPC SVPWM control method.

According to the FFT analysis, the harmonic content of the grid currents shown in Figure 6.14 was 12.22%, which is quite high and appears to violate the IEEE 519™-2014 standard. Furthermore, the effect of this unbalance is likely to negatively impact the long-term health of the battery when charged in such environments.

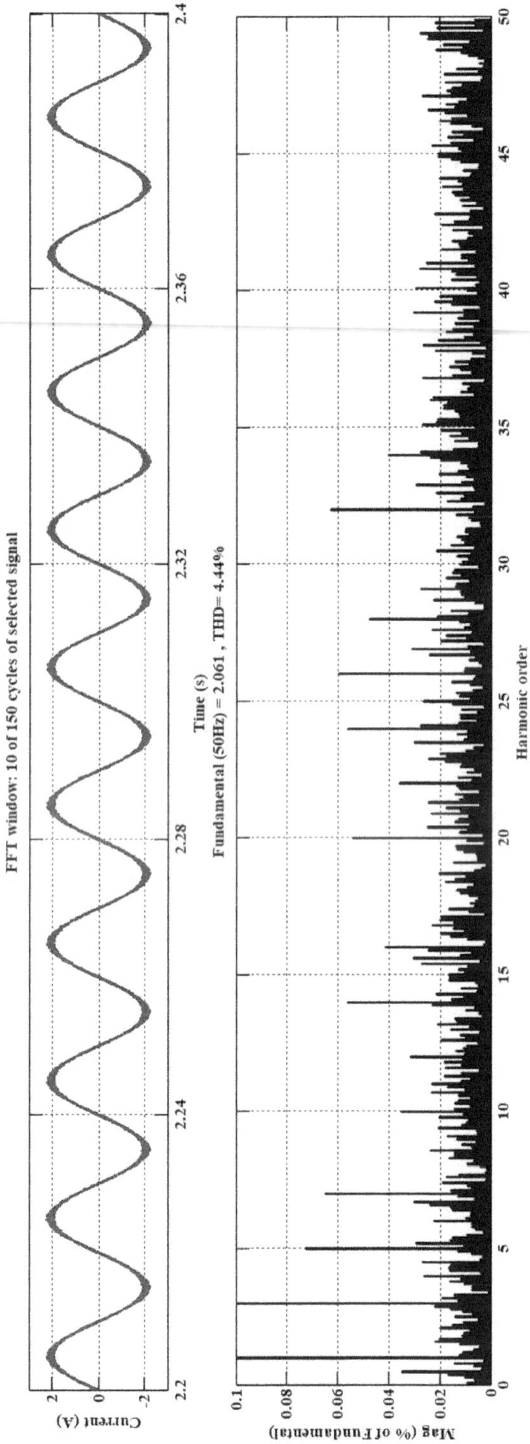

Figure 6.12 FFT analysis of current waveform for finite set MPC control-based algorithm during single-phase unbalance (20% A sag).

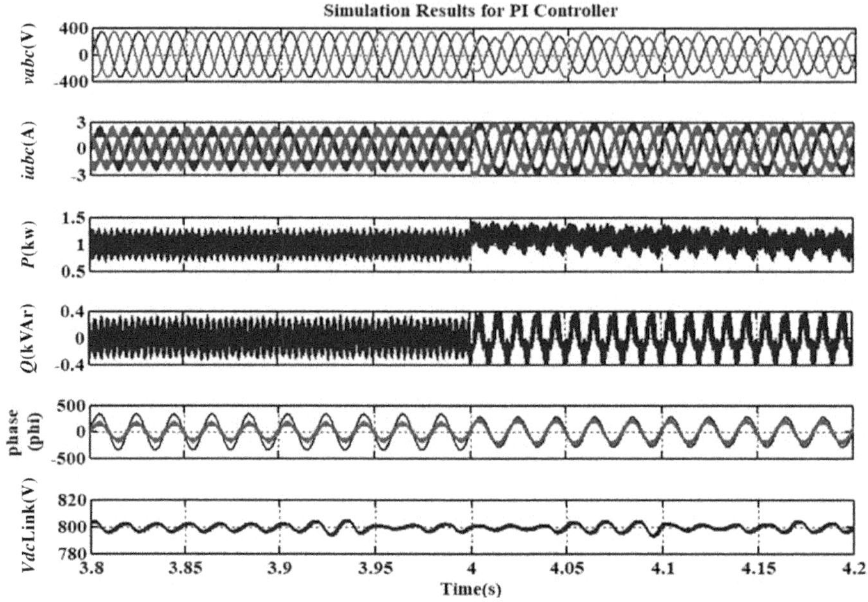

Figure 6.13 Simulation results for PI SVPWM control-based algorithm during multiple phase unbalance (20% A and 30% B sag) grid operations.

The FS-MPC SVPWM algorithm shows much better results in terms of harmonic limit violations compared to previous, as shown in Figure 6.15 When applied to the MSZS level 3 charger, the FSMPC SVPWM algorithm shows notable benefits over the more intricate and challenging-to-tune PI SVPWM control algorithm in terms of DC link voltage changes and active/reactive power oscillations. Furthermore, the FSMPC SVPWM approach shows much better performance in terms of harmonic limit violations than the previous method, as illustrated in Figure 6.16.

The results show that the grid current distortion level is reduced from 12.76% to 4.44% with single-phase unbalance. The higher THD values represent the PI SVPWM control, while the lower values represent the FS-MPC SVPWM algorithm. A similar outcome is seen with multiple-phase unbalance—the grid current distortion level is reduced from 12.22% with PI SVPWM control to 4.3% with FSMPC SVPWM control. Table 6.2 provides a comprehensive overview of the improvements in DC link variations and distortion levels.

6.4 CONCLUSION

The results depicted in Figures 6.12 and 6.16 demonstrate the significant improvements in grid current distortion levels achievable through advanced control algorithms. Specifically, the data in Table 6.1 shows that grid current

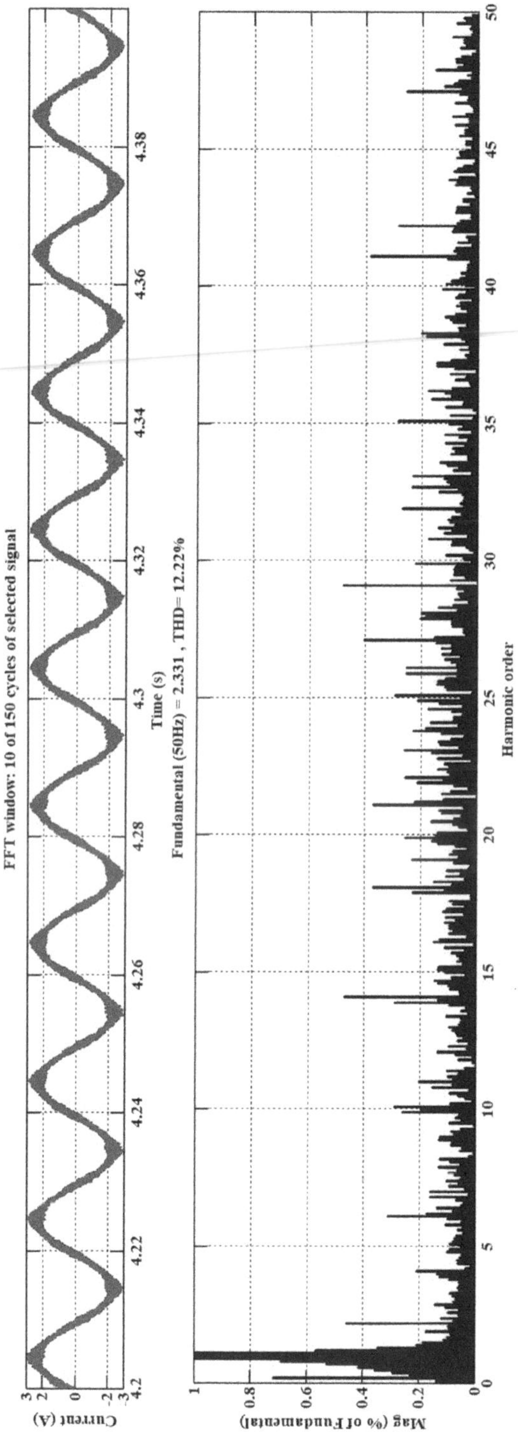

Figure 6.14 FFT analysis of current waveform for PI SVPWM control-based algorithm during multiple phase unbalance (20% A and 30% B sag).

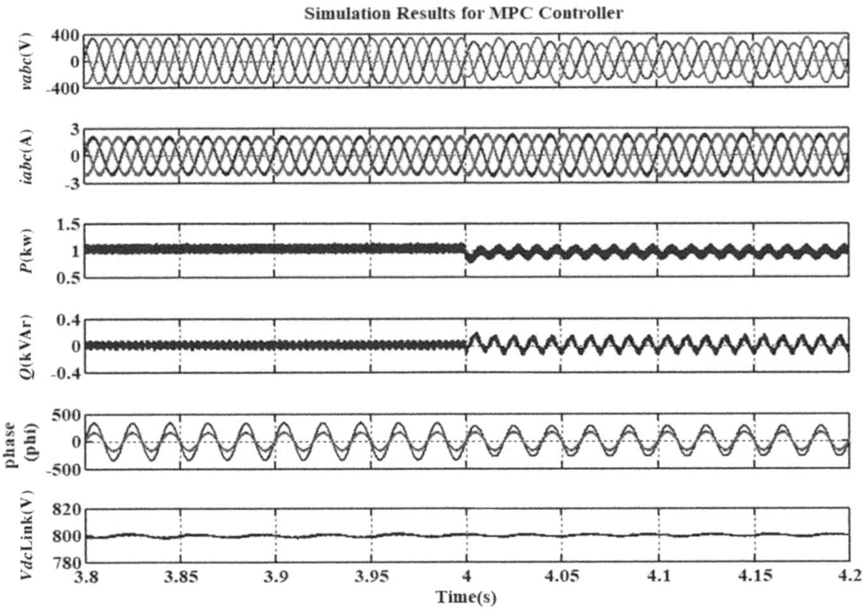

Figure 6.15 Simulation results for FS-MPC SVPWM control-based algorithm during multiple phase unbalance (20% A and 30% B sag) grid operations.

distortion was reduced from 12.76% under traditional PI SVPWM control to just 4.44% with the more sophisticated FSMPC SVPWM algorithm, representing a substantial 65% reduction. This trend is also observed under conditions of multiple-phase unbalance, where grid current distortion dropped from 12.22% with PI SVPWM to only 4.3% using the FSMPC SVPWM approach, again a 65% improvement. These significant decreases in grid current distortion levels significantly impact modern power networks' power quality and stability, which face increasing challenges due to integrating renewable energy sources and the rapid growth of EVs. As the chapter outlines, the influx of EVs connecting to distribution microgrids is causing considerable voltage distortion and imbalance, worsened by the shift toward faster level 3 charging, which places greater demands on the grid. Maintaining grid stability and power quality is crucial for reliable power system operation and ensuring maximum battery longevity and performance for EV owners. The comprehensive data in Table 6.2 further underscores the advantages of the FSMPC SVPWM control algorithm, highlighting its superiority over traditional PI control in mitigating the impacts of DC link variations and distortion levels. Developing autonomous, self-regulating microgrid systems with advanced control capabilities is crucial in enabling the seamless integration of renewable energy and EV charging infrastructure within modern power networks.

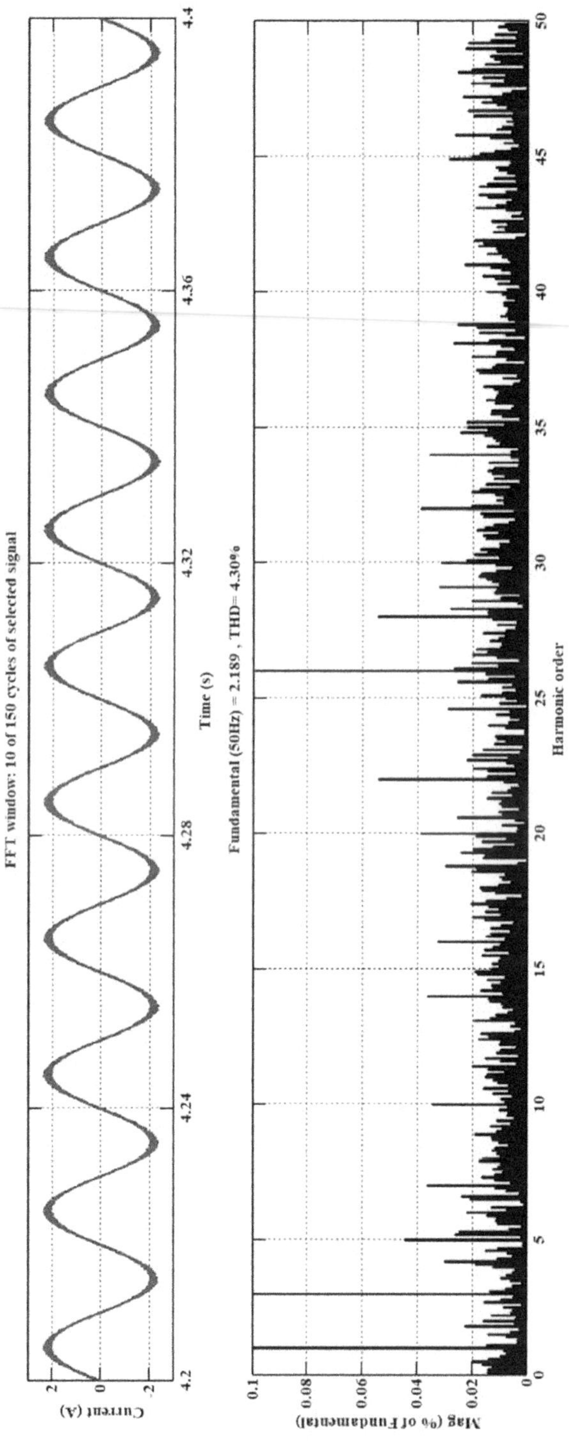

Figure 6.16 FFT analysis of current waveform for FS-MPC SVPWM control-based algorithm during multiple phase unbalance (20% A and 30% B sag).

Table 6.2 Summary of simulation results for different control algorithm

	PI SVPWM control	*FS-MPC SVPWM control*
Single-phase unbalance	12.76%	4.44%
Multiple phase unbalance	12.22%	4.30%

ACKNOWLEDGMENT

This chapter presents experiments conducted at Sardar Vallabhbhai National Institute of Technology (SVNIT) Electrical Engineering Department Surat, Gujarat, India, under the guidance and supervision of Professor Dr. Anandita Chowdhury.

REFERENCES

1. X. Li, and S. Wang, "Energy management and operational control methods for grid battery energy storage systems," in *CSEE Journal of Power and Energy Systems*, vol. 7, no. 5, pp. 1026–1040, 2019.
2. Z. Yang, C. Shen, L. Zhang, M. L. Crow, and S. Atcitty, "Integration of a StatCom and battery energy storage," in *IEEE Transactions on Power Systems*, vol. 16, no. 2, pp. 254–260, 2001.
3. A. Chakraborty, S. K. Musunuri, A. K. Srivastava, and A. K. Kondabathini, "Integrating STATCOM and battery energy storage system for power system transient stability: A review and application," in *Advances in Power Electronics*, vol. 2012, no. 1, p. 676010, 2012.
4. A. Mohd, E. Ortjohann, A. Schmelter, N. Hamsic, and D. Morton, "Challenges in integrating distributed energy storage systems into future smart grid," *2008 IEEE International Symposium on Industrial Electronics*, IEEE, pp. 1627–1632, 2008.
5. K. Kim, T. Yoon, G. Byeon, H. Jung, H. Kim, and G. Jang, "Power demand and power quality analysis of EV charging station using BESS in MicroGrid," *2012 IEEE Vehicle Power and Propulsion Conference*, IEEE, pp. 996–1001, 2012.
6. M. Moghbel, S. Masoum, A. Fereidouni, and S. Deilami, "Optimal sizing, siting and operation of custom power devices with STATCOM and APLC functions for real-time reactive power and network voltage quality control of smart grid," in *IEEE Trans Smart Grid*, vol. 9, no. 6, pp. 5564–5575, 2017.
7. A. Camacho, M. Castilla, J. Miret, A. Borrell, and L. G. de Vicuña, "Active and reactive power strategies with peak current limitation for distributed generation inverters during unbalanced grid faults," in *IEEE Transactions on Industrial Electronics*, vol. 62, no. 3, pp. 1515–1525, 2014.
8. A. Tiwari, A. Tiwari, A. Chowdhury, A. Chowdhury, and A. Chowdhury, "Modified switched Z-source topology for inverter applications," in *Lecture Notes in Electrical Engineering*, vol. 616, Springer, 2021. doi: 10.1007/978-981-16-1978-6_3.
9. A. Tiwari and A. Chowdhury, "Bi-q MSZSI topology for grid-tied inverter under ideal grid conditions," vol. 939, Springer, 2022, doi: 10.1007/978-981-19-4364-5_7.
10. A. Tiwari and A. Chowdhury, "Finite set-model predictive control for modified quasi switched Z source level 3 off-board charger for active power oscillations damping in weak grids," in *ELECTRICA*, vol. 24, no. 3, pp. 336–345, 2024. doi: 10.5152/electrica.2024.23081.

11. A. Tiwari and A. Chowdhury, “An indirect PI-MPC hybrid method-based bi-quasi-modified switched Z source inverters for vehicle to grid applications,” in *ELECTRICA*, vol. 23, no. 1, pp. 20–27, 2023.
12. A. Renault, J. Pacher, L. Comparatore, M. Ayala, J. Rodas, and R. Gregor, “MPC with space vector phase-shift PWM (SV-PSPWM) technique with harmonic mitigation strategy for shunt active power filters based on H-bridge multilevel converter,” in *Frontiers in Energy Research*, vol. 10, 2022. doi: 10.3389/fenrg.2022.779108.
13. M. W. Qaisar, J. Fang, and W. Li, “Switching harmonic attenuation of grid-tied converters using LLCL filters and MPC-SVPWM,” *Proceedings of 2024 IEEE 7th International Electrical and Energy Conference, CIEEC 2024*, Institute of Electrical and Electronics Engineers Inc., pp. 1566–1572, 2024. doi: 10.1109/CIEEC60922.2024.10583736.
14. V. F. Pires, D. Foito, and A. Cordeiro, “Three-phase qZ-source inverter with fault tolerant capability,” in *IET Power Electronics*, vol. 10, no. 14, pp. 1852–1858, 2017.
15. P. T. Staats, W. M. Grady, A. Arapostathis, and R. S. Thallam, “A statistical method for predicting the net harmonic currents generated by a concentration of electric vehicle battery chargers,” in *IEEE Transactions on Power Delivery*, vol. 12, no. 3, pp. 1258–1266, 1997.
16. A. Tiwari and A. Chowdhury, “Impact of NL and EV penetration on IEEE 13 Node Test Feeder and q-MSZS BESS Charger,” *2024 IEEE 3rd International Conference on Electrical Power and Energy Systems (ICEPES)*, IEEE, pp. 1–5, 2024.
17 A. Tiwari, and A. Chowdhury, “Optimizing residential energy systems: a comprehensive study of quasi-modified switched Z-source DC–DC converters with MPPT and storage integration,” in *Electrical Engineering*, 1–22, 2025.
18 X. Guo, W. Liu, and Z. Lu, “Flexible power regulation and current-limited control of the grid-connected inverter under unbalanced grid voltage faults,” in *IEEE Transactions on Industrial Electronics*, vol. 64, no. 9, pp. 7425–7432, 2017.
19 N. Jin, C. Gan, and L. Guo, “Predictive control of bidirectional voltage source converter with reduced current harmonics and flexible power regulation under unbalanced grid,” in *IEEE Transactions on Energy Conversion*, vol. 33, no. 3, pp. 1118–1131, 2017.

Chapter 7

Design and implementation of efficient EV charger for CC and CV mode of operation while maintaining grid code and quality

Dibyendu Mandal and Md Waseem Ahmad

7.1 INTRODUCTION

In general, plug-in hybrid electric vehicles (PHEV) and electric vehicles (EVs) are powered by an on-board charger (OBC) or an off-board charger connected to a grid [1]. A variety of charging technologies for EVs are discussed in the literature and used in practice. All these methodologies have some merits and demerits in terms of efficiency, cost, weight, charging time, and availability. The most easily available and cost-effective EV charger is the single-phase OBC, which uses home infrastructure to charge the Li-ion battery. The basic architecture of all the topologies for high-voltage (HV) battery charging is almost similar, which is shown in Figure 7.1. It comprises an AC–DC power factor correction (PFC) stage followed by a DC–DC converter. In this chapter, a PFC controller with dual-control loops is employed for a full-bridge rectifier to charge a HV battery, with a DC–DC buck converter integrated between the rectifier and the battery. The specifications for the system considered are listed in Table 7.1.

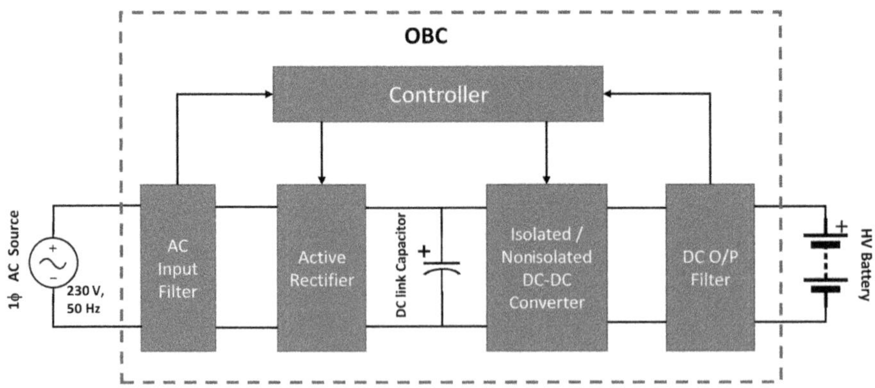

Figure 7.1 Single-phase HV battery charging architecture.

DOI: 10.1201/9781003515807-7

Table 7.1 Specifications for the EV charger design

Category	*Symbol*	*Specification*
Battery nominal voltage	V_{batt_nom}	285 V
Battery fully charged voltage	V_{batt_full}	330 V
Battery rated capacity	Q_{batt}	110 Ah
Battery maximum energy	E_{batt}	36.3 kWh
Charging power	$P_{in} = P_{out}$	3.6 kW
Charging time	–	10 hours
Grid voltage (RMS)	V_g	230 V
Grid current (RMS)	I_g	16 A
Grid frequency	f	50 Hz
DC-link voltage	V_{DC}	400 V

In single-phase grid-connected systems, the DC-link voltage exhibits second-harmonic oscillations at twice the grid frequency, leading to performance degradation. This challenge has prompted extensive research aimed at mitigating such oscillations. Grid-connected EV chargers must comply with standards such as IEEE 519 and IEC 61000-3-2, which impose limits on total harmonic distortion (THD) and require a high power factor (PF) to ensure grid compatibility and operational efficiency.

Full-bridge rectifier

Rectifiers are used to convert an AC power to a DC power. The full-bridge rectifier circuit is the most efficient rectifier circuit among the rectifiers as it converts both the half cycles of an AC component to a pulsating DC component. In active full-bridge rectifier, four insulated gate bipolar transistors (IGBTs) are used, of which two IGBTs in the same leg will not operate at the same time. The switching on of the IGBTs are controlled by PWM gate pulses. Since the output of the rectifier is pulsating DC, a DC-link capacitor is being used to smoothen the pulsation for getting low ripple DC quantity.

7.1.1 Working principle

In the rectifier for the positive half cycle of the AC signal, IG1 and IG4 will gets switched on and for negative half cycle IG3 and IG2 gets switched on. The current flow path is same through the capacitive load during the positive and negative half cycles, as shown in Figure 7.2. In the output we get only the positive pulsating DC signal, which has frequency two times the input signal frequency (Figure 7.3).

Since the battery receives constant power delivery and a single-phase power grid cannot provide constant power, an intermediary energy storage device, a DC-link capacitor, must be used. This element's size is critical for the

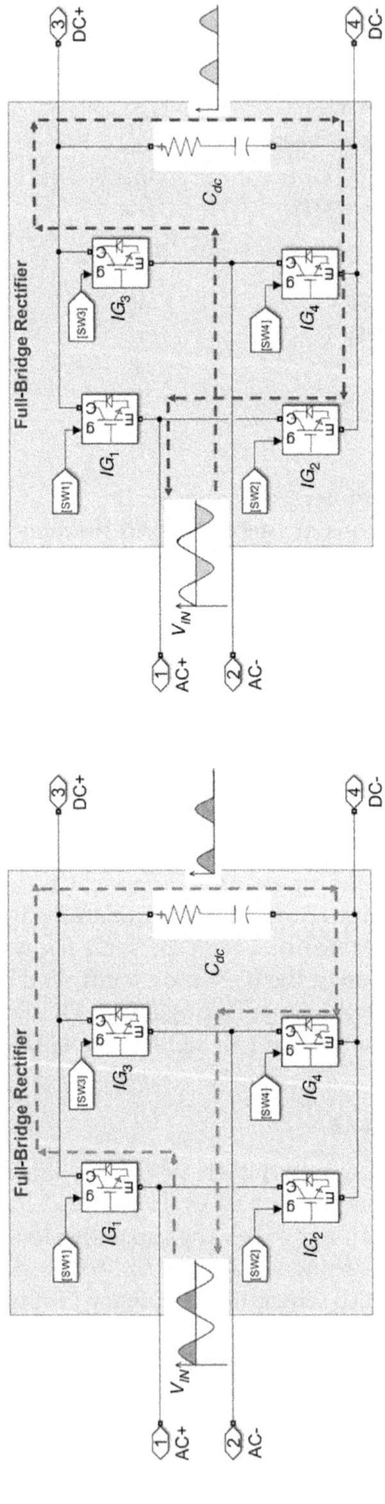

Figure 7.2 Full-bridge rectifier positive and negative half cycle current flow diagram.

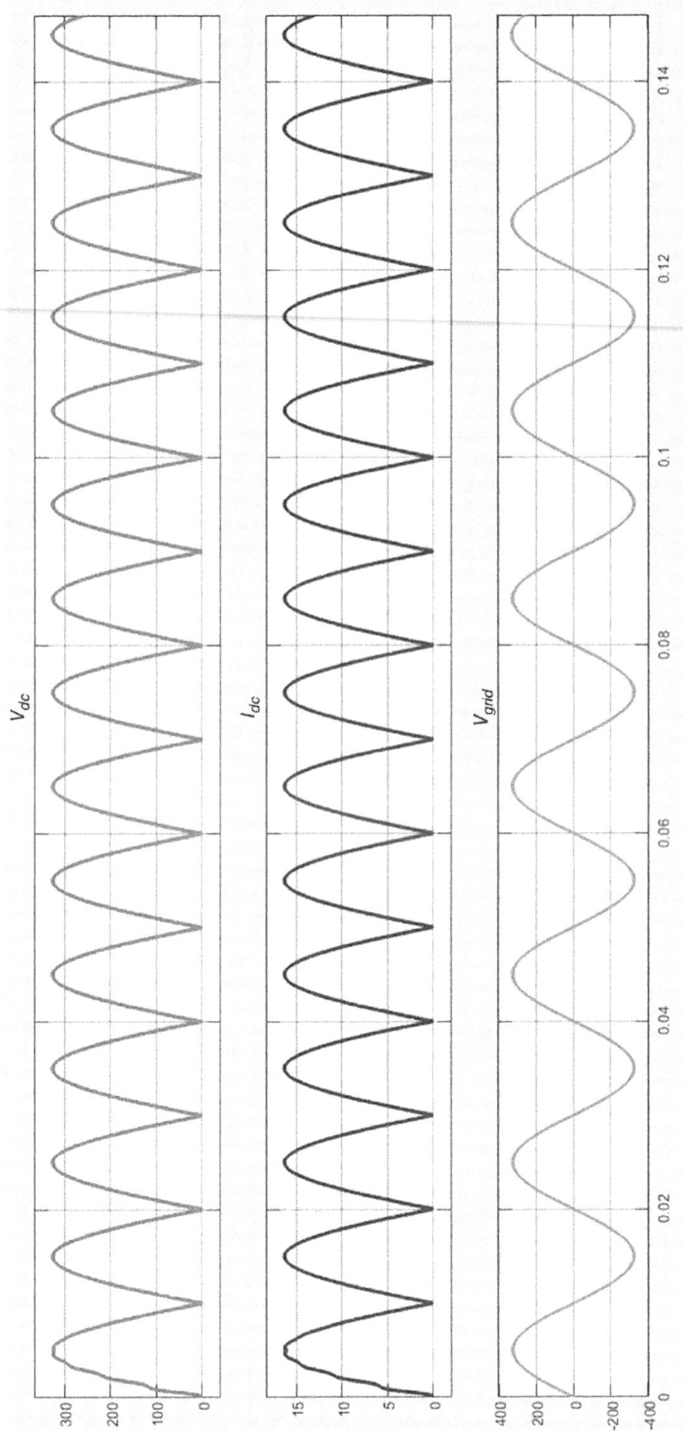

Figure 7.3 Full-wave rectified pulsating DC waveform.

full-bridge AC–DC bidirectional converter to function properly. Capacitor is used to limit or filter ripple voltage across DC link, absorb harmonics.

Instantaneous voltage and current equation without any harmonics for a capacitive load can be expressed as follows:

$$\left.\begin{aligned} v(t) &= \sqrt{2}\ V_g \sin\omega t \\ i(t) &= \sqrt{2}\ I_g \sin(\omega t + \varnothing) \end{aligned}\right\} \tag{7.1}$$

Now the instantaneous power can be expressed as follows:

$$\left.\begin{aligned} p(t) &= v(t).i(t) = V_g\, I_g[2\sin\omega t \sin(\omega t + \varnothing)] \\ &= V_g\, I_g \cos\varnothing(1-\cos 2\omega t) + V_g\, I_g \sin\varnothing \sin 2\omega t \\ &= P(1-\cos 2\omega t) + Q\sin 2\omega t \\ &= \bar{p} + \tilde{p} + \tilde{q} \end{aligned}\right\} \tag{7.2}$$

where $\tilde{p} = (-P\cos 2\omega t)$ will contribute to the energy consumption in the DC-link capacitor. Now, from power balance, it is possible to derive the DC-link capacitor value as shown below:

$$\frac{1}{2}C_{dc}\left(v_{dc\_max}^2 - v_{dc\_min}^2\right) = \int_{-\frac{T}{8}}^{\frac{T}{8}} \tilde{p}\, dt$$

$$\begin{aligned} C_{dc} &= \left| \frac{2\displaystyle\int_{-\frac{T}{8}}^{\frac{T}{8}} (-P\cos 2\omega t)}{\left(v_{dc\_max}^2 - v_{dc\_min}^2\right)} \right| \\ &= \frac{P/_{\omega}\left[2\sin\left(2 * 2\pi f * \frac{T}{8}\right)\right]}{2V_{dc} * 2\Delta V_{dc}} \qquad [as\, f = 1/T] \\ &= \frac{P}{2V_{dc} * \omega * \Delta V_{dc}} \end{aligned} \tag{7.3}$$

where V_{dc} = rated DC voltage and ΔV_{dc} = ripple component of DC voltage.

In high-frequency applications, the ESR becomes more important. High ESR can result in increased ripple voltage and poor regulation. Furthermore, the resistive component of the ESR causes power dissipation, which causes the capacitor to heat up. Hence, selecting a capacitor with low ESR is essential to ensure proper performance and longevity of electronic devices. It is calculated by Equation (7.4).

$$ESR = (\tan\delta)/(2 * \pi * f * C_{dc}) \tag{7.4}$$

The output DC voltage across DC-link capacitor is shown in Figure 7.4.

PFC Controller Design for Full-Bridge Rectifier

In order to optimize the real power extracted from the grid, PFC modifies the power supply's input current to align with the grid voltage. In PFC circuit, grid current follows the grid voltage as pure resistive load without any harmonics. PFC converters are used to increase efficiency and power quality in power electronics applications. In the study of different literature, it has been found that the PFC controller for AC voltage rectification is done with two control loops:

- Outer loop is voltage control.
- Inner loop is current control.

In Ref. [2], a predictive current control is used. In Refs. [3, 4], both the control loops are used with PI compensator. In Ref. [5], the controllers are designed with fuzzy logic. In Ref. [6], the current control loop is designed with PR compensator. In Ref. [7], an adaptive PI compensator is designed for voltage control loop to find the damping ratio (ξ).

Here, PFC controller with two control loops is used for full-bridge rectifier to charge HV battery. To design the current and voltage controller, two things have been considered.

a. Inner current control loop is ten times slower than the PWM switching frequency, to accumulate all the small sampled changes.
b. The outer voltage control loop is further considered ten times slower than the current control loop, so that the change in voltage control can be followed by current controller.

7.1.2 Current controller design

To design any controller, the circuit has to be linearized. The simple linear circuit for rectifier with load (R_{Load}), DC-link capacitor (C_{dc}), line inductor (L_{in}), and ESR of line inductor (R) is shown in Figure 7.5.

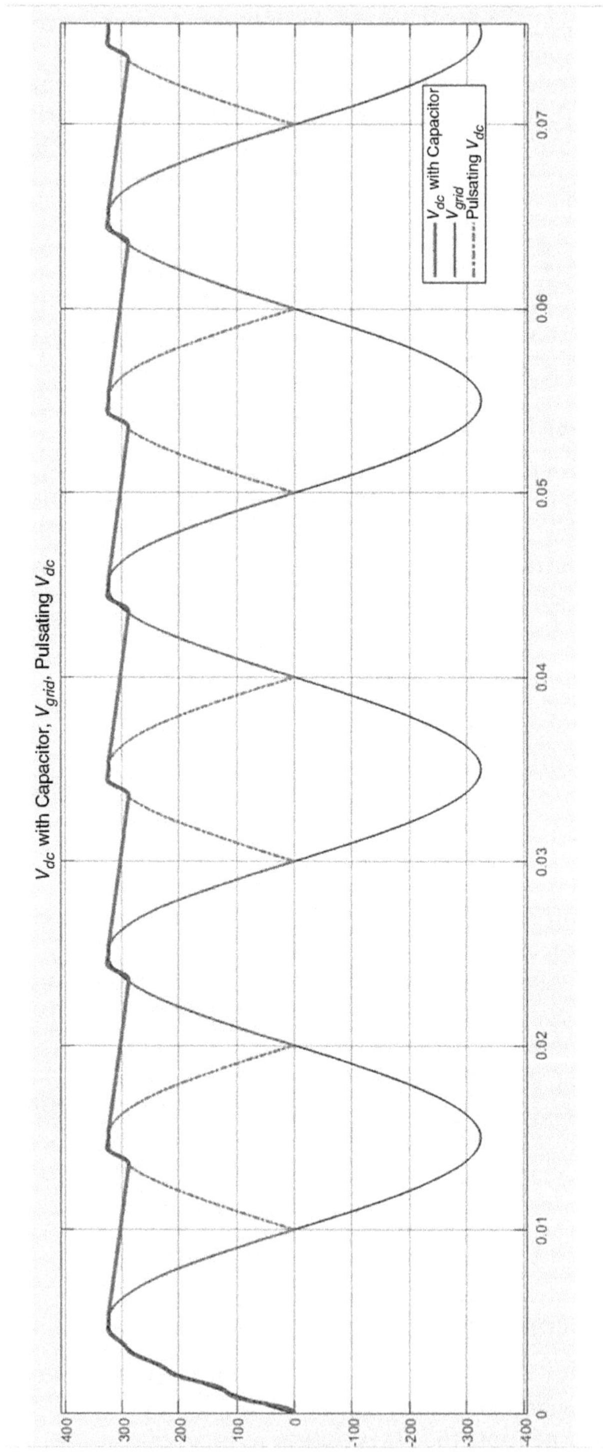

Figure 7.4 Impact of DC-link capacitor in full-wave-rectified DC.

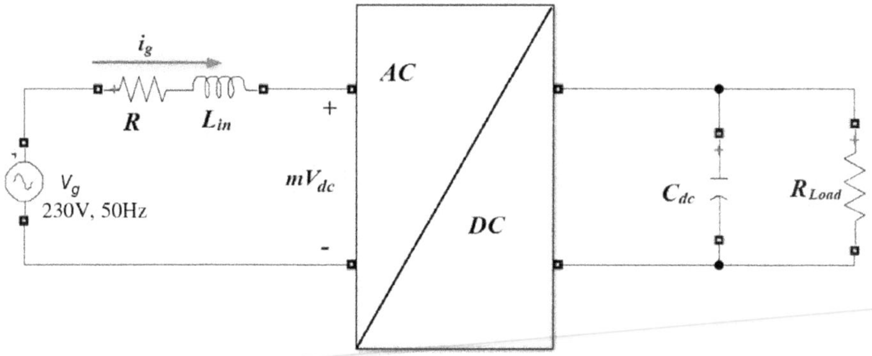

Figure 7.5 Rectifier-linearized model.

From the above circuit,

$$L\frac{di_g}{dt}+Ri_g=-V_g+mV_{dc} \tag{7.5}$$

As there is no control of the grid voltage and its disturbance, it has been neglected in further calculation. Taking Laplace transform of equation (7.5):

$$\begin{aligned} &sLI_g+RI_g=mV_{dc} \\ &\Rightarrow \frac{I_g}{mV_{dc}}=\frac{1}{sL+R} \end{aligned} \tag{7.6}$$

If rectifier switch resistance is considered, then Equation (7.6) will become

$$\frac{I_g}{mV_{dc}}=\frac{1}{sL+(R+r_{on})} \tag{7.7}$$

Equation (7.6) or (7.7) is plant transfer function. The representation of Equations (7.5) and (7.7) with feedback control, PI compensator, and feedforward grid voltage is shown in Figure 7.6.

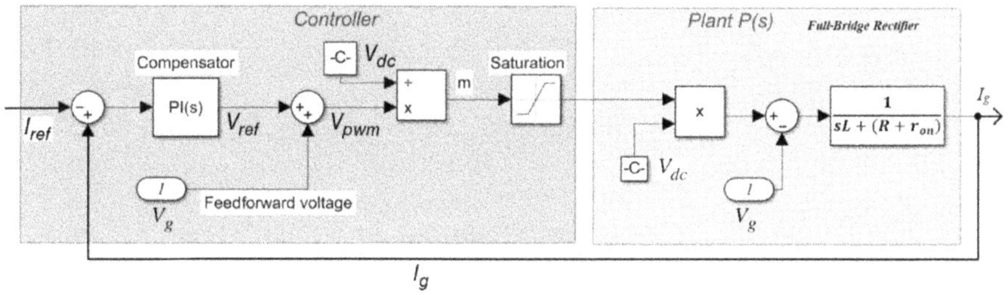

Figure 7.6 Current controller with feedback and feedforward loop.

Now the open-loop gain with PI compensator:

$$G(s) = \left(K_p + \frac{K_i}{s}\right) * \left(\frac{1}{sL + (R + r_{on})}\right)$$

$$\Rightarrow G(s) = \frac{K_p}{sL} \frac{\left(s + \frac{K_i}{K_p}\right)}{\left(s + \frac{R + r_{on}}{L}\right)} \tag{7.8}$$

Based on the block diagram of Figure 7.6 and Equation (7.8), the open-loop full-bridge rectifier has a stable pole at $p = -\frac{R+r_{on}}{L}$. Typically, this pole is fairly close to the origin and corresponds to a slow natural response. To improve the open-loop frequency response, the pole can be canceled by the zero of the PI compensators.

$$\frac{K_i}{K_p} = \frac{R + r_{on}}{L} \tag{7.9}$$

$$\frac{K_p}{L} = \frac{1}{\tau_i} \tag{7.10}$$

Thus, $\frac{K_i}{K_p} = \frac{R+r_{on}}{L}$ and $\frac{K_p}{L} = \frac{1}{\tau_i}$, where τ_i is the desired time constant of the closed-loop system and the obtained closed-loop transfer function is as shown in Equation (7.11):

$$\frac{I_g}{I_{ref}} = \frac{1}{1 + \tau_i s} \tag{7.11}$$

This time constant should be small for faster current-control response and $\frac{1}{\tau_i}$ should be ten times smaller than the switching frequency.

7.1.3 Voltage controller design

The DC bus current i_{dc} is represented as follows:

$$i_{dc} = i_C + i_{Load} = C\frac{dV_C}{dt} + i_{Load} \tag{7.12}$$

AC side active and reactive powers are represented by Equation (7.13)

$$\left.\begin{aligned} P_{ac} &= \frac{1}{2} V_{gM}\ I_{gM} \cos\varnothing \\ Q_{ac} &= \frac{1}{2} V_{gM}\ I_{gM} \sin\varnothing \end{aligned}\right\} \tag{7.13}$$

When $\cos\varnothing = 1$

$$P_{ac} = \frac{1}{2} V_{gM}\ I_{gM}$$

$$Q_{ac} = 0$$

The DC side active power P_{dc} is sum of

i. active power P_{Load} consumed by DC side load, and
ii. active power P_C required to maintain the DC-link average voltage V_{dc}^{avg} equal to its ref value V^*.

$$\left.\begin{aligned} P_{dc} &= V_{dc}^{avg} \cdot I_{dc}^{avg} = P_{Load} + P_C \\ P_{Load} &= V_{dc}^{avg} \cdot I_{Load}^{avg} \\ P_C &= V_{dc}^{avg} \cdot I_C^{avg} \end{aligned}\right\} \tag{7.14}$$

Neglecting the power loss in the converter switches and L filter, the resultant DC side active power P_{dc} equals to the AC side active power P_{ac}.

$$V_{dc}^{avg} \cdot I_{dc}^{avg} \approx \frac{1}{2} V_{gM} \cdot I_{gM}$$

$$\Rightarrow I_{dc}^{avg} = \frac{V_{gM} \cdot I_{gM}}{2\ V_{dc}^{avg}} = G \cdot I_{gM}$$

where

$$G = \frac{V_{gM}}{2\ V_{dc}^{avg}} \tag{7.15}$$

To simplify the grid-connected voltage controller model, system nonlinearity is neglected, and current control loop is ten times faster than voltage control loop. In this way, the current control loop can be represented as constant gain G inside the DC-link voltage control loop. The block diagram of voltage controller is shown in Figure 7.7.

From Figure 7.7, it can be seen that the DC-link voltage control has two inputs: V_{dc}^* and I_{Load}^{avg}. Since V_{dc}^* is constant, in the following the load current input is only considered as external disturbance for computation.

Loop gain or open-loop transfer function of the above control system block is

$$L(s) = \left(K_p + \frac{K_i}{s}\right) * \left(\frac{V_{gM}}{2\ V_{dc}^{avg}}\right) * \frac{1}{Cs} \tag{7.16}$$

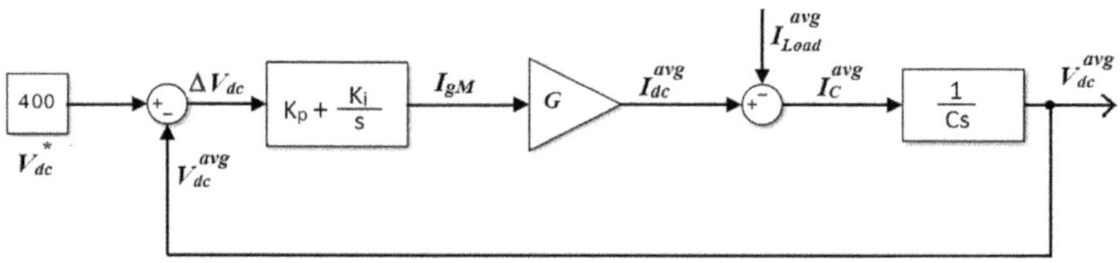

Figure 7.7 DC-link voltage controller.

The closed-loop transfer function for output voltage V_{dc}^{avg} to the input voltage V_{dc}^* is

$$\frac{V_{dc}^{avg}}{V_{dc}^*} = \frac{\left(K_p + \frac{K_i}{s}\right) * \frac{G}{Cs}}{1 + \left(K_p + \frac{K_i}{s}\right) * \frac{G}{Cs}}$$

$$\frac{V_{dc}^{avg}}{V_{dc}^*} = \frac{\frac{GK_p}{C}s + \frac{GK_i}{C}}{s^2 + \frac{GK_p}{C}s + \frac{GK_i}{C}} \approx \frac{2\xi\omega_n s + \omega_n^2}{s^2 + 2\xi\omega_n s + \omega_n^2} \tag{7.17}$$

The closed-loop transfer function for output voltage V_{dc}^{avg} to the input load current disturbance I_{Load}^{avg} is

$$\frac{V_{dc}^{avg}}{I_{Load}^{avg}} = \frac{-\frac{1}{Cs}}{s^2 + \frac{GK_p}{C}s + \frac{GK_i}{C}} \approx \frac{-\frac{1}{Cs}}{s^2 + 2\xi\omega_n s + \omega_n^2} \tag{7.18}$$

By comparing denominator of Equations (7.17) and (7.18) with characteristic equation $s^2 + 2\xi\omega_n s + \omega_n^2$, the derived K_p and K_i value can be found as follows:

$$\left. \begin{aligned} K_p &= \frac{2C}{G}\xi\omega_n \\ K_i &= \left(\frac{C}{G}\right)\omega_n^2 \end{aligned} \right\} \tag{7.19}$$

The pole of the closed-loop transfer function from Equations (7.17) and (7.18) is as follows:

$$P_{(1,2)} = -\xi\omega_n \pm j\omega_n\sqrt{1-\xi^2}, \text{ for } 0 \le \xi \le 1$$

In order to have a stable system, the K_p and K_i values should be positive. The time required for the DC-link voltage controller to be equal to the ref voltage is as follows:

$$\tau_{V_{dc}} = \frac{\pi}{Im\left(P_{(1,2)}\right)}$$
$$\Rightarrow \tau_{V_{dc}} = \frac{1}{\omega_n} * \frac{\pi}{\sqrt{1-\xi^2}} \tag{7.20}$$

Voltage controller has to be ten times slower than the current controller. That means

$$\tau_{V_{dc}} \geq 10\tau_i$$

To design the proper K_p and K_i along with Equations (7.19) and (7.20), the following points need to be considered.

- Damping factor ξ should be chosen based on two conditions.
  i. When ξ is high, the system is overdamped and there is lower dynamic response.
  ii. When ξ is close to zero, this will result in oscillatory response.
- The maximum value of ω_n can be found from Equation (7.20) and $\omega_n \ll \omega_n^{max}$.
- DC bus voltage fluctuation should be less than 10%.
- The magnitude of the third harmonics must be lower than 3% during steady-state condition.

7.1.4 Single-phase PLL

PLL is a closed-loop system that locks the phase and frequency of the output to the input. It is basically an internal oscillator that is controlled such that its phase and frequency are locked to the external periodic signal. PLL consists of three fundamental blocks, as shown in Figure 7.8.

- *Phase Detector:* The PD block generates an output that is proportional to the phase difference between the input signal with frequency f_{in} and output signal with frequency f_{out}.

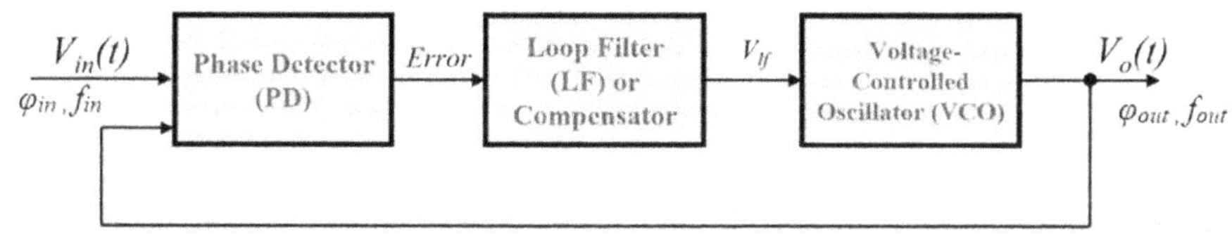

Figure 7.8 PLL inside blocks generalize structure.

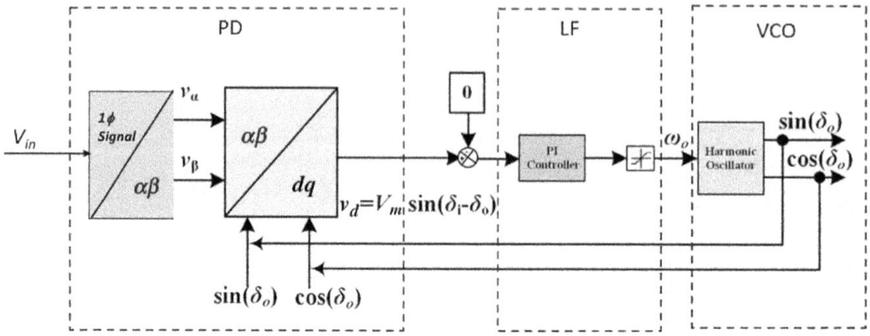

Figure 7.9 Single-phase PLL block.

- *Loop Filter:* This block presents a low-pass filtering characteristic to attenuate the high frequency AC component from PD output.
- *Voltage-Controlled Oscillator (VCO):* VCO block generates at its output an AC signal whose frequency is a function of the input provided by the LF. VCO generates the o/p signal whose frequency is proportional to the input signal amplitude of the VCO block.

The main objective of PLL is to track if the output of a system connected to grid follow the grid frequency and phase. There is always fluctuation of grid voltage from 49 Hz to 51 Hz. This is tracked dynamically and very fast. The PD block is implemented with reference frame transformation. Single-phase PLL block diagram is shown in Figure 7.9.

7.1.4.1 PD block design

From Clark transformation, we have

$$\overline{V} = V_\alpha + jV_\beta = V_m e^{+j(\omega t - 90)} \tag{7.21}$$

For a single-phase system,

$$\begin{aligned} V_\alpha &= V_m \sin(\omega t) \\ V_\beta &= V_m \sin(\omega t - 90^\circ) \end{aligned} \tag{7.22}$$

7.1.4.1.1 *β-component generation using SOGI*

Second-order generalized integrator (SOGI) structures are notch filters that are readily adjusted to the grid frequency. Two first-order integrators are used by the SOGI-PLL to build the orthogonal signal generator (OSG) as shown in Figure 7.12.

The transfer functions from V_{in} to V_α and V_β can be inferred using the Mason formula:

$$G_\alpha(s) = \frac{V_\alpha(s)}{V_{in}(s)} = \frac{K\omega_e s}{s^2 + K\omega_e s + \omega_e^2}$$
$$G_\beta(s) = \frac{V_\beta(s)}{V_{in}(s)} = \frac{K\omega_e^2}{s^2 + K\omega_e s + \omega_e^2} \tag{7.23}$$

where $\omega_e = 2\pi \cdot 50$ rad/s and K is the loop gain. A comparative analysis of the TF of Equation (7.23) is done for $K = \sqrt{2}$ and $K = 10$ in Figures 7.10 and 7.11. With $K = \sqrt{2}$, we get much better slope. Also, from the Bode plot, it is evident that β-component is with 90° phase lag from α-component at any value of K.

The angular frequency ω_e in Figure 7.12 is the feedback frequency signal from LF module which makes SOGI a frequency adaptive OSG.

From Park transform, we know

$$\overline{V_{dq}} = \left[v_\alpha + jv_\beta\right] * e^{-j\omega t} = v_d + jv_q$$
$$\Rightarrow \begin{bmatrix} v_d \\ v_q \end{bmatrix} = \begin{bmatrix} \cos(\omega t)\sin(\omega t) \\ -\sin(\omega t)\cos(\omega t) \end{bmatrix} \begin{bmatrix} v_\alpha \\ v_\beta \end{bmatrix} \tag{7.24}$$

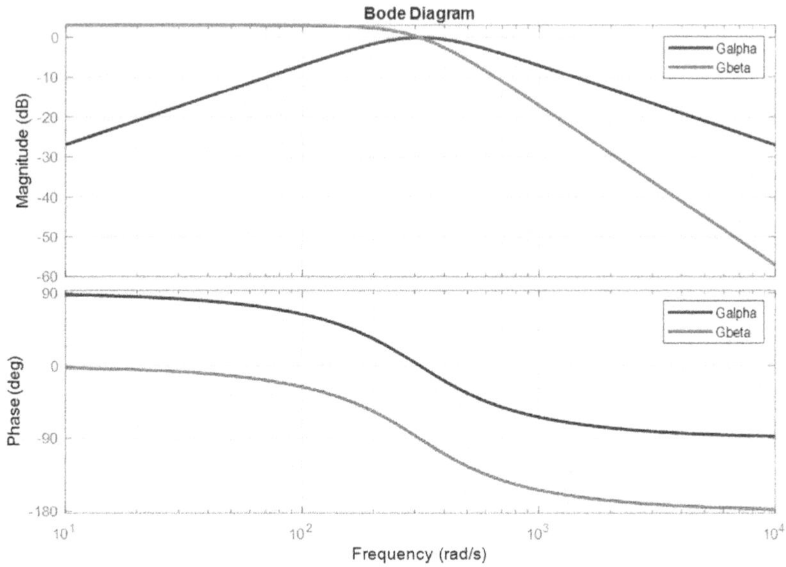

Figure 7.10 Bode plot for V_α and V_β at $\omega_e = 2\pi \cdot 50$ rad/s and $K = \sqrt{2}$.

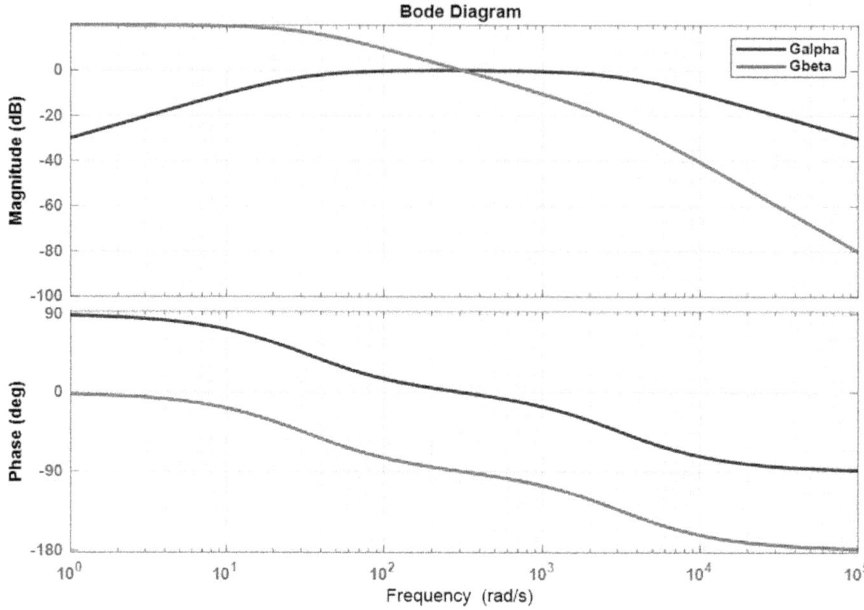

Figure 7.11 Bode plot for V_α and V_β at $\omega_e = 2\pi \cdot 50$ rad/s and $K = 10$.

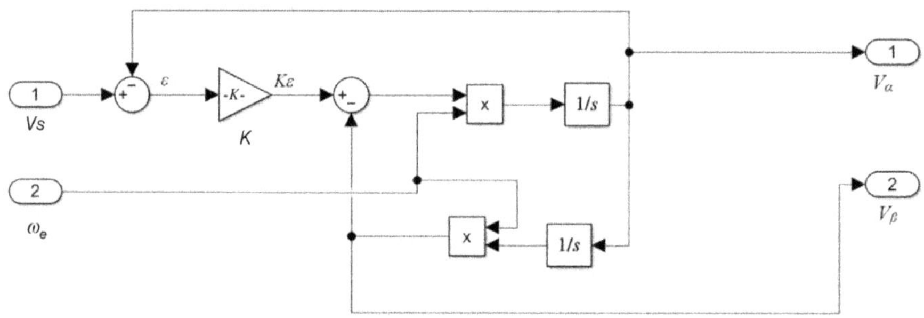

Figure 7.12 Orthogonal signal generator based on SOGI.

When $\alpha\beta$-components are multiplied with proper $\sin(\omega t)$ and $\cos(\omega t)$ signal from VCO, we get pure DC. If we consider the input signal has phase ($\delta_i = \omega_i t$) and output signal has ($\delta_o = \omega_o t + \theta_o$), then

$$\therefore\ V_{dq} = V_m e^{j\left(\delta_i - \delta_o - 90^\circ\right)}$$
$$V_d = V_m \sin\left(\delta_i - \delta_o\right) \tag{7.25}$$

When $\delta_i = \delta_o$, $V_d = 0$. If $V_d \neq 0$, then there is phase error. That is why output of PD is compared with 0.

7.1.4.2 Loop filter design and control block

In Refs. [8–10], loop filter has been designed with a PI compensator. In this design loop, the filter is designed with a Type 2 compensator. Type 2 compensator is generally used for its simpler design and has a phase margin greater than 45° to reduce overshoot.

Now the output of PD is nonlinear, as shown in Equation (7.25). To design a compensator, this has to be linearized. Lower value of $(\delta_i - \delta_0)$, we can say

$$\sin(\delta_i - \delta_0) \cong (\delta_i - \delta_0) \tag{7.26}$$

The main design criteria for compensator:

- Open-loop gain $G(s) \to \infty$ at steady state.
- At any frequency, $G(s) \neq -1$, since system will be unstable at $G(s) = -1$.
  - This means when gain = 1, phase stay away from 180°.
  - When phase is 180°, gain stay away from 1.

  Type 2 compensator transfer function is

$$C(s) = \frac{A}{s}\left(\frac{s+\omega_z}{s+\omega_p}\right) \tag{7.27}$$

where $\omega_z \ll \omega_p$, $\omega_z = P/\alpha$ and $\omega_p = P$.

The phase margin and crossover frequency for Type 2 compensator:

$$\delta_m = \sin^{-1}\left(\frac{\alpha-1}{\alpha+1}\right), \ \omega_c = \frac{P}{\sqrt{\alpha}} \tag{7.28}$$

Plant transfer function is

$$P(s) = \frac{V_m}{s} \tag{7.29}$$

So, from Equations (2.18) and (2.20), the open-loop transfer function is as follows;

$$G(s) = C(s) \cdot P(s) = \left(\frac{A}{s}\right)\left(\frac{s+\omega_z}{s+\omega_p}\right)\left(\frac{V_m}{s}\right) \tag{7.30}$$

The linearized model is shown in Figure 7.13.

If a phase margin $\delta_m = 52°$ and cross-over frequency $\omega_c = 200$ rad/s is chosen, then $A = 359.52$, $\omega_z = 68.4\frac{\text{rad}}{\text{s}}$, $\omega_p = 584.7\frac{\text{rad}}{\text{s}}$, and $V_m = 230\sqrt{2}$. With the above value, the open-loop transfer Bode plot is shown in Figure 7.14.

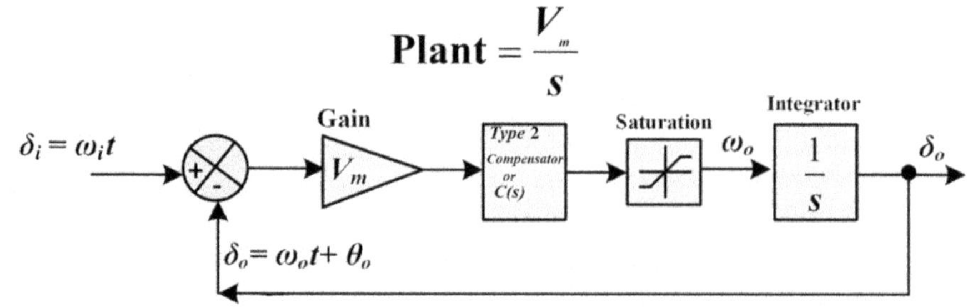

Figure 7.13 Linearized PLL control module with Type 2 compensator.

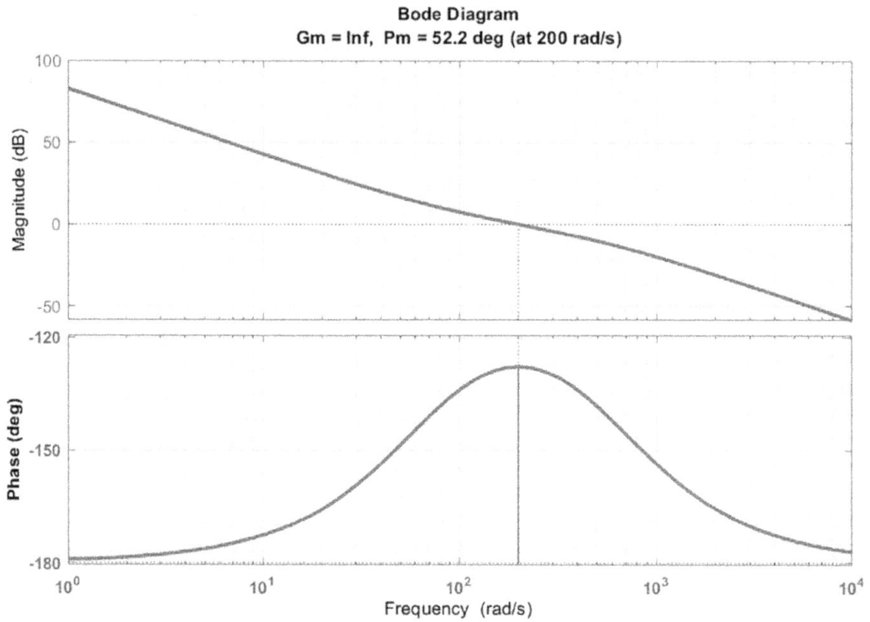

Figure 7.14 Bode plot of open-loop TF of PLL control.

To reduce the start-up transient, a feedforward loop with angular frequency $(\omega^* = 2\pi.50 \text{ rad/s})$ shall be used. This helps to track the frequency change very fast without much overshoot.

MATLAB implementation of the control block is shown in Figure 7.15.

7.1.4.3 Voltage controlled oscillator (VCO) design

The term "voltage-controlled oscillator" indicates that the input voltage controls the oscillator's instantaneous frequency of output. This type of oscillator can generate an output signal frequency within a broad range, contingent upon the DC voltage applied to it.

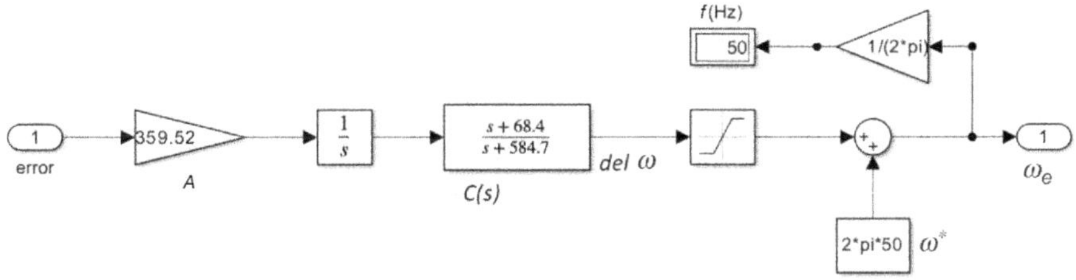

Figure 7.15 Loop filter implementation in MATLAB with Type 2 compensator.

Generally, a variety of VCO forms are utilized. It may be an LC or crystal oscillator, an RC oscillator, or a multivibrator. When an LC oscillator is used, the output signal's oscillation frequency will be

$$f = \frac{1}{\left(2\pi\sqrt{LC}\right)} \tag{7.31}$$

Thus, we can state that capacitance decreases with increasing input or control voltage. As a result, there is a direct proportionality between the oscillation frequency and control voltage. In other words, as one rises, so will the other. Figure 7.16 shows the imperial relation of input voltage and output frequency in VCO.

As in single-phase PLL, output signal of VCO is required to be sinusoid with unity gain, a harmonic oscillator can be used. A harmonic oscillator is designed with first-order differential equation, keeping future prospects in mind to be implemented digitally. Any differential equation can be implemented digitally using the trapezoidal method or backward–forward Euler's method with less consumption of resources.

We need to generate two orthogonal signals (sine and cosine) with unity gain and angular frequency ω.

Assume $x(t) = \sin(\omega t)$ and $y(t) = \cos(\omega t)$, where $x(0) = 0$ and $y(0) = 1$.

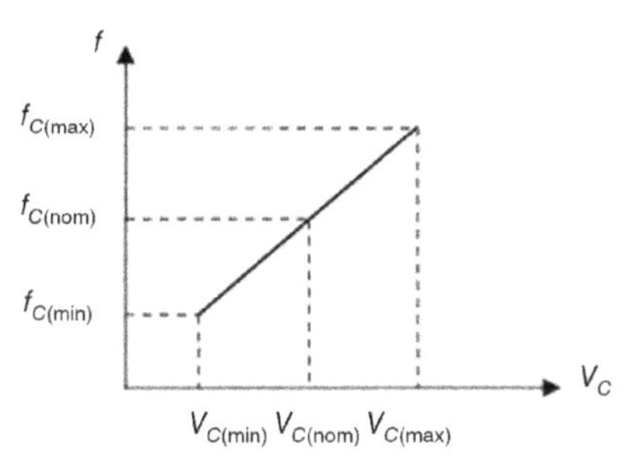

Figure 7.16 Output frequency versus input voltage.

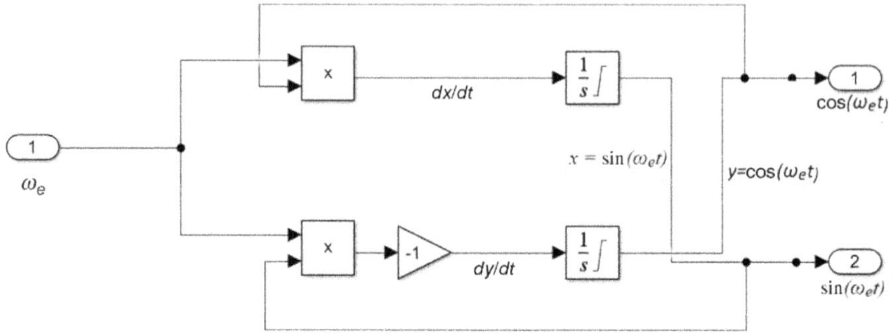

Figure 7.17 MATLAB implementation of harmonic oscillator.

These signals can be generated with a simple differential equation, which is shown in Equation (7.32)

$$\frac{dx}{dt} = \omega\cos(\omega t) = \omega y$$
$$\frac{dy}{dt} = -\omega\sin(\omega t) = -\omega x \tag{7.32}$$

MATLAB model of the above equation is shown in Figure 7.17. The initial value of the integrators in sin(ωt) path is set to 0 and cos(ωt) path is set to 1 with both paths limited to (–1,1).

The gate pulses for the rectifier switches are generated by a PWM generator having switching frequency f_{sw} = 10 kHz with variable duty cycle. The duty cycle of the PWM generator is decided by the voltage and current controller discussed in Sections 7.1.2 and 7.1.3. The complete MATLAB design for the PWM generator is shown in Figure 7.18.

Buck converter design

The buck converter output voltage, which is battery voltage V_{bat}, is less than the input voltage of the converter, which is DC-link voltage V_{dc}. The circuit diagram of the buck converter is shown in Figure 7.19.

In the circuit in Figure 7.19, IG1 is working as a switch to regulate the voltage and IG2 is working as a freewheeling diode. Usually, buck converter works in both continuous conduction mode (CCM) and discontinuous conduction mode (DCM). Here CCM approach is discussed. In CCM inductor current, i_L will never cross 0 A. Due to the switching operation, input current i_{dc} is discontinuous but inductor current i_L and battery charging current i_{bat} are continuous.

The operation of buck converter is divided into two modes when switch IG1 is ON and when switch IG1 is OFF. The switching ON and OFF of the

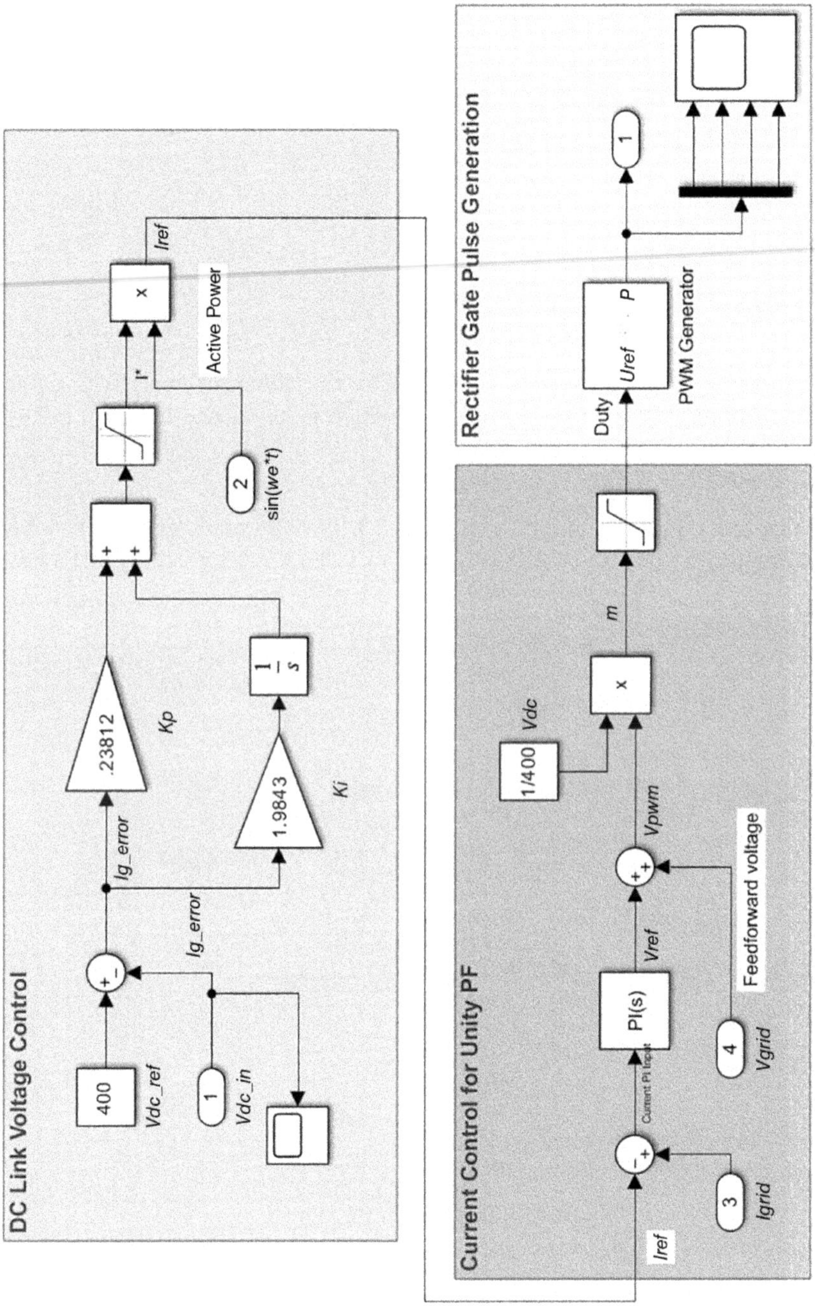

Figure 7.18 Gate pulse generation for full-bridge rectifier in MATLAB.

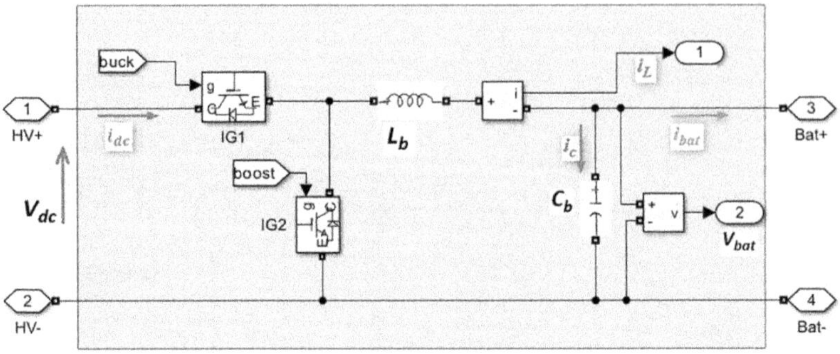

Figure 7.19 Buck converter circuit.

IG1 is done by PWM gate pulses decided by the constant current (CC) and constant voltage (CV) mode of charging, which is explained in Section 7.4.1.

7.1.4.3.1 Switch IG1 Is ON

The equivalent circuit is shown in Figure 7.20. The switch is on for DT_s, where T_s is the switching period and D is the duty cycle. Applying Kirchhoff's law,

$$V_L = V_{dc} - V_{bat} = L_b \frac{di_L}{dt}$$
$$\Rightarrow (\Delta i_L)_{on} = \left(\frac{V_{dc} - V_{bat}}{L_b}\right) DT_s \tag{7.33}$$

Δi_L is the ripple component of inductor current.

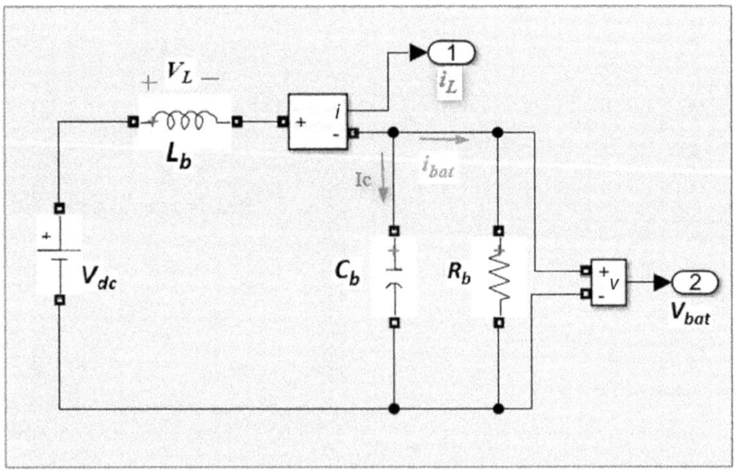

Figure 7.20 Buck converter equivalent circuit for switch ON operation.

7.1.4.3.2 Switch IGI Is OFF

The equivalent circuit for switch operation is shown in Figure 7.21. The switch is off for $(1-D)T_s$, where the inductor will get discharged and inductor current i_L will flow through load (i_{bat}) and through capacitor (i_c).

The voltage across inductor is as follows:

$$V_L = -V_{bat} = L_b \frac{di_L}{dt}$$
$$\Rightarrow (\Delta i_L)_{off} = \left(\frac{-V_{bat}}{L_b}\right)(1-D)T_s \tag{7.34}$$

In steady-state applying volt–second balance with Equations (7.33) and (7.34), the derived relation between input and output voltage with duty cycle D is as follows:

$$(\Delta i_L)_{on} + (\Delta i_L)_{off} = 0$$
$$\Rightarrow \left(\frac{V_{dc} - V_{bat}}{L_b}\right) DT_s - \left(\frac{V_{bat}}{L_b}\right)(1-D)T_s = 0 \tag{7.35}$$
$$\Rightarrow V_{bat} = DV_{dc} \text{ (where } D < 1)$$

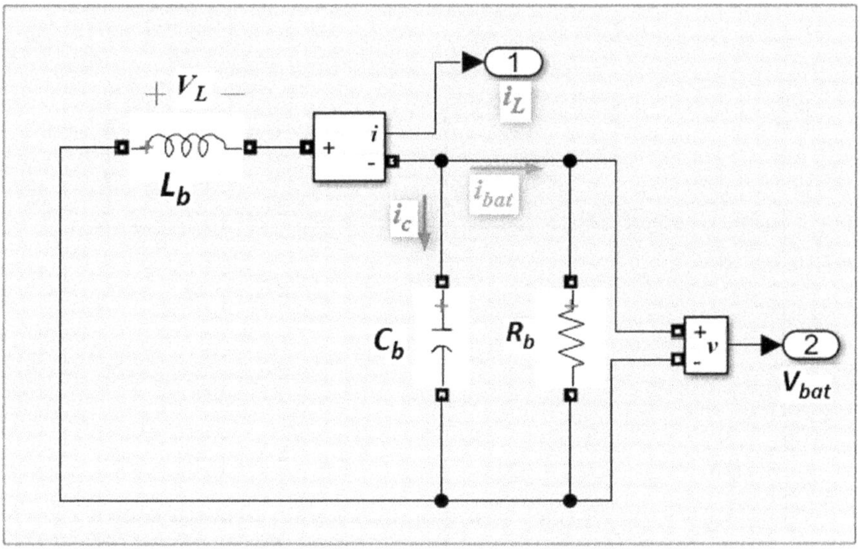

Figure 7.21 Buck converter equivalent circuit for switch OFF operation.

From Equation (7.33), the equation for inductor calculation can be written as follows:

$$L_b = D\left(\frac{V_{dc} - V_{bat}}{\Delta i_L \ f_s}\right), \left[\text{where } f_s = \frac{1}{T_s}\right] \tag{7.36}$$

The graph of inductor current and capacitor current during switch-on and switch-off operation is shown in Figure 7.22. From the graph, charge stored in the capacitor is ΔQ.

$$\Delta Q = C_b * \Delta V_{bat}$$

$$\Rightarrow \frac{1}{2}\left(\frac{\Delta i_L}{2}\right)\left(\frac{T_s}{2}\right) = C_b * \Delta V_{bat}$$

$$\Rightarrow \Delta V_{bat} = \frac{T_s * \Delta i_L}{8C_b} = \frac{T_s}{8C_b} * \left(\frac{V_{bat}}{L_b}\right)(1-D)T_s \tag{7.37}$$

$$\Rightarrow C_b = \frac{(1-D)}{8 * L_b\left(\frac{\Delta V_{bat}}{V_{bat}}\right) f_s^2}$$

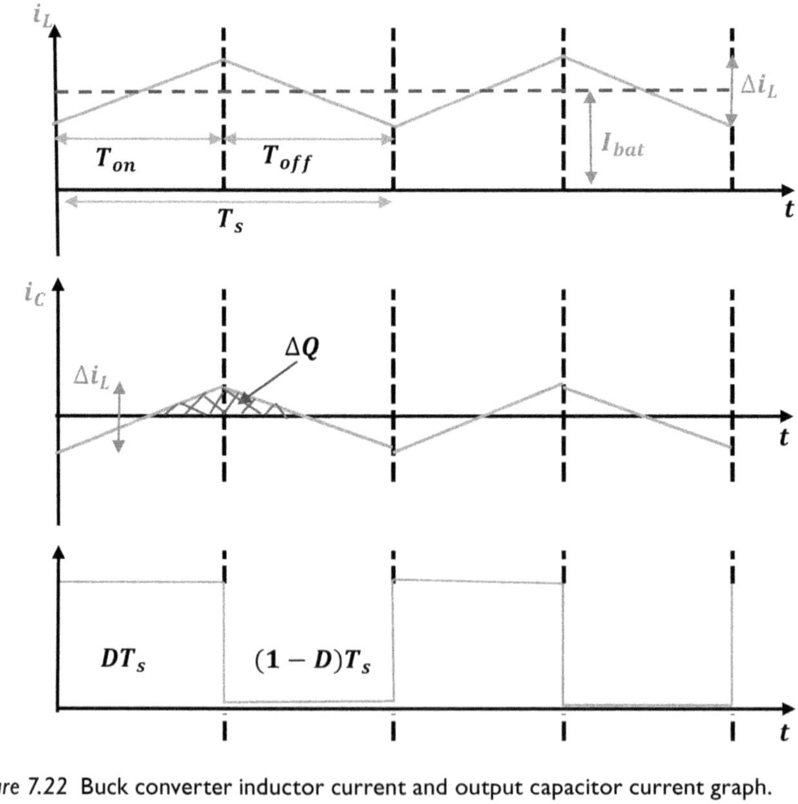

Figure 7.22 Buck converter inductor current and output capacitor current graph.

7.1.5 CC and CV mode of charging

There are several literatures where CC and CV mode of charging is discussed. This is one of the most important criteria for Li-ion battery charging, which is critical for maximizing battery health and performance.

In CC mode of charging, battery is charged with a CC but voltage and SOC of the battery will increase. Typically, in CC mode, a DC fast-charging Li-ion battery charged with max charging current is not more than 1 C. CC mode charging duration varies from manufacturer to manufacturer, and it is decided either based on SOC or fully charged voltage.

When the battery reaches that threshold SOC or threshold voltage value, it starts charging in CV mode till it reaches the final SOC value. The charging current slowly decreases to its low current threshold (cutoff current), and it will cease to 0 after it reaches final SOC. This phenomenon is shown in Figure 7.23.

In this discussion, the CC and CV modes of charging are decided based on the battery voltage V_{bat}.

- When $V_{bat} < 330$ V, it is CC mode of charging.
- When $V_{bat} \geq 330$ V, it is CV mode of charging.

7.1.5.1 CC mode of charging

The block diagram for generating PWM gate pulses for the DC–DC converter to work in a CC mode is shown in Figure 7.24.

I_{ref} is the maximum charging current being compared with the charging current to generate the error quantity. This error amount is being controlled

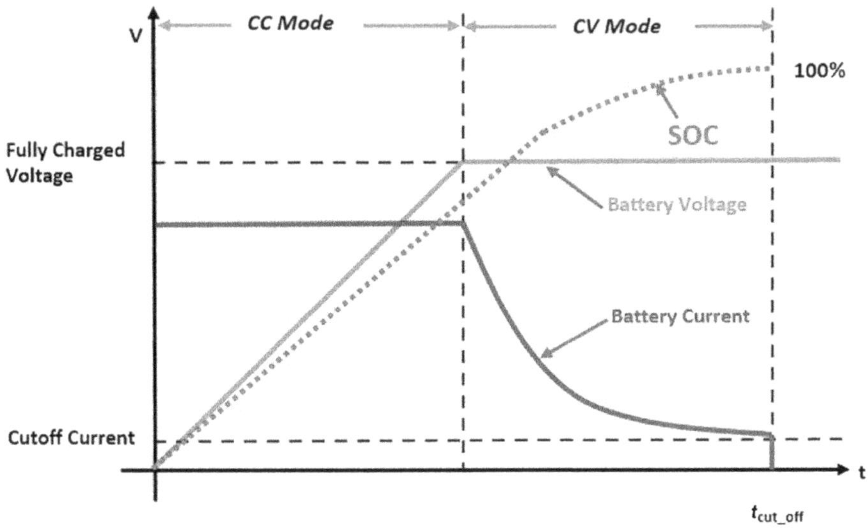

Figure 7.23 CC and CV modes of charging graph.

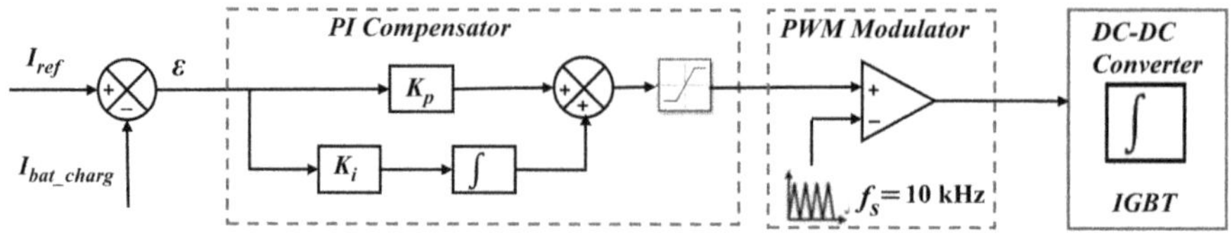

Figure 7.24 Control system for CC mode of charging.

by the PI compensator gain so that it reaches final steady-state reference value. The output of the PI compensator is the duty cycle to generate the gate pulses for the DC–DC converter. The switch of the DC–DC converter operates in such a way by the controllable duty that the converter output current, that is, the battery charging current I_{bat}, reaches steady-state value I_{ref} and stays in that way till battery voltage V_{bat} reaches 330 V.

The PI compensator gain K_p and K_i value for the CC mode of the charging is calculated based on Ziegler–Nichols method and manual tuning.

1. Set the compensator gain K_c value to a minimum and make integral component 0. This makes the compensator work as proportional only.
2. Slowly increase the K_c till the output of the converter reaches an oscillating steady-state value.
3. When the value of K_c causes a sustained periodic oscillation or near to it, make this compensator gain as ultimate gain $K_u = K_c$.
4. Now measure the oscillation period and mark it as ultimate oscillation period P_u.
5. From K_u and P_u values, as shown in Table 7.2, K_p, τ_i, and τ_d are calculated. Further, a manual fine-tune is done.
6. Finally, $K_i = K_p/\tau_i$ and $K_d = K_p * \tau_d$. In this case, K_d is not calculated because of the PI controller.

7.1.5.2 CV mode of charging

The general block diagram for generating PWM gate pulses for the DC–DC converter to work in a CV mode is shown in Figure 7.25.

Table 7.2 K_p, τ_i, τ_d calculation table in Ziegler–Nichols method

	K_p	τ_i	τ_d
P	$K_u/2$	–	–
PI	$K_u/2.2$	$P_u/1.2$	–
PID	$K_u/1.7$	$P_u/2$	$P_u/8$

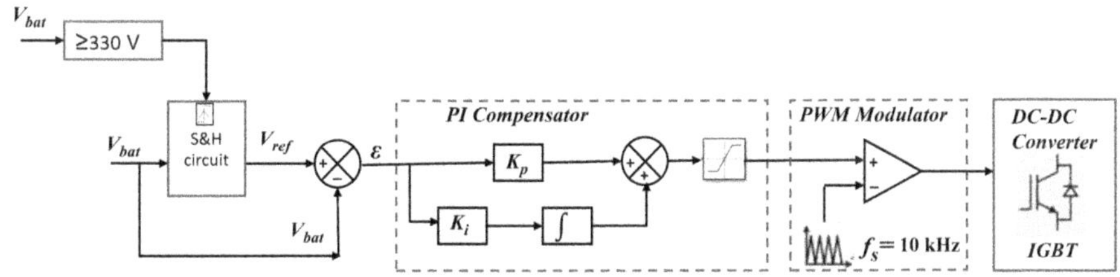

Figure 7.25 Control system for CV mode of charging.

In the CC mode, when battery voltage V_{bat} reaches maximum charging voltage 330 V, the battery voltage on that time is taken as reference voltage for CV mode of charging. Due to the fluctuation of output voltage, the final stabilized averaged voltage is $V_{bat} = 329.999\ V$. This is done by "sample and hold" circuit with "edge rise" functionality. A sample-and-hold (S/H) circuit's main function is to take samples of its input signal and hold those samples for some time. It gives the last hold value output as soon as there is edge rise detected in the enable condition [11].

The detection of $V_{bat} \geq 330\ V$ and producing an edge rise signal is done by MATLAB program. The average battery voltage $V_{bat} = 329.999\ V$ is taken as reference voltage for voltage control loop. This voltage is again compared with the actual battery voltage and the error value will act as an input to the CV mode PI compensator module. The PI compensator will generate the required duty for CV mode of operation and generate the necessary switching pulses for IG1 of Buck converter.

Here also the PI compensator value is being set by Ziegler–Nichols method. Fine-tuning of the PI compensator gain (K_p and K_i) is done manually.

During simulation, it was found that due to battery characteristics, the switchover of charging method from CC to CV should be decided based on *battery fully charged voltage*. The fully charged voltage of the battery is 330 V. So, the controller is designed with that perspective and this switchover is shown in Figure 7.27. The charging current is negative, which can be seen in the figure. As per the Li-ion battery characteristics during CC mode of charging, the voltage and SOC will rise and the charging current will be constant. From the simulation graph shown in Figure 7.27, we can see that during CC mode, the charging current is constant at almost 10.9 A and the battery voltage rises until 330 V. Also, SOC rises around 95%.

On the other hand, in CV mode of charging, the charging voltage should be constant and the charging current should decrease over time and SOC should increase. From Figure 7.27, we can see that battery voltage is being constant to the fully charged voltage 330 V and the current has decreased to around 8 A. But SOC is rising.

If we further expand the current graph for the CV mode section, we will see that the current is actually decreasing slowly from 8 A as the SOC increases. But voltage is constant. This behavior is shown in Figure 7.28.

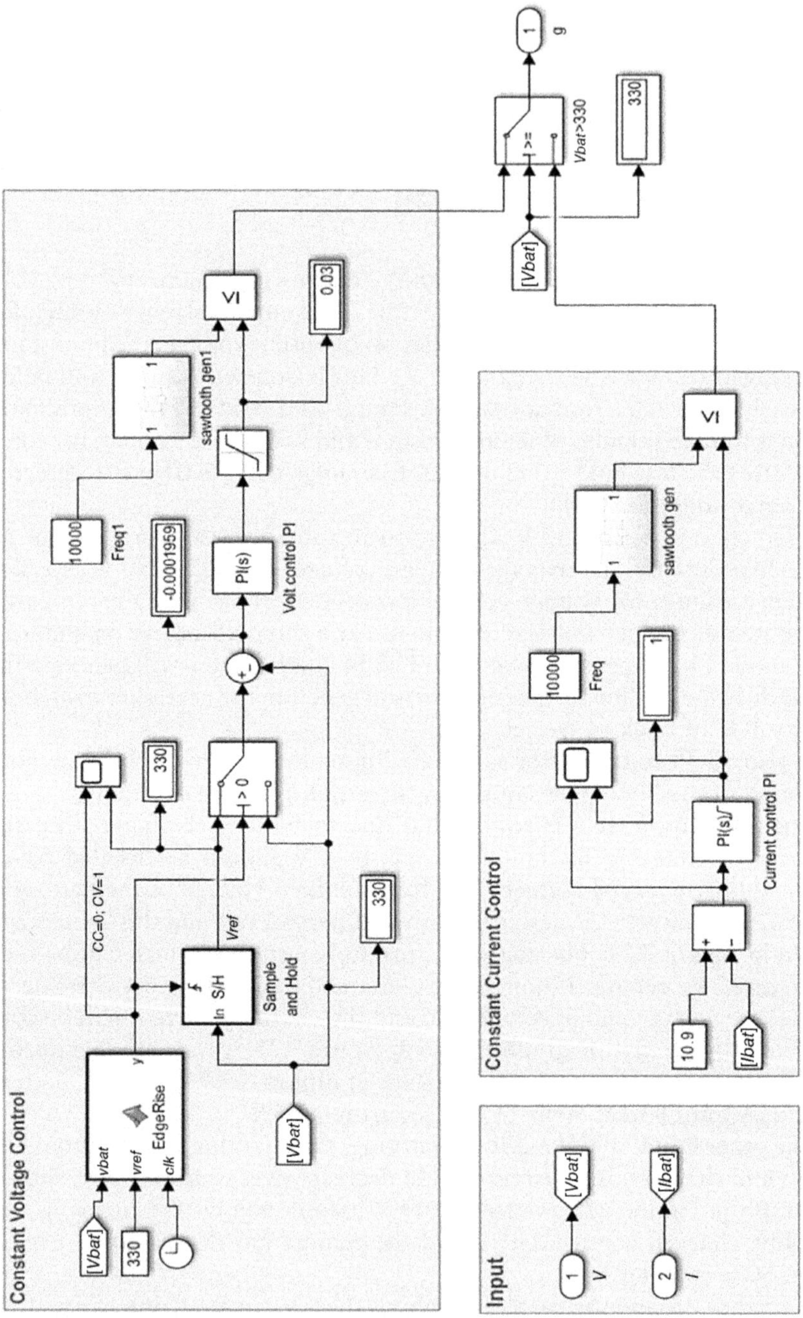

Figure 7.26 MATLAB implementation of CC and CV controllers.

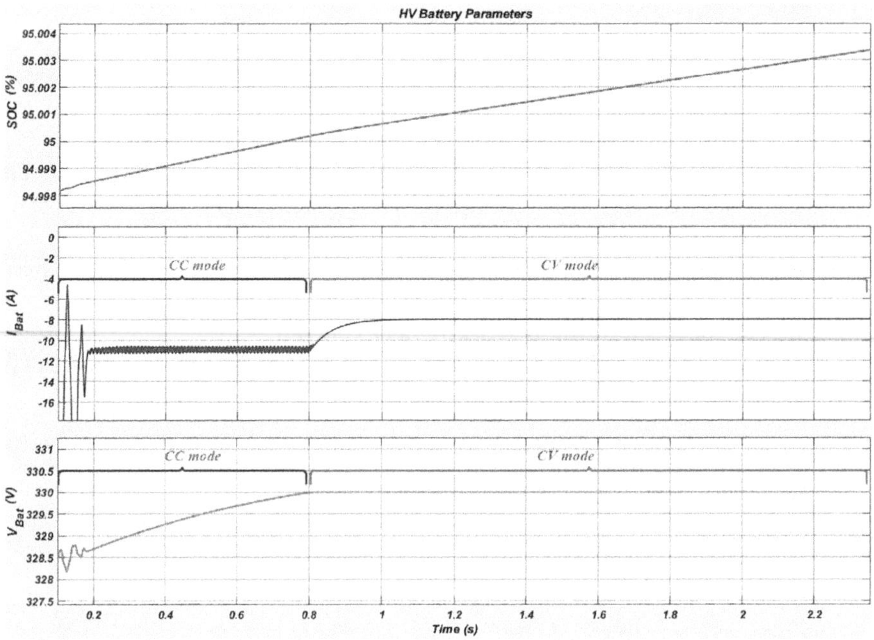

Figure 7.27 HV battery parameter with CC and CV charging.

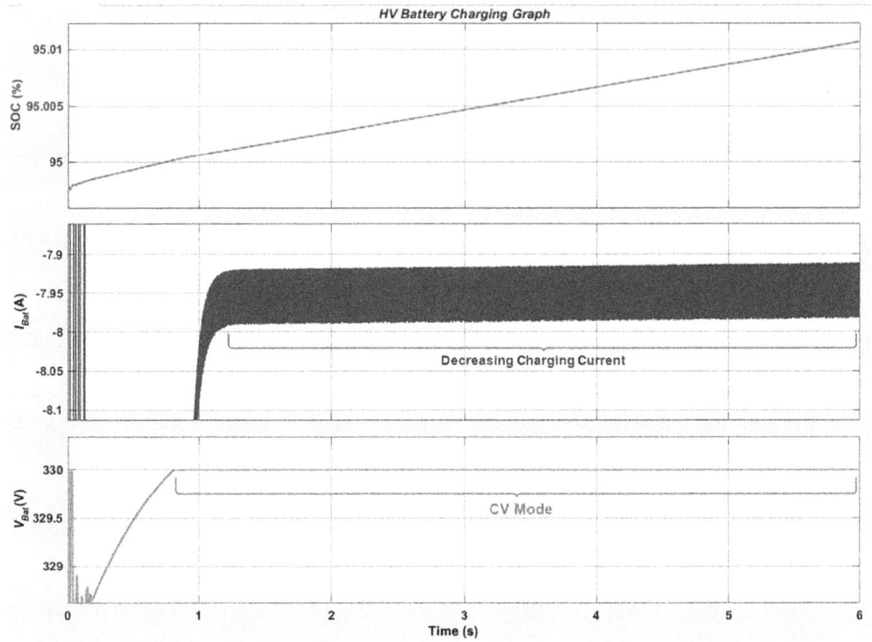

Figure 7.28 Charging current decreases in CV mode.

7.2 CONCLUSION

From the above simulation experiment, we are able to achieve the following:

1. As per IEEE 519 and IEEE 1547, the grid code is PF >0.95, THD_i <5%, and contribution of the third harmonic should be less than 3%. The designed PFC-controlled charger is able to achieve PF = 0.986, THD_i = 2.38%, and third harmonic = 2.21%, which is pretty good.
2. The efficiency of the converter is found to be very good with a conversion ratio of 99.1%.
3. The CC mode charging is working as expected.
4. CV mode of charging is also working as per battery characteristics. The charging current is decreasing with time.

All these aspects have been validated via simulation.

REFERENCES

1. Ali Emadi, Young Joo Lee, and Kaushik Rajashekara, "Power electronics and motor drives in electric, hybrid electric, and plug-in hybrid electric vehicles", in IEEE Transactions on Industrial Electronics, vol. 55, no. 6, pp. 2237–2245, 2008.
2. J. G. Pinto, V. Monteiro, H. Gonçalves, and J. L. Afonso, "Onboard reconfigurable battery charger for electric vehicles with traction-to-auxiliary mode," in IEEE Transactions on Vehicular Technology, vol. 63, no. 3, pp. 1104–1116, 2014. doi: 10.1109/TVT.2013.2283531.
3. C. Shi, Y. Tang, and A. Khaligh, "A single-phase integrated onboard battery charger using propulsion system for plug-in electric vehicles," in IEEE Transactions on Vehicular Technology, vol. 66, no. 12, pp. 10899–10910, 2017. doi: 10.1109/TVT.2017.2729345.
4. K. Kim, S. Park, S. Lee, T. Lee, and C. Won, "Battery charging system for PHEV and EV using single phase AC/DC PWM buck converter", 2010 IEEE Vehicle Power and Propulsion Conference, pp. 1–6, 2010. doi: 10.1109/VPPC.2010.5729195.
5. R. Metidji, B. Metidji, and B. Mendil, "Design and implementation of a unity power factor fuzzy battery charger using an ultrasparse matrix rectifier," in IEEE Transactions on Power Electronics, vol. 28, no. 5, pp. 2269–2276, 2013. doi: 10.1109/TPEL.2012.2211107.
6. H. V. Nguyen and D.-C. Lee, "Single-phase multifunctional onboard battery chargers with active power decoupling capability," 2018 IEEE Applied Power Electronics Conference and Exposition (APEC), San Antonio, TX, USA, 2018, pp. 3434–3439.
7. M. Merai, M. W. Naouar, I. Slama-Belkhodja, and E. Monmasson, "An Adaptive PI controller design for DC-link voltage control of single-phase grid-connected converters," in IEEE Transactions on Industrial Electronics, vol. 66, no. 8, pp. 6241–6249, 2019. doi: 10.1109/TIE.2018.2871796.
8. L. G. Barbosa Rolim, D. Rodrigues da Costa and M. Aredes, "Analysis and software implementation of a robust synchronizing PLL circuit based on the pq theory," in IEEE Transactions on Industrial Electronics, vol. 53, no. 6, pp. 1919–1926, 2006. doi: 10.1109/TIE.2006.885483.

9. Q. Guan, Y. Zhang, Y. Kang, and J. M. Guerrero, "Single-phase phase-locked loop based on derivative elements," in IEEE Transactions on Power Electronics, vol. 32, no. 6, pp. 4411–4420, 2017. doi: 10.1109/TPEL.2016.2602229.
10. M. Ciobotaru, R. Teodorescu, and F. Blaabjerg, "A new single-phase PLL structure based on second order generalized integrator," 2006 37th IEEE Power Electronics Specialists Conference, Jeju, Korea (South), 2006, pp. 1–6. doi: 10.1109/pesc.2006.1711988.
11. M. E. Waltari and K. A. I. Halonen, Sample-and-Hold Operation, in: Circuit Techniques for Low-Voltage and High-Speed A/D Converters, The International Series in Engineering and Computer Science, vol. 709. Springer, Boston, MA. https://doi.org/10.1007/0-306-47979-6_3

Chapter 8

Mitigating circulating currents in electric vehicle charging stations with multiple parallel-connected nonidentical three-phase Vienna rectifiers

Chia-Chi Chu

8.1 INTRODUCTION

Recent advancements in environmental considerations have catalyzed significant progress in electric vehicles (EVs). These vehicles utilize high-power DC charging systems sourced from AC grids to efficiently drive their electric motors. Among the available converter options, the three-level Vienna rectifier has gained popularity due to its practical simplicity, offering fewer active power switches compared to other multilevel converters [1–6]. However, this class of Vienna rectifiers still encounters several technical difficulties [7]. Overcoming these challenges requires sophisticated control strategies distinct from other multilevel converters. Innovative approaches such as zero-sequence injection (ZSI) modulation [8] and other advanced compensation methods [9] have already been proposed to enhance the performance and reliability of Vienna rectifiers, particularly in managing DC-link voltages under varied conditions.

Expanding the power capacity of charging stations often involves paralleling multiple lower-rated Vienna rectifiers. However, this method can introduce zero-sequence circulating currents (ZSCC), especially when phase parameters differ across system phases. Under worst-case scenario, such as interleaved operations, ZSCC can manifest as (i) low-frequency (LF) ZSCC, and (ii) high-frequency components (HF) ZSCC, resulting in significant increases of power losses. Effective mitigation strategies include advanced control methods aimed at minimizing current ripple through precise phase angle manipulation in switch cycle patterns.

Research efforts have extensively addressed circulating currents especially for multiple parallel-connected converters [10–24]. On the other hand, Vienna rectifiers have received comparatively less attention [7, 25–29], prompting recent studies to propose integrated control schemes specifically tailored for multiple parallel-interleaved Vienna rectifiers [1].

This work builds upon previous findings by introducing a novel control scheme designed to optimize ZSCC without altering circuit topologies. Key contributions include the following:

1. Develop an average ZSCC model for two parallel-interleaved Vienna rectifiers.

DOI: 10.1201/9781003515807-8

2. Implement a two-layer control scheme integrating proportional-integral-resonant (PIR) control for LF-ZSCC mitigation and discontinuous pulse width modulation (DPWM) techniques for HF-ZSCC reduction.
3. Provide a guideline for DPWM scheme design in both single and multiple Vienna rectifier configurations.
4. Validate through comprehensive simulations and hardware experiments, confirming efficacy under diverse operating conditions, including interleaved modes with nonidentical inductors.

Subsequent sections of this chapter provide a detailed analysis and validation of the proposed control scheme. Section 8.2 first discusses space vector of the conventional Vienna rectifier. Then, an average ZSCC model for multiple parallel-connected Vienna rectifiers is presented. Section 8.3 focuses on employing PIR control to mitigate LF-ZSCC, while Section 8.4 explores DPWM techniques aimed at reducing HF-ZSCC. Both computer simulations and laboratory experiments are presented in Section 8.5 to demonstrate the efficiency of the proposed control method. Finally, Section 8.6 provides some conclusions.

8.1.1 Background

8.1.1.1 Operating principles

Figure 8.1 depicts a schematic of two parallel-connected three-phase Vienna rectifiers. These AC input terminals (a1, b1, c1; a2, b2, c2) will be linked to the positive point (P), the neutral point (O), or the negative point (N) of the DC bus. This configuration results in three operational states based on the

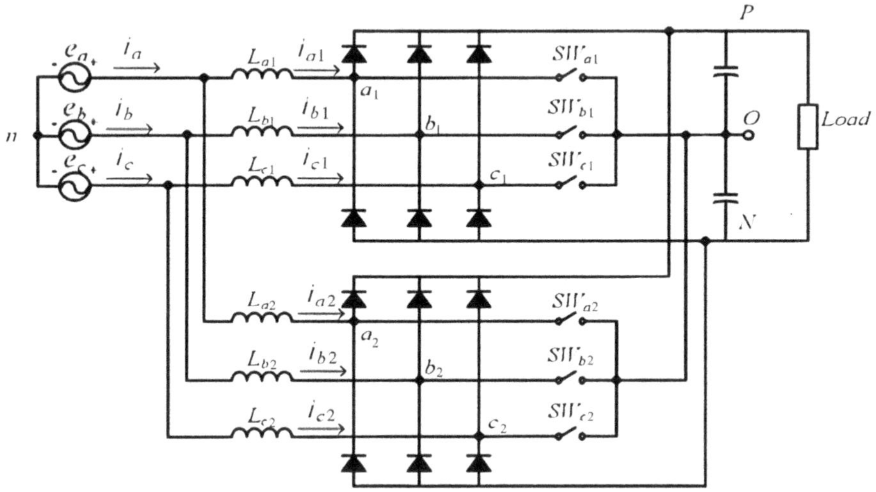

Figure 8.1 Schematic of two parallel-connected three-phase Vienna rectifiers.

switch settings and the polarity of AC currents. To define these states, the notion of the switching function will be utilized as follows:

$$S_{xk} = \begin{cases} 1, \text{ if } SW_{xk} \text{ turns off and } i_{xk} > 0 \\ 0, \text{ if } SW_{xk} \text{ turns on,} \\ -1, \text{ if } SW_{xk} \text{ turns off and } i_{xk} < 0. \end{cases} \tag{8.1}$$

Here, x is the phase index, k is the index of the rectifier, and S_{xk} represents the corresponding switching function. Thus, the voltage v_{xkO} can be determined using

$$v_{xkO} = \frac{V_{dc}}{2} S_{xk} \tag{8.2}$$

Thus, each phase (P, O, N) defines three possible states, influencing the input voltages that regulate AC currents and the corresponding ZSCC.

8.2 SPACE VECTOR MODULATION

In the α–β plane, one three-phase Vienna rectifier can be represented by 27 space vectors, categorized into three states P, O, and N, as shown in Figure 8.2. These vectors are crucial for generating controllable AC input voltages necessary for practical applications. This control is achieved by combining different vectors with specific dwell times [17], employing various modulation techniques described in the literature. Among these techniques,

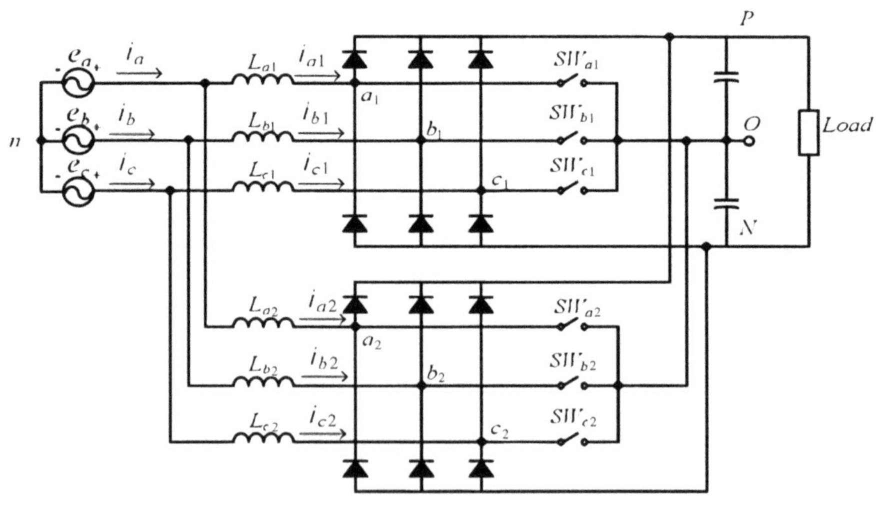

Figure 8.2 Twenty-seven space vectors used in a Vienna rectifier.

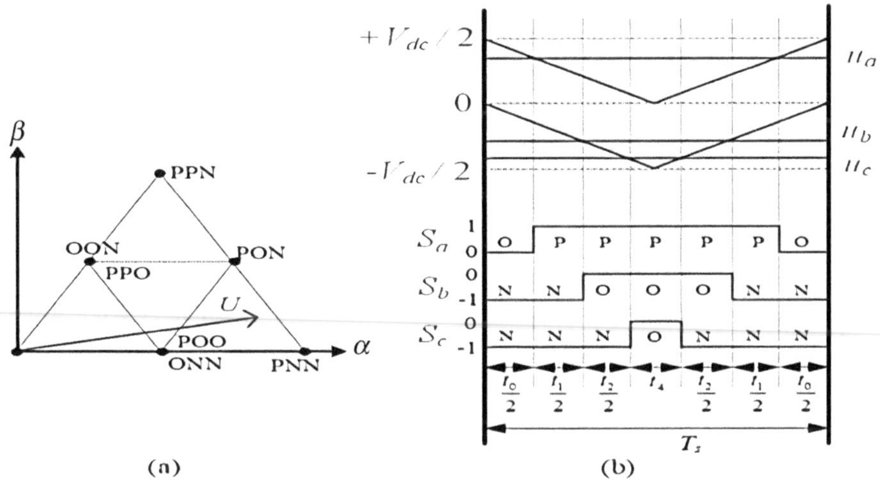

Figure 8.3 Voltage vectors utilized in PDPWM. (a) Reference vectors. (b) Carrier waveforms and corresponding switching sequences.

carrier-based modulation (CBM) is noteworthy for its computational efficiency and extensive use. We will apply one special CBM, called phase disposition pulse width modulation (PDPWM), for subsequent study.

When the PDPWM is applied, the voltage reference vectors (u_a, u_b, u_c) are considered to compare with two level-shifted carriers, as shown in Figure 8.3. The vector U represents the three-phase references in the α–β plane.

All possible space vectors are detailed in Table 8.1 [20]. It is worth mentioning that two vectors [PPP] and [NNN] will not particulate for a single Vienna rectifier. Nevertheless, these vectors still may appear in other applications.

When employing PDPWM, the corresponding voltage reference vector can be described as follows:

$$U = \frac{1}{T_s}\left(u_{\alpha\beta 1}t_1 + u_{\alpha\beta 2}t_2 + u_{\alpha\beta 3}t_3 + u_{\alpha\beta 4}t_4\right) \tag{8.3}$$

Table 8.1 Classification of space vectors

Name	*Type*	*Space vectors*
Zero vector	N-type	[N N N]
	O-type	[O O O]
	P-type	[P P P]
Small vector	N-type	[O N N], [N O O], [O O N], [N N O], [N O N], [O N O]
	P-type	[P O O], [O P P], [P P O], [O O P], [O P O], [P O P]
Median vector		[P O N], [O P N], [N P O], [N O P], [O N P], [P N O]
Large vector		[P N N], [P P N], [N P N], [N P P], [N N P], [P N P]

where $(u_{\alpha\beta 1}, u_{\alpha\beta 2}, u_{\alpha\beta 3})$ and $u_{\alpha\beta 4}$ represent the space vectors located at t_1, t_2, t_3, and t_4, respectively. In practice, there are two potential combinations of $u_{\alpha\beta 1}$ and $u_{\alpha\beta 4}$:

1. "Zero vector" → "Small vector" → "Small vector"→ a new "Zero vector"
2. "Small vector"→ Large vector" → "Median vector" → a new "Small vector"

$u_{\alpha\beta 1}$ and $u_{\alpha\beta 4}$ will occupy the same location, which will be either "Zero vectors" or "Small vectors," as depicted in Figure 8.2. Despite locating the same position, $u_{\alpha\beta 1}$ and $u_{\alpha\beta 4}$ differ in their vector definitions, as outlined in Table 8.1. Therefore, adjusting the ratio of t_1 and t_4 can modify the corresponding zero-sequence.

8.3 LF-ZSCC MITIGATION

First, the circulating current, which is defined by the difference in those current components between any two different rectifiers in the system, will be elucidated. Specifically, in the context of two parallel-connected rectifiers, it can be expressed as follows:

$$\left[i_{cir,a}, i_{cir,b}, i_{cir,c}\right]^T = \frac{1}{2}\left[i_{a1} - i_{a2}, i_{b1} - i_{b2}, i_{c1} - i_{c2}\right]^T \tag{8.4}$$

Under the d–q axis synchronous reference frame, these circulating currents in Equation (8.4) will be reformulated as follows [30, 33]:

$$\left[i_{cir,d}, i_{cir,q}, i_{cir,z}\right]^T = \frac{1}{2}\left[i_{d1} - i_{d2}, i_{q1} - i_{q2}, i_{z1} - i_{z2}\right]^T \tag{8.5}$$

Here, $i_{cir,d}$ and $i_{cir,q}$ denote the d–q axis circulating currents. These components will be mitigated using the conventional current control method already implemented in each individual rectifier. However, $i_{cir,z}$, referred to as the ZSCC, requires some designated control schemes.

To formulate an appropriate mathematical model for reducing ZSCC, we can apply the Kirchhoff's voltage law (KVL) to the system shown in Figure 8.1. Given that the resistance R_s is sufficiently small, its dynamic behavior can be disregarded. Hence, we can derive the following set of differential equations:

$$\begin{bmatrix} L_{ak}\dfrac{di_{ak}}{dt} \\ L_{bk}\dfrac{di_{bk}}{dt} \\ L_{ck}\dfrac{di_{ck}}{dt} \end{bmatrix} = \begin{bmatrix} e_a \\ e_b \\ e_c \end{bmatrix} - \begin{bmatrix} v_{akO} \\ v_{bkO} \\ v_{ckO} \end{bmatrix} - \begin{bmatrix} v_{nO} \\ v_{nO} \\ v_{nO} \end{bmatrix} \tag{8.6}$$

After applying the dqz-axis coordinate transformation, Equation (8.6) can be converted into the following set of differential equations [22]:

$$\left(\overline{L_k}+\frac{L_{c2\omega k}}{3}\right)\frac{di_{dk}}{dt}+\frac{L_{s2\omega k}}{3}\frac{di_{qx}}{dt}-\frac{\omega L_{s2\omega k}}{3}i_{dk}-\omega\left(\overline{L_k}+\frac{L_{c2\omega k}}{3}\right)i_{qk}$$
$$+\frac{2}{3}L_{c\omega k}\frac{di_{zk}}{dt}=e_d-v_{dk} \tag{8.7}$$

$$-\frac{L_{s2\omega k}}{3}\frac{di_{dk}}{dt}+\left(\overline{L_k}-\frac{L_{c2\omega k}}{3}\right)\frac{di_{qk}}{dt}+\omega\left(\overline{L_k}-\frac{L_{c2\omega k}}{3}\right)i_{dk}+\frac{\omega L_{s2\omega k}}{3}i_{qk}$$
$$-\frac{2}{3}L_{s\omega k}\frac{di_{zk}}{dt}=e_q-v_{qk} \tag{8.8}$$

$$\frac{L_{c\omega k}}{3}\frac{di_{dk}}{dt}-\frac{L_{s\omega k}}{3}\frac{di_{qk}}{dt}-\frac{\omega L_{s\omega k}}{3}i_{dk}+\frac{\omega L_{c\omega k}}{3}i_{qk}+\overline{L_k}\frac{di_{zk}}{dt}=e_z-v_{zk}+v_{n0} \tag{8.9}$$

where e_d, e_q, and e_z represent the d-axis, q-axis, and z-axis entries of the utility grid vector $\vec{e}$, respectively. Similarly, v_{dx}, v_{qx}, and v_{zx} denote the d-axis, q-axis, and z-axis entries, respectively, of the input voltage in the kth Vienna rectifier. Constants $L_{c\omega k}, L_{s\omega k}, L_{c2\omega k}, L_{s2\omega k}$, and $\overline{L_k}$ are defined as follows:

$$\begin{aligned}
L_{c\omega k} &= L_{ak}\cos(\omega t)+L_{bk}\cos(\omega t-2\pi/3)+L_{ck}\cos(\omega t+2\pi/3)\\
L_{s\omega k} &= L_{ak}\sin(\omega t)+L_{bk}\sin(\omega t-2\pi/3)+L_{ck}\sin(\omega t+2\pi/3)\\
L_{c2\omega k} &= L_{ak}\cos(2\omega t)+L_{bk}\cos(2\omega t-2\pi/3)+L_{ck}\cos(2\omega t+2\pi/3)\\
L_{s2\omega k} &= L_{ak}\sin(2\omega t)+L_{bk}\sin(2\omega t-2\pi/3)+L_{ck}\sin(2\omega t+2\pi/3)\\
\overline{L_k} &= \frac{1}{3}(L_{ak}+L_{bk}+L_{ck})
\end{aligned} \tag{8.10}$$

Prior to considering the ZSCC, it is essential to effectively control the i_{dk} and i_{qk}, particularly in the presence of unbalanced inductors. According to Equation (8.10), when the three-phase filter inductors L_{ak}, L_{bk}, and L_{ck} are identical, $L_{c\omega k}$, $L_{s\omega k}$, $L_{c2\omega k}$, and $L_{s2\omega k}$ become zero. Thus, Equations (8.7) and (8.8) can be simplified as follows with DC components only:

$$L_k\frac{di_{dk}}{dt}=e_d-v_{dk}-\omega L_k i_{qk} \tag{8.11a}$$

$$L_k\frac{di_{qk}}{dt}=e_q-v_{qk}+\omega L_k i_{dk} \tag{8.11b}$$

As discussed in Ref. [30], under balanced conditions, a conventional PI controller suffices for control purposes. However, when these filter inductors are not the same, nonzero values of $L_{c2\omega k}$ and $L_{s2\omega k}$ will introduce a 2ω frequency component into i_{dk} and i_{qk}. To attenuate 2ω component, the PI control is substituted by the following PIR control [31–34]:

$$K_{dq}(s) = k_p + \frac{k_i}{s} + \frac{k_r \cdot 2\omega_c s}{s^2 + 2\omega_c s + (2\omega_0)^2} = PI(s) + F(s) \tag{8.12}$$

In Equation (8.12), the conventional PI compensator is expressed as $PI(s) = k_p + \frac{k_i}{s}$. The resonant regulator $F(s) = \frac{k_r \cdot 2\omega_c s}{s^2 + 2\omega_c s + (2\omega_0)^2}$ included in Equation (8.12) incorporates the resonant gain k_r, the angular frequency of the grid voltage ω_0, and the cutoff frequency ω_c. In this PIR control, $PI(s)$ primarily regulates the DC components, while $F(s)$ effectively controls these double-frequency AC signals [35].

Since the Vienna rectifier is usually operated under the unity power factor, i_{qk} tends to zero. Therefore, it can be simplified to

$$\bar{L}_k \frac{di_{z_k}}{dt} = e_z + v_{nO} - v_{z_k} + \frac{\omega L_{sw_k}}{3} i_{d_k} \tag{8.13}$$

Since $i_{z1} = -i_{z2}$, Equation (8.13) can be rewritten as follows:

$$\left(\bar{L}_1 + \bar{L}_2\right)\frac{di_{z_1}}{dt} = -\left(v_{z_1} - v_{z_2}\right) + \frac{\omega L_{s\omega 1}}{3} i_{d_1} + \frac{\omega L_{s\omega 2}}{3} i_{d_2} \tag{8.14}$$

Thus, the dynamical behavior of the ZSCC in a parallel-connected Vienna rectifier can be described by Equation (8.14). When $L_{c2\omega k}$ and $L_{s2\omega k}$ are nonzero, and $v_{z1} \neq v_{z2}$, the ZSCC is generated, including a fundamental frequency component when $L_{s\omega k}$ is involved. To mitigate effects of LF-ZSCC, a PIR control will be used. Figure 8.4 illustrates the control scheme for mitigating this LF-ZSCC. By employing the internal model principle again, the corresponding current controller can be stated as follows:

$$K_z(s) = k_{pz} + \frac{k_{iz}}{s} + \frac{k_{rz} \cdot 2\omega_c s}{s^2 + \omega_c s + (2\omega_0)^2} = PI_z(s) + F_z(s) \tag{8.15}$$

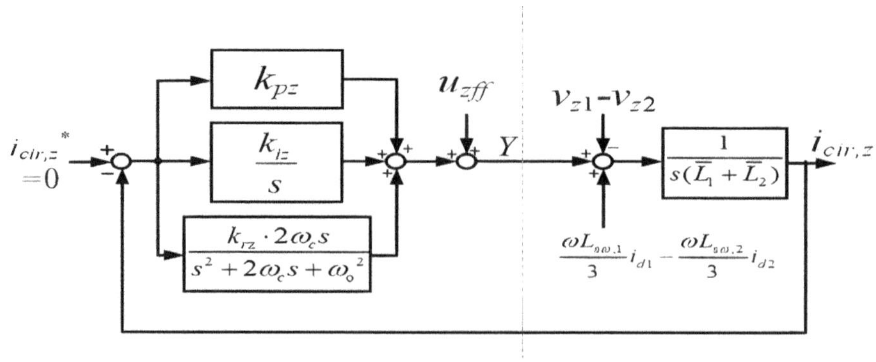

Figure 8.4 The proposed control diagram to alleviate $i_{cir,z}$.

where $PI_z(s) = k_{pz} + \frac{k_{iz}}{s}$ is the PI compensator and $F_z(s) = \frac{k_{rz} \cdot 2\omega_c s}{s^2 + 2\omega_c s + (2\omega_0)^2}$ is the z-axis circulating current compensator.

The output of the ZSCC control Y can be obtained by Equation (8.16) as the difference of u_{z1} and u_{z2}. Equation (8.5) specifies the ZSCC, where u_{zff} represents the feedforward control corresponding to the initial difference between u_{z1} and u_{z2}.

$$u'_{x1} = u_{x1} + Y/2, \text{ and } u'_{x2} = u_{x2} - Y/2 \tag{8.16}$$

8.4 HF-ZSCC MITIGATION

Interleaved operations are a commonly adopted method in parallel-connected converters to efficiently reduce passive component requirements, particularly in systems with multiple parallel-connected converters. Nevertheless, these interleaved operations may involve shifted switching patterns, leading to some voltage drops occurring at staggered times within the switching period. This phenomenon results in the generation of substantial HF-ZSCC.

To illuminate this phenomenon, let us consider two parallel-connected Vienna rectifiers with closely spaced input voltage references in phase a. These switching patterns under conventional interleaved operations can be shown in Figure 8.5. Under this situation, $V_{a1} - V_{a2}$ can be treated as the voltage source of the circulating current i_{cir}. These fluctuations will constitute the HF-ZSCC. Notably, $i_{cir,a}$ reaches its maximum when $u_a = V_{dc}/4$, which coincides with the

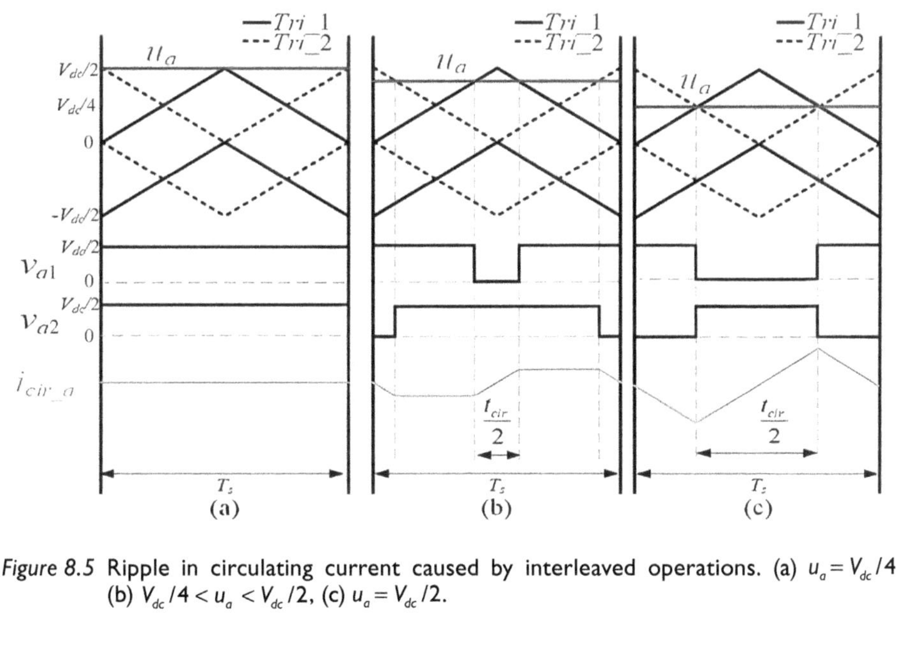

Figure 8.5 Ripple in circulating current caused by interleaved operations. (a) $u_a = V_{dc}/4$ (b) $V_{dc}/4 < u_a < V_{dc}/2$, (c) $u_a = V_{dc}/2$.

longest circulation period t_{cir}. Conversely, $i_{cir,a}=0$ if $u_a=\pm V_{dc}/2$ or $u_a=0$, since the voltage waveform of phase a in these two Vienna rectifiers is exactly the same. Their relationship can be stated as follows:

$$i_{cir,x,peak} \propto \min\left(\left|u_x + V_{dc}/2\right|, \left|u_x\right|, \left|u_x - V_{dc}/2\right|\right) \tag{8.17}$$

Eliminating HF-ZSCC can be challenging. In the past, various technological solutions have been considered, including (i) adding interphase reactors [36, 37] or (ii) modifying these LCL filters [16]. However, integrating these additional components inevitably increases the overall system cost. An alternative approach is to employ a DPWM control scheme to address HF-ZSCC without modifying the conventional circuit topology, thereby providing a more cost-effective solution.

To reduce the HF-ZSCC, it is important to position the voltage references away from $\pm V_{dc}/4$ and closer to $\pm V_{dc}/2$ or 0. A DPWM can achieve this goal by utilizing the zero sequence injection (ZSI) method, which can be stated as follows:

$$u''_{xk} = u'_{xk} + u_0 \tag{8.18}$$

Here, u'_{xk} and u''_{xk} represent the voltage reference before and after activating the ZSI, respectively. u_0 is the command signal to generate the zero-sequence. To minimize these ripples, the following optimization problem can be formulated as stated in Equation (8.17) for determining $u_0 = \arg\min Z(u_0)$:

$$Z(u_0) = \sum_{x=a,b,c,k=1,2} \min\left(\left|u''_{xk} + V_{dc}/2\right|, \left|u''_{xk}\right|, \left|u''_{xk} - V_{dc}/2\right|\right) \tag{8.19}$$

To delve into DPWM, we begin by applying it to a single Vienna rectifier. Let u_{max}, u_{mid}, and u_{min} represent the maximum value, the middle value, and the minimum value among three-phase variables u_a, u_b, and u_c, respectively. $d_{top} = V_{dc}/2 - u_{max}$ is the distance from u_{max} to $V_{dc}/2$, $d_{zero} = \left|u_{mid}\right|$ as the distance from u_{mid} to zero, and $d_{bottom} = V_{dc}/2 + u_{min}$ as the distance from u_{min} to $-V_{dc}/2$.

When $u_{mid} > 0$, as shown in Figure 8.6, three cases can be categorized by the number of phases with voltage references exceeding $\pm V_{dc}/4$:

1. All three-phase voltage references are less than or equal to $\pm V_{dc}/4$, as depicted in Figure 8.6a. Here, applying a negative u_0 reduces the ripples of $i_{cir,max}$ and $i_{cir,mid}$ since $i_{cir,max}$ and $i_{cir,mid}$ move farther from $V_{dc}/4$. However, the circulating current $i_{cir,mid}$ increases as u_{min} approaches $-V_{dc}/4$. All three voltages will vary linearly, resulting in decreased circulating currents for two phases and an increased current for one phase, thereby mitigating the HF-ZSCC.

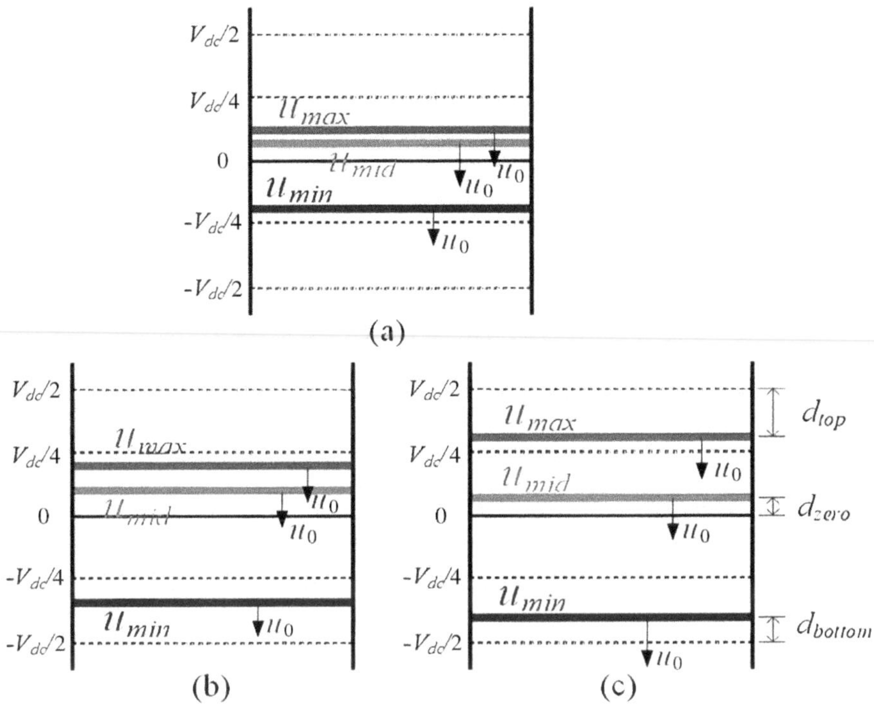

Figure 8.6 Scenarios for determining u_0 in a Vienna rectifier.

2. One-phase voltage reference is larger than $\pm V_{dc}/4$, shown in Figure 8.6b. Under this situation, enabling a negative u_0 will decrease these ripples of circulating currents.
3. Two-phase voltage references will surpass $\pm V_{dc}/4$, illustrated in Figure 8.6c. This result is similar to case (1). For brevity, we omit the details.

It is important to note that in practical implementations the value of u_0 is confined by d_{zero} and d_{bottom} due to these modulation constraints. The strategy for reducing HF-ZSCC by choosing u_0 when $u_{mid} < 0$ can be further illustrated by expanding upon the scenarios outlined earlier, as summarized in Figure 8.7. Essentially, u_0 can be chosen to *drive the nearest phase towards* $\pm V_{dc}/2$ *or 0*. When a modulation index M is set to 0.8, simulation waveforms in Figure 8.8 verify this rule. This figure can clearly demonstrate that when employing the DPWM scheme, the periods during which voltage references reach $\pm V_{dc}/2$ and 0 are extended.

However, directly implementing the DPWM scheme from Figure 8.7 appears unsuitable for parallel multiple Vienna rectifiers due to potential ZSCC arising from differing zero-sequence voltage commands. Therefore, modifications to Figure 8.7 are needed. Specific details are omitted here. In summary, Table 8.2 outlines our DPWM scheme tailored for multiple

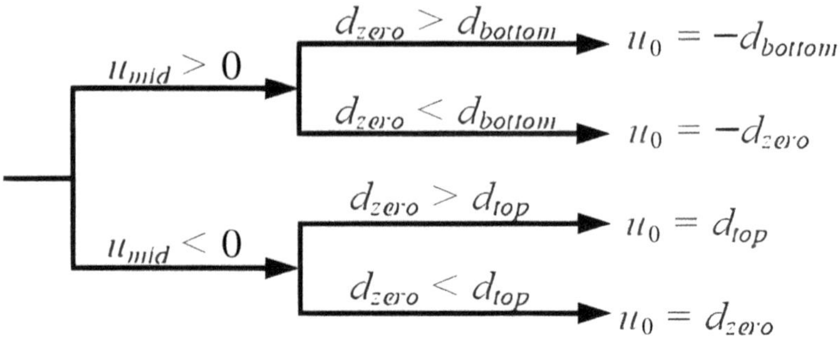

Figure 8.7 Approach to determine u_0 for alleviating the HF-ZSCC.

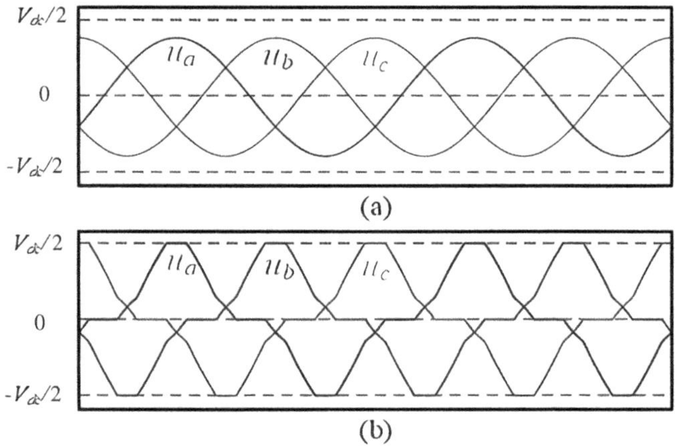

Figure 8.8 Reference waveform with a modulation index of 0.8. (a) Without applying DPWM. (b) Applying DPWM.

Table 8.2 DPWM for two parallel-connected Vienna rectifiers

Case 1	$u_{mid1} > 0, u_{mid2} > 0$			
Minimum term	d_{zero1}	$d_{bottom1}$	d_{zero2}	$d_{bottom2}$
u_0	$-d_{zero1}$	$-d_{bottom1}$	d_{zero2}	$d_{bottom2}$
Case 2	$u_{mid1} > 0, u_{mid2} < 0$			
Minimum term	d_{zero1}	$d_{bottom1}$	d_{zero2}	d_{top2}
u_0	$-d_{zero1}$	$-d_{bottom1}$	d_{zero2}	d_{top2}
Case 3	$u_{mid1} < 0, u_{mid2} > 0$			
Minimum term	d_{zero1}	d_{top1}	d_{zero2}	$d_{bottom2}$
u_0	$-d_{zero1}$	d_{top1}	$-d_{zero2}$	$-d_{bottom2}$
Case 4	$u_{mid1} < 0, u_{mid2} < 0$			
Minimum term	d_{zero1}	d_{top1}	d_{zero2}	d_{top2}
u_0	d_{zero1}	d_{top1}	d_{zero2}	d_{top2}

parallel-connected Vienna rectifiers. This table involves identifying the shortest distance among $d_{top,k}$, $d_{zero,k}$, and $d_{bottom,k}$, and then adjusting the nearest phase to $\pm V_{dc}/2$ or 0, which can be intuitively understood.

8.4.1 Extensions to multiple parallel-connected rectifiers

From our extensive review of the literature, there is no coherent definition about the ZSCC in n parallel-connected Vienna rectifiers. Nonetheless, by carefully controlling i_{dk}, i_{qk}, and i_{zk} as described in Equation (8.5) using the DPWM, it is still possible to achieve the acceptable dynamical performance by following the design guideline below:

1. *Management of Circulating Current*: For the kth rectifier, i_{dk} and i_{qk} are regulated by u_{dk} and u_{qk}, respectively.
2. *LF-ZSCC Mitigation:* A sequential approach is proposed. Initially, the first two parallel-connected rectifiers employ a PIR controller, depicted in Figure 8.4. Once the ZSCC between the first and second rectifiers is alleviated, the control action of the third rectifier will be enabled. This iterative process continues until all rectifiers are addressed, as illustrated in Figure 8.9. If n rectifiers are considered, n – 1 PIR control are

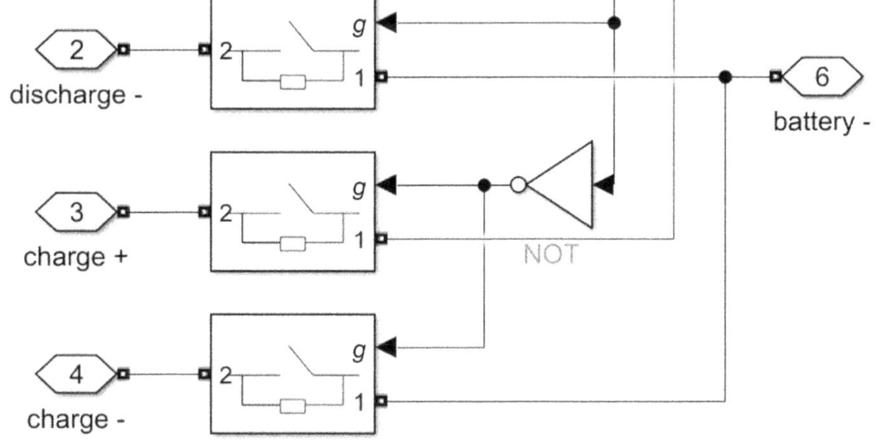

Figure 8.9 ZSCCs for *n* parallel-connected Vienna rectifiers.

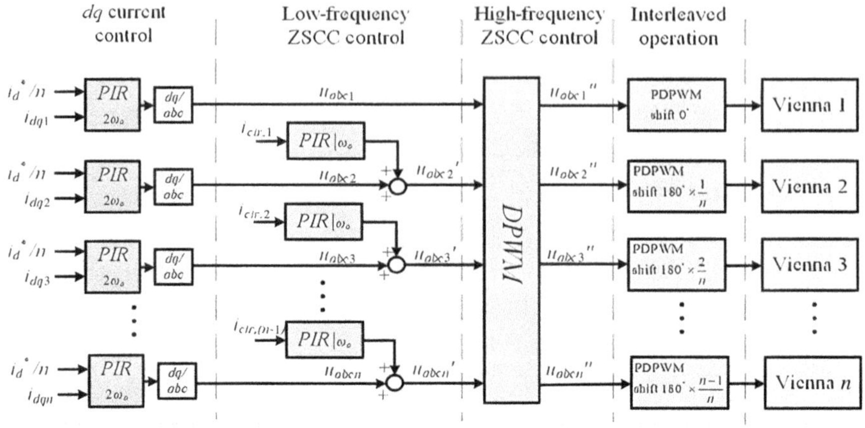

Figure 8.10 The proposed control block diagram for *n* parallel-connected Vienna rectifiers.

required to nullify all i_{zk}. The formulation for ZSCC in the *k*th PIR control can be stated in Equation (8.20), and these outputs are designated in Equation (8.21):

$$i_{cir,k} = \left(\sum_{n=1}^{k} i_{zn}\right) - i_{z(k+1)},\; v_{zff,k} = \sum_{n=1}^{k} u_{zn} \tag{8.20}$$

$$u'_{x(k+1)} = u_{x(k+1)} + Y_k \tag{8.21}$$

3. *HF-ZSCC Mitigation:* By following the principles outlined in Table 8.2, u_0 can be determined.

In summary, Figure 8.10 illustrates the integrate control diagram for *n* parallel-connected three-phase Vienna rectifiers.

8.5 SIMULATION VALIDATIONS

Extensive simulations were performed by using PLECS to analyze the system as shown below.

8.5.1 Linear loads

All parameters utilized under study are outlined in Table 8.3. An internal resistance $R_s = 0.1\ \Omega$ was included to model the internal resistance of the reactor. With a load rating $P_{load} = 30$ kW, each phase of the system corresponds to a resistive load of 21.33 Ω. Notably, the inductors of the input filter in the first

Table 8.3 System parameters under study.

V_{dc}	800 V
E_{abc} (RMS)	380 V
L_{a1}, L_{b1}, L_{c1}	0.7 mH, 1 mH, 1.3 mH
L_{a2}, L_{b2}, L_{c2}	1 mH, 1 mH, 1 mH
L_{a3}, L_{b2}, L_{c2}	1 mH, 1 mH, 1 mH
C	1,250 uF
f_o	60 Hz
f_s	30 kHz
Current control loop	$k_p = 10, k_i = 50, k_r = 50, \omega_c = 7.5$
ZSCC control loop	$k_{pz} = 1, k_{iz} = 5, k_{r1} = 5, \omega_c = 7.5$
Digitization	Zero-order hold (ZOH)

rectifier exhibit significant imbalance, differing by 30%. The design proceeds as follows:

1. *Current Control in d–q Axis*: The time constant of the current closed-loop system was selected as $\tau = 0.5$ ms. This choice balances rapid current control response with a sufficiently narrow bandwidth for stable closed-loop control. To enhance the frequency response, the pole-zero cancellation technique will be employed [30, p. 51]. Therefore, the parameters k_p and k_i for the PI compensator $PI(s)$ are chosen as follows:

$$k_p = \overline{L_k}/\tau = 10 \text{ and } k_i = R_s/\tau = 200 \tag{8.22}$$

   To compensate for these double-frequency current components in the rotating synchronous frame, a resonant control described by Equation (8.12) is recommended. Considering AC grid voltage frequency fluctuations and digital control numerical errors [31], the cutoff frequency ω_c is set to ±2% of the grid frequency ω_0, resulting in ω_0 = 7.536 rad/s. When $k_r = 50$, the Bode plot of the d–q axis current control is depicted in Figure 8.11, illustrating phase margin (PM) of 85.1° and gain margin (GM) of 35.6 dB.
2. *Current Control in z-Axis*: k_{pz}, k_{iz}, and k_{rz} in $K_z(s)$ adhere to the principles outlined earlier. We chose $\omega_c = 7.5$, $k_{pz} = 1$, $k_{iz} = 5$, and $k_{rz} = 5$. To validate the stability margins of this design, the Bode plot of its open-loop z-axis circulating current is presented in Figure 8.12. Thus, this closed-loop system has phase margin (PM) of 73.3° and gain margin (GM) of 41.6 dB.

To validate the proposed control scheme, extensive simulations were conducted. Figure 8.13 illustrates time-domain waveforms of i_{z1}, i_{z2}, and i_{z3} along with the three-phase input AC current. In Figure 8.13a, without ZSCC

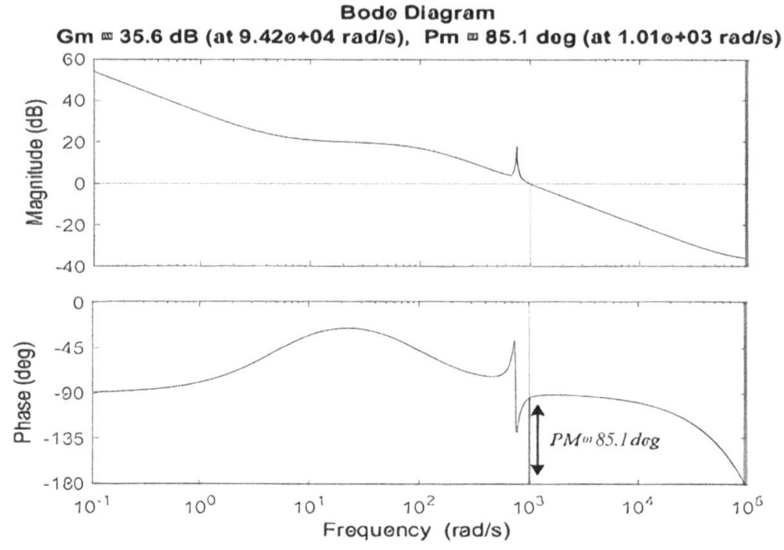

Figure 8.11 Bode plot of the open-loop current in *d–q* axis.

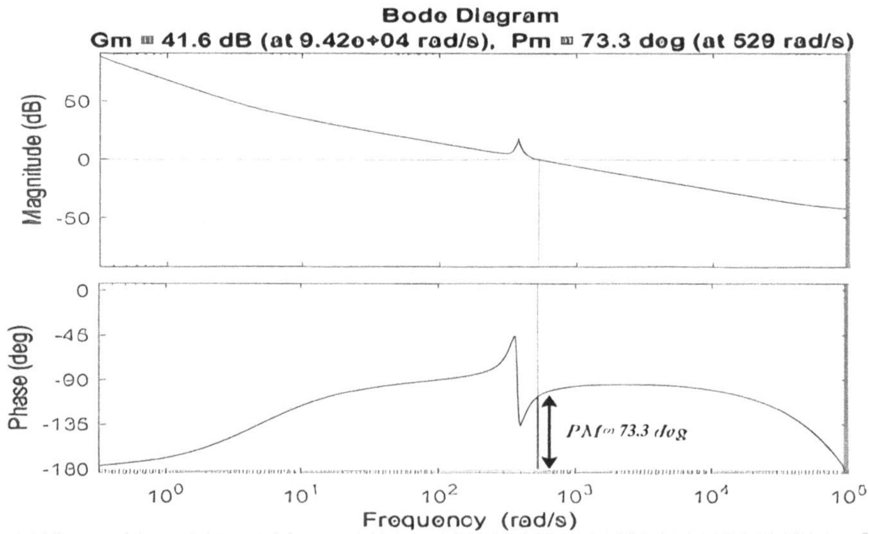

Figure 8.12 Bode plot of the open-loop current in z-axis.

control, i_{z1}, i_{z2}, and i_{z3} exhibit a fundamental frequency component due to the existence of the unequal inductors used in input filters. Additionally, interleaved operation introduces HF-ZSCC, evident from the increased amplitude of i_{z1}, i_{z2}, and i_{z3} due to high-frequency components. Figure 8.13b demonstrates the mitigation effect when the LF-ZSCC control is enabled, even though the

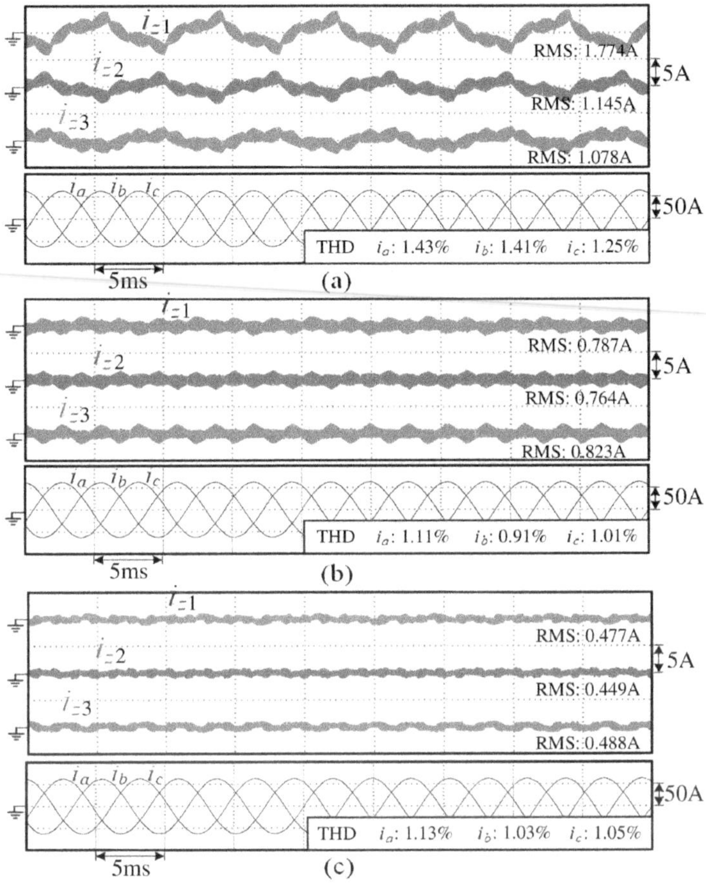

Figure 8.13 Simulation waveforms. (a) *d–q* axis current control only. (b) *d–q* axis current control + LF-ZSCC control. (c) The proposed integrated control scheme.

HF-ZSCC remains noticeable. With the proposed method, as shown in Figure 8.13c, HF-ZSCC is effectively mitigated, resulting in narrower waveforms for i_{z1}, i_{z2}, and i_{z3} compared to Figure 8.13b. Furthermore, significant improvements of total harmonic distortion (THD) in the input AC current waveforms can be observed.

Further analysis about the frequency spectrum of i_{z1} is presented in Figure 8.14. In Figure 8.14a, both LF-ZSCC and HF-ZSCC are evident. When the LF-ZSCC control is activated, as shown in Figure 8.14b, these LF components are reduced. However, these HF components remain present. In contrast, employing the proposed DPWM scheme, Figure 8.14c clearly demonstrates mitigation of the HF component.

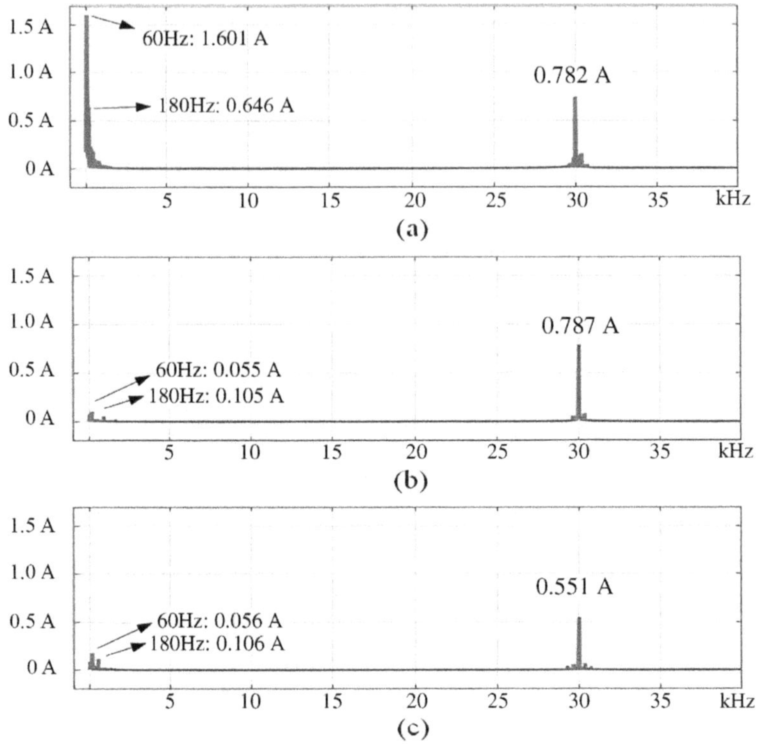

Figure 8.14 Frequency spectrum of i_{z1}. (a) *d–q* axis current control only. (b) *d–q*-axis current control + LF-ZSCC control. (c) The proposed integrated control.

To confirm the efficacy of the PIR control, we compare its output waveforms with the PI control. Figure 8.15 illustrates the frequency spectrum of i_{z1} using both controllers. While both PI and PIR controls notably can reduce ZSCC, the PIR control demonstrates more effective suppression of the ω component. Similar benefits are observed in the d–q axis current, where PIR control effectively manages the 2ω component in i_{dk} and i_{qk}.

8.5.2 Nonlinear loads

To further illustrate the practical application of the proposed scheme in EV charging stations, we examine a buck converter with a resistor simulating a nonlinear load in EV charging stations.

It is worth noting that since the circulating currents generated by nonlinear loads are treated as disturbance sources in the proposed control scheme, they do not impact the dynamic performance of the method. Their attenuation rates can be determined by the transfer function of the closed-loop system.

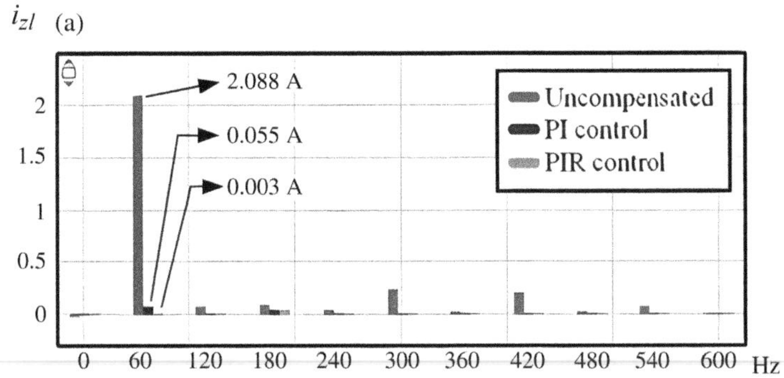

Figure 8.15 Frequency spectrum of i_{zl}—without any control actions, PI control, and PIR control.

Figure 8.16 shows the circuit diagram. A DC voltage source V_o = 400 V is considered [38]. The nonlinear DC load is rated at 30 kW, corresponding to a resistor value of 5.33 Ω, as indicated in Figure 8.17. Other circuit parameters such as inductors and capacitors are outlined in Ref. [39, p. 260].

Figure 8.18 illustrates the time-domain waveforms. In Figure 8.18a, LF-ZSCC as well as HF-ZSCC are noticeable when *d–q* axis current control is employed. Conversely, Figure 8.18b demonstrates the efficacy of the proposed integrated control. Consequently, the input AC current exhibits reduced THD.

We also study the frequency spectrum of the ZSC. Figure 8.19a shows that if *d–q* axis current control is activated, both LF-ZSCC HF-ZSCC are clearly visible. By contrast, Figure 8.19b clearly shows that if the proposed integrated control method is applied, both LF-ZSCC and HF-ZSCC are significantly mitigated.

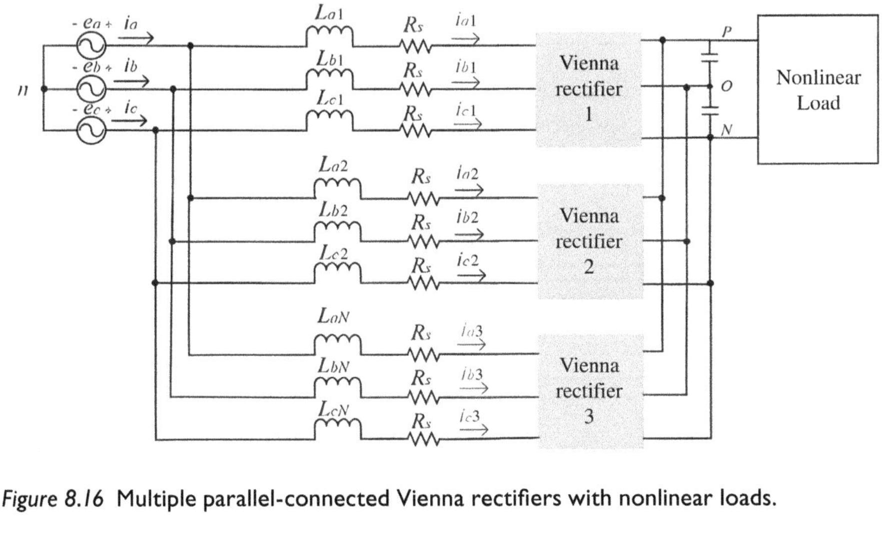

Figure 8.16 Multiple parallel-connected Vienna rectifiers with nonlinear loads.

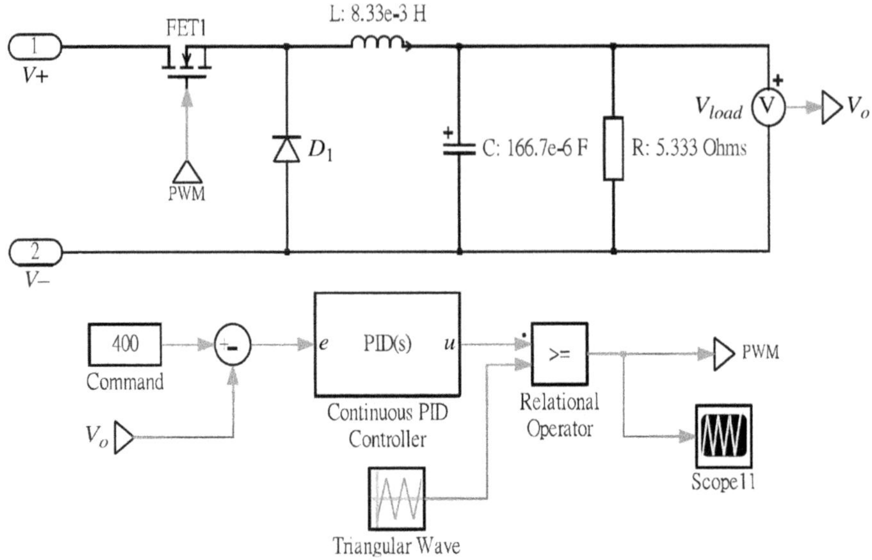

Figure 8.17 The nonlinear DC load used in our study.

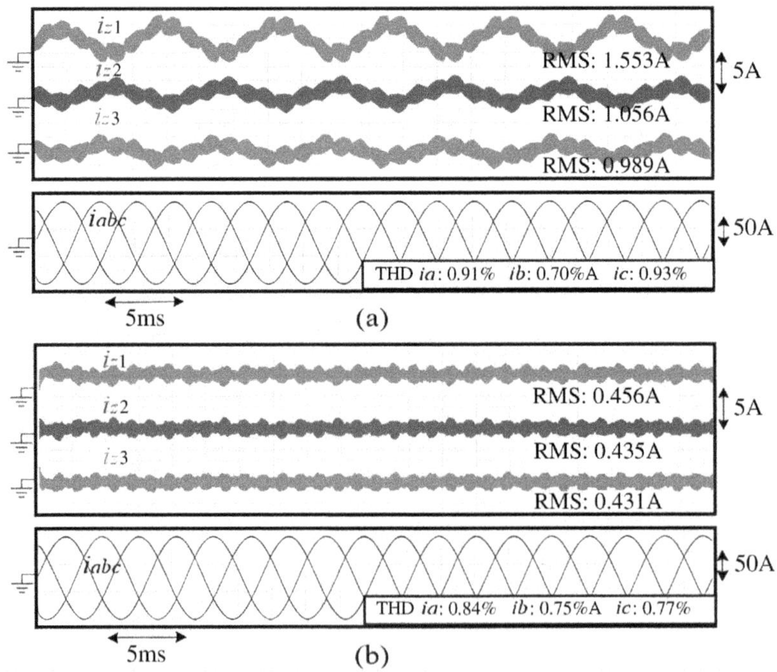

Figure 8.18 Time-domain waveforms of multiple parallel-connected Vienna rectifiers with nonlinear loads. (a) *d–q* axis current control only. (b) The proposed integrated control scheme.

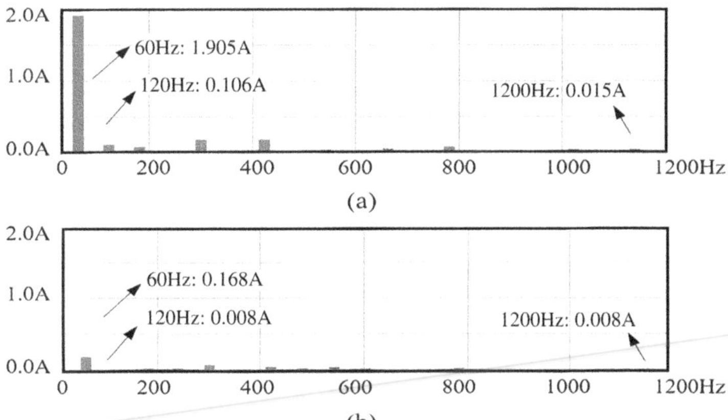

Figure 8.19 Frequency spectrum of i_{z1} with nonlinear loads. (a) *d–q* axis current control only. (b) The proposed integrated control scheme.

8.5.3 Experiment validation

Hardware laboratory experiments were also conducted to validate the performance of the proposed control scheme. In our laboratory, a system with two parallel-connected three-phase Vienna rectifiers are developed. Two rectifiers are controlled only by one digital signal processing processor TMS320F28335 [40–41].

The hardware setup of the entire system is illustrated in Figure 8.20. Their system parameters are detailed in Table 8.4. Controller parameters such

Figure 8.20 Hardware setup of the system in our laboratory.

Table 8.4 Parameters of parallel-connected Vienna rectifiers in laboratory experiments

V_{dc}	800 V
e_{abc} (RMS)	380 V
P_{load}	6 kW
R_s	0.1 Ω
L_{a1}, L_{b1}, L_{c1}	5 mH, 8 mH, 10 mH
L_{a2}, L_{b2}, L_{c2}	1 mH, 1 mH, 1 mH
L_{a3}, L_{b2}, L_{c2}	5 mH, 5 mH, 5 mH
C	2,240 uF
f_o	60 Hz
f_s	15 kHz
Current control loop	$k_p = 15, k_i = 100, k_r = 100, \omega_c = 7.5$
ZSCC control loop	$k_{pz} = 15, k_{iz} = 100, k_{r1} = 100, \omega_c = 7.5$
Digitization	Zero-order hold (ZOH)

as $k_p, k_i, k_r, k_{pz}, k_{iz}, k_{rz}$, and ω_c used in these PIR control adhere to the design guideline outlined earlier.

In our experiments, the first rectifier was equipped with highly unequal filter inductors with 100% difference. Figure 8.21 illustrates time-domain waveforms with and without the proposed integrated control scheme. It is evident that these results can demonstrate that the improvement of the proposed method is significant since the RMS of the ZSCC can be reduced. Moreover, i_{z1} becomes more stable and exhibits reduced fluctuations.

While the THD of the input current in Figure 8.21b shows a slight increase compared to Figure 8.21a, the reduction in losses appears more pronounced.

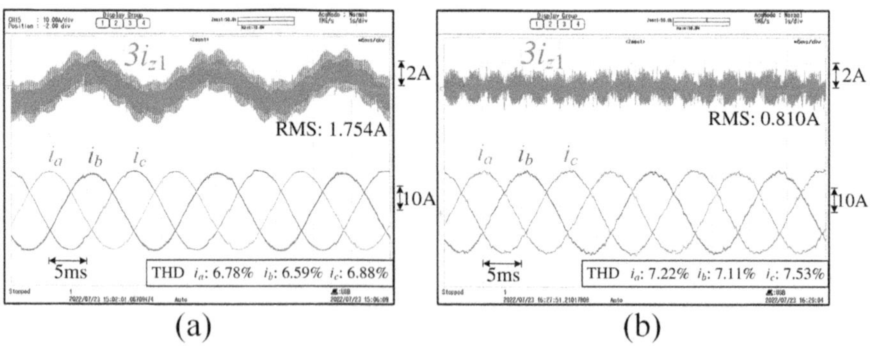

Figure 8.21 Experiment results under different control schemes. (a) *d–q* axis current control only. (b) The proposed integrated control scheme.

8.6 CONCLUSION

This chapter introduces an innovative control method tailored for parallel-connected three-phase Vienna rectifiers equipped with uneven filter inductors under interleaved operations. The primary objective is to alleviate these circulating currents. To achieve this, the chapter explores the PDPWM scheme to reduce computational complexity and utilizes space vectors to generate an adjustable ZSC.

Additionally, the chapter develops an average model and examines the dynamics of LF-ZSCC. PIR control is employed in the inner loop to alleviate the LF-ZSCC. A DPWM scheme is suggested to find ZSI for mitigating the HF-ZSCC. Possible extensions to multiple parallel-connected rectifiers are also discussed. Computer simulations as well as laboratory experiments have been conducted to confirm the efficacy of the proposed integrated control method, demonstrating its practical applicability and performance improvements in mitigating ZSCC in parallel Vienna rectifiers.

The main challenge in scaling the proposed ZSCC mitigation techniques for large-scale EV charging systems with hundreds of Vienna rectifiers is the increased complexity of the control scheme shown in Figure 8.9, making analytical study more difficult. Fortunately, significant progress has been made in microgrid control, which enables us to address this scaling issue through a distributed control scheme. Moreover, further investigation could address the following issues:

1. Integrating Vienna rectifiers with various types of renewable energy sources.
2. Examining the effect of temperature variations in the power electronic components of Vienna rectifiers on ZSCC and the efficiency of the proposed mitigation method.
3. Assessing the impact of high-frequency electromagnetic interference (EMI) on control schemes used to mitigate ZSCC in Vienna rectifiers.
4. Analyzing the effects of communication delays and synchronization challenges between parallel-connected Vienna rectifiers in real-time control environments.
5. Evaluating the performance of the proposed control schemes when Vienna rectifiers are used with fast or ultrafast EV chargers.

We hope to report further progress on these unresolved issues in the near future.

REFERENCES

1. S.-C. Peng, Z.-L. Li, S. C. Gulipalli, W.-C. Chen, and C.-C. Chu, "Mitigating circulating currents of parallel three-phase Vienna rectifiers with unbalanced filter inductors," in *IEEE Trans. Ind. Appl.*, vol. 59, no. 5, pp. 5804–5817, 2023.

2. J. W. Kolar, U. Drofenik, and F. C. Zach, "VIENNA Rectifier II: A novel single-stage high-frequency isolated three-phase PWM rectifier system," in *IEEE Trans. Ind. Electron.*, vol. 46, no. 4, pp. 674–691, 1999.
3. M. Hartmann, S. D. Round, H. Ertl, and J. W. Kolar, "Digital current controller for a 1 MHz, 10 kW three-phase VIENNA rectifier," in *IEEE Trans. Power Electron.*, vol. 24, no. 11, pp. 2496–2508, 2009.
4. J.-S. Lee, K.-B. Lee, and F. Blaabjerg, "Predictive control with discrete space-vector modulation of Vienna rectifier for driving PMSG of wind turbine systems," in *IEEE Trans. Power Electron.*, vol. 34, no. 12, pp. 12368–12383, 2019.
5. X. Xing, X. Li, C. Qin, Z. Liu, and C. Zhang, "Two-layer pulse width modulation strategy for common-mode voltage and current harmonic distortion reduction in Vienna rectifier," in *IEEE Trans. Ind. Electron.*, vol. 67, no. 9, pp. 7470–7483, 2020.
6. M. A. H. Rafi and J. Bauman, "A comprehensive review of DC fast-charging stations with energy storage: Architectures, power converters, and analysis," in *IEEE Trans. Transp. Electrific*, vol. 7, no. 2, pp. 345–368, 2021.
7. W. Han, X. Li, C. Qin, X. Pang, and C. Zhang, "Decoupling control of circulating current suppression and current distortion elimination for the paralleled Vienna-type rectifiers," in *2022 IEEE Energy Conversion Congress* and Exposition (ECCE), IEEE, pp. 1–6, 2022.
8. W. Ding, C. Zhang, F. Gao, B. Duan, and H. Qiu, "A zero-sequence component injection modulation method with compensation for current harmonic mitigation of a Vienna rectifier," in *IEEE Trans. Power Electron.*, vol. 34, no. 1, pp. 801–814, 2019.
9. Z. Miao, H. Tong, X. Jin, W. Yao, Z. Lu, and Z. Ma, "DQ-frame zero-crossing effect modeling and current distortion compensation method for Vienna rectifier," in *IEEE Trans. Power Electron.*, vol. 35, no. 7, pp. 7612–7623, 2020.
10. C.-T. Pan and Y.-H. Liao, "Modeling and coordinate control of circulating currents in parallel three-phase boost rectifiers," in *IEEE Trans. Ind. Electron.*, vol. 54, no. 2, pp. 825–838, 2007.
11. T.-P. Chen, "Circulating zero-sequence current control of parallel three-phase inverters," in *IEE Proc. Electric Power Appl.*, vol. 153, no. 2, pp. 282–288, 2006.
12. T.-P. Chen, "Dual-modulator compensation technique for parallel inverters using space-vector modulation," in *IEEE Trans. Ind. Electron.*, vol. 56, no. 8, pp. 3004–3012, 2009.
13. T.-P. Chen, "Zero-sequence circulating current reduction method for parallel HEPWM inverters between AC bus and DC bus," in *IEEE Trans. Ind. Electron.*, vol. 59, no. 1, pp. 290–300, 2012.
14. H. Chen, and X. Xing, "Circulating current analysis and suppression for module grid-connected inverters under unbalanced conditions," in *IEEE Access*, vol. 6, pp. 69120–69129, 2018.
15. Z. Xueguang, W. Li, Y. Xiao, G. Wang, and D. Xu, "Analysis and suppression of circulating current caused by carrier phase difference in parallel voltage source inverters with SVPWM," in *IEEE Trans. Power Electron.*, vol. 33, no. 12, pp. 11007–11020, 2018.
16. Z. Shao, X. Zhang, F. Wang, and R. Cao, "Modeling and elimination of zero-sequence circulating currents in parallel three-level *t*-type grid-connected inverters," in *IEEE Trans. Power Electron.*, vol. 30, no. 2, pp. 1050–1063, 2015.
17. K. Sun, X. Lin, Y. Li, Y. Gao, and L. Zhang, "Improved modulation mechanism of parallel-operated *t*-type three-level PWM rectifiers for neutral-point potential balancing and circulating current suppression," in *IEEE Trans. Power Electron.*, vol. 33, no. 9, pp. 7466–7479, 2018.

18. X. Xing, X. Li, C. Qin, J. Chen, and C. Zhang, "An optimized zero-sequence voltage injection method for eliminating circulating current and reducing common mode voltage of parallel-connected three-level converters," in *IEEE Trans. Ind. Electron.*, vol. 67, no. 8, pp. 6583–6596, 2020.
19. X. Liu, T. Liu, A. Chen, X. Xing, and C. Zhang, "Circulating current suppression for paralleled three-level *t*-type inverters with online inductance identification," in *IEEE Trans. Ind. Appl*, vol. 57, no. 5, pp. 5052–5062, 2021.
20. A. Chen, Z. Zhang, X. Xing, K. Li, C. Du, and C. Zhang, "Modeling and suppression of circulating currents for multi-paralleled three-level *t*-type inverters," in *IEEE Trans. Ind. Appl.*, vol. 55, no. 4, pp. 3978–3988, 2019.
21. X. Liu, T. Liu, A. Chen, X. Xing, and C. Zhang, "Suppression of circulating currents for paralleled three-level *t*-type inverters under unbalanced operating conditions," in *2020 IEEE* Applied Power Electronics Conference and Exposition (APEC), IEEE, pp. 3003–3007, 2020.
22. C. Qin, C. Zhang, A. Chen, X. Xing, and G. Zhang, "Circulating current suppression for parallel three-level inverters under unbalanced operating conditions," in *IEEE J. Emerg. Sel. Topics Power Electron.*, vol. 7, no. 1, pp. 480–492, 2019.
23. B. Ren, X. Sun, M. Yu, J. Liu, and Q. Zhang, "Circulating current analysis and the improved Δ–Σ digital control strategy for multi-paralleled three-level *t*-type grid-connected inverters," in *IEEE Trans. Ind. Electron.*, vol. 67, no. 4, pp. 2810–2821, 2020.
24. Y. Fu, N. Cui, Z. Chen, and C. Zhang, "An improved finite-time control strategy for zero-sequence circulating current suppression of parallel three-level rectifiers," in *IEEE J. Emerg. Sel. Topics Power Electron.*, vol. 8, no. 4, pp. 3933–3943, 2020.
25. L. Hang, B. Li, M. Zhang, Y. Wang, and L. M. Tolbert, "Equivalence of SVM and carrier-based PWM in three-phase/wire/level Vienna rectifier and capability of unbalanced-load control," in *IEEE Trans. Ind. Electron.*, vol. 61, no. 1, pp. 20–28, 2014.
26. J.-S. Lee and K.-B. Lee, "Carrier-based discontinuous PWM method for Vienna rectifiers," in *IEEE Trans. Power Electron.*, vol. 30, no. 6, pp. 2896–2900, 2015.
27. Q. Wang, X. Zhang, R. Burgos, D. Boroyevich, A. M. White, and M. Kheraluwala, "Design and implementation of a two-channel interleaved Vienna-type rectifier with >99% efficiency," in *IEEE Trans. Power Electron.*, vol. 33, no. 1, pp. 226–239, 2018.
28. T. Wang, C. Chen, P. Liu, T. Liu, Z. Chao, and S. Duan, "A hybrid space-vector modulation method for harmonics and current ripple reduction of interleaved Vienna rectifier," in *IEEE Trans. Ind. Electron.*, vol. 67, no. 10, pp. 8088–8099, 2020.
29. L. Song, S. Duan, R. Li, X. Liu, and B. Ji, "A hybrid discontinuous PWM strategy for current ripple and neutral-point fluctuation reduction of parallel Vienna rectifier," in *IEEE Trans. Ind. Electron.*, vol. 70, no. 3, pp. 2531–2542, 2023.
30. A. Yazdani and R. Iravani, *Voltage-Sourced Converters in Power Systems: Modeling, Control, and Applications*, Hoboken, NJ: Wiley, 2010.
31. M. Liserre, R. Teodorescu, and F. Blaabjerg, "Multiple harmonics control for three-phase grid converter systems with the use of PI-RES current controller in a rotating frame," in *IEEE Trans. Power Electron.*, vol. 21, no. 3, pp. 836–841, 2006.
32. S. Buso and P. Mattavelli, *Digital Control in Power Electronics*, Synthesis Lectures on Power Electronics, vol. 2, San Rafael, CA: Morgan & Claypool Publisher, 2006.

33. R. Teodorescu, M. Liserre, and P. Rodriguez, *Grid Converters for Photo-Voltaic and Wind Power Systems*. Hoboken, NJ: Wiley, 2011.
34. S. N. Vukosavic, *Grid-Side Converters Control and Design*, Cham, Switzerland: Springer, 2018.
35. J. Hu, Y. He, L. Xu, and B. W. Williams, "Improved control of DFIG systems during network unbalance using PI-r current regulators," in *IEEE Trans. Ind. Electron.*, vol. 56, no. 2, pp. 39–451, 2009.
36. D. Zhang, F. Wang, R. Burgos, and D. Boroyevich, "Common-mode circulating current control of paralleled interleaved three-phase two-level voltage-source converters with discontinuous space-vector modulation," in *IEEE Trans. Power Electron.*, vol. 26, no. 12, pp. 3925–3935, 2011.
37. D. Zhang, F. Wang, R. Burgos, and D. Boroyevich, "Total flux minimization control for integrated inter-phase inductors in paralleled, inter-leaved three-phase two-level voltage-source converters with discontinuous space-vector modulation," in *IEEE Trans. Power Electron.*, vol. 27, no. 4, pp. 1679–1688, 2012.
38. A. Arancibia and K. Strunz, "Modeling of an electric vehicle charging station for fast DC charging," in *2012 IEEE* International Electric Vehicle Conference, IEEE, pp. 1–6, 2012.
39. M. H. Rashid, *Power Electronics: Devices, Circuits, and Applications*, 3rd ed., Upper Saddle River, NJ: Pearson Prentice Hall, 2004.
40. K. Sozanski, *Digital Signal Processing in Power Electronics Control Circuits*, 2nd ed., London, UK: Springer, 2017.
41. J. W. Kimball and M. Zawodniok, "Reducing common-mode voltage in three-phase sine-triangle PWM with interleaved carriers," in *IEEE Trans. Power Electron.*, vol. 26, no. 8, pp. 2229–2236, 2011.

Chapter 9

Analysis of vehicle-to-grid as an auxiliary support for grid faults

D. Suchitra, R. Rajarajeswari, D. Anitha, R. Uthra, Pradeep Vishnuram, Mohit Bajaj, and Arvind R. Singh

9.1 INTRODUCTION

Electric vehicle charging stations (EVCS) can permit either a unidirectional or bidirectional power flow. Both the vehicle-to-grid mode (V2G) and the grid-to-vehicle (G2V) modes are enhanced by the bidirectional power flow. EVs can be used as backup generation, supplying the energy back to the grid when needed, in the event of a system failure, peak load demand, or other unforeseen scenarios, which increase the overall reliability of the distribution grid.

Profiles for G2V and V2G were studied in several scenarios when plug-in hybrid electric vehicles (PHEVs) require full charging. Time-coordinated optimal power flow (TCOPF) formulation was presented to provide an optimal method for analyzing the impacts of PHEVs on load profiling, peak shaving, and other operational issues [1]. Rutherford and Yousefzadeh conducted modeling and detailed analysis of the impact of EV on distribution transformers which are the essential part of the network that mitigates distribution network loss and offers a battery load profiling of EVs [2]. Furthermore, the modeling and optimization process that outlines the construction of objective function for load profiling for optimized charging/discharging of PEVs at different scenarios is proposed [3]. A single-phase full-bridge AC–DC converter and a dual active bridge (DAB) converter are optimized for usage in V2G and G2V environment for achieving bidirectional energy fluidity [4].

The current state scenario of operational cities that employs the use of V2G technology and their pricing strategies were detailed in Ref. [5]. EVCS on voltage power equilibrium, peak load, power quality, transformer loss, and their impacts were examined [6]. Load-side management/demand-side management approach with V2G technology was implemented in EVs [7]. The utilization of batteries as energy storage systems in microgrids is examined to facilitate power management by storing excess energy and returning it to the source when required. Additionally, a comprehensive overview of charging station design architecture to ensure optimal dynamic performance regarding DC bus voltage stability is presented [8].

A V2G system connected to the microgrid consisting of solar, wind, and diesel power generation along with residential and industrial load is

DOI: 10.1201/9781003515807-9

considered where V2G acts as a frequency regulator in microgrid. In this research, the effect of V2G on the system frequency is observed [9]. Detailed analysis on V2G charging techniques, its different charging methods and architecture, demand-side load view, and how the trends affect the growth of EV is reported in Ref. [10]. V2G projects that are going on in the world, and their success and failure stories as case studies, are presented in Ref. [11]. The charging technologies that exist in the industry in terms of converter topologies, and strategies for charging control to highlight fast-charging techniques, are detailed. The charging industry is incorporating genetic algorithms that are aligned with contemporary requirements [12]. The principal trends in the automobile sector, particularly within the EV domain and its prospective developments alongside alternative fuel sources, are delineated in Ref. [13].

The power usage optimization in the V2G service is presented by the predictive model. The mechanism's uniqueness as a sort of model predictive control (MPC) is that it seeks to develop a cost-effective way to manage electric energy in an EV. The model allows for a highly precise computation of the vehicle's demand for energy related with progressive movement, considering the vehicle's characteristics, suspension structure, and aerodynamics [14]. This chapter utilizes South Africa as a case study to model various scenarios that assess the technological and environmental impacts of increasing EV adoption. EV adoption can promote the transportation and power sectors in Nigeria despite the country's natural gas-dominated energy infrastructure, although at a rather high cost [15].

Goel et al. examined the different modeling approaches and optimization techniques used in research on the market infiltration rate of EVs, PHEVs, and EVs [16]. There are many potential financial and environmental advantages of integrating EVs into a smart electric grid, according to Noori et al. [17]. EV regulation services can be used to balance out renewable energy sources like solar and wind in virtual power plants and microgrids. As V2G technology acts as a boon permitting the bidirectional energy exchange between EVs and the power grid, the valuable services such as regulation of main grid, spinning reserve, and compensation of reactive power in power system can be accomplished [18]. This chapter highlights the primary optimization strategies for achieving specific V2G targets under various constraints. Furthermore, it aids the grid by utilizing the surplus energy from idle EV chargers.

Ravi and Aziz [19] attempted to illustrate some of the possible auxiliary services that could be provided by such a system and investigate the challenges, effects, and potential market penetration of V2G technology. During off-peak hours, EV integration with the grid seems to offer a variety of financial and environmental advantages. EVs can function as an additional unit to reduce the issues associated with renewable energy systems and so offer ancillary services to the grid in terms of controlling voltage and frequency, according to Mohammad et al. [20].

An overview of the existing converter topologies for EV charging stations, comprising power levels, power flow direction, and various charging control levels, was provided in Refs. [21–23]. A detailed description of the charging facility and its capacity for EV charging stations connected to the power grid has also been provided. Vijayakumar et al. [24] suggested integrating grid and plug-in EVs with the distribution network to satisfy the demands of the additional load. Additionally, an independent wind energy system was installed into the EV station which charges the vehicles as its source.

From the literature review, the research gaps identified are the EV battery as an auxiliary source is not analyzed for the V2G and G2V environment [25–30]; the deployment of low-voltage ride-through (LVRT) and auxiliary sources like electric car batteries is not regarded as an auxiliary source under overloading conditions in EV charging stations [31–33]; and under the above scenario, the system is not considered to operate on the critical loads at the background operation even at the time of grid failure.

Thus, the work focuses on analyzing the charging and discharging schedules of V2G and G2V operations, both with and without grid faults, considering two levels of charging architecture. In the context of Level 1 EV charging architecture, trickle charging, a very low rate charging condition, makes the battery's charge level to be slow and causes lesser energy efficiency of the battery. To overcome this, Level 2 charging system is also proposed to facilitate a faster and smarter charging system due to the shorter trickle phase and leads to a higher efficiency. A MATLAB simulation of the grid-to-vehicle-to-grid operation for Levels 1 and 2 charging is performed employing an auxiliary power source such as the EV battery to meet critical loads during grid fault conditions.

The organization of this chapter is as follows: Section 9.1 starts with an introduction, followed by the literature survey and the identification of research gaps. Section 9.2 discusses the utilization of EV batteries for voltage support, accompanied by a power flow diagram. Section 9.3 outlines battery parameters, followed by Section 9.4, which presents the simulation results with discussion for both the charging levels. Section 9.5 concludes the chapter with a summary.

9.2 EV BATTERY AS VOLTAGE SUPPORT WITH POWER FLOW DIAGRAM

Various articles were reviewed for gaining thorough knowledge on charging and discharging, bidirectional power transfer from charging stations to the source of energy, and electric car batteries as an auxiliary power source. This survey was with a moto of modeling the charging design for Level 1 and Level 2 electrical chargers with the residential load connected to charger when acting as a voltage source. This analysis was made with an aim to investigate the impact on the battery, load and preferred mode of operation (V2G, G2V) during the injection of a fault into the grid. Accordingly, the architecture of battery charging system and its circuits making for the various charging levels, including loads and faults, with the verification of output voltage, current, and SoC for both the instances can be reported.

9.2.1 Low-voltage ride-through

The capacity of operational loads to stay in service during a voltage dip produced by any disturbance is known as LVRT capability. The transmission system operators examine certain rigorous criteria on renewable energy sources such as wind, solar, geothermal, and other renewable energy sources in order to include reactive power regulation and ride-through capabilities. Transmission system operators in most countries have published tight grid-supporting standards for these renewable energy islands experiencing grid failures. In the case of a grid voltage sag, the auxiliary power source should also offer reactive power (up to 100% current capacity) to assist with voltage recovery. The amount of reactive current needed as a function of grid voltage amplitude is shown in the German and Danish grid codes. As a result, STATCOM is frequently used in LVRT applications to fulfill grid code requirements. Using the LVRT idea as inspiration, it is quite feasible that during a failure in a grid-connected system, an electrical charging station may give power to important loads using the V2G concept and deliver power for the duration of the problem.

9.2.2 EV battery as voltage support

EVs possess fewer moving components that transform electrical energy into mechanical and/or translational energy with greater efficiency. In comparison to gas-powered engines, the EVs exhibit less energy loss and have the added advantage of serving as battery banks when connected to an electric charging station integrated with the grid. The increasing demand for a reliable and stable energy supply for electric automobiles facilitates the growth of renewable energy sources like solar and wind. An electric car battery provides voltage support to mitigate system faults, electrical surges, and power distribution during peak hours. The EV charging station, and the efficient battery management system of the vehicle, can function cohesively, maintaining a consistent voltage output to the grid or load side. They can respond within milliseconds to variations in voltage or current. Research is being conducted on the intelligent power electronic devices utilized in EV charging topologies to develop an efficient EV charger.

9.2.3 Level 1 charging architecture

Level 1 EV charger uses a 120 V and 15–20 A of current. Level 1 charger is the basic architecture for charging that employs the use of trickle charging of the battery to ensure the vehicle is being charged and minimal damage is caused to the battery. The operating cost of such an architecture is minimal and uses a simple linear charger topology consisting of power electronics devices.

9.2.4 Level 2 charging architecture

A typical Level 2 EV charger accepts a continuous voltage supply of 200–240 V and will charge a common EV at a steady rate of 20–100 km/h, depending on how much power the charger can offer and how much power the EV can accept. Most residential dwellings in the United States use a constant supply of 240 V for equipment like the washing machine and dryer, which is basically two 120-V circuits combined. Alternatively, most commercial premises have a constant 208-V three-phase AC supply. Figure 9.1 shows the structure of Level 1 and Level 2 EV chargers, which ends with power electronic devices. Furthermore, Table 9.1 shows constraints for modeling both Level 1 and Level 2 chargers.

9.2.5 EV power flow diagram

In general, an EV charger has three basic components—connection controller, rectifier, and buck–boost converter. They allow the battery to be charged from the grid. To allow bidirectional flow of power, the same sequence is

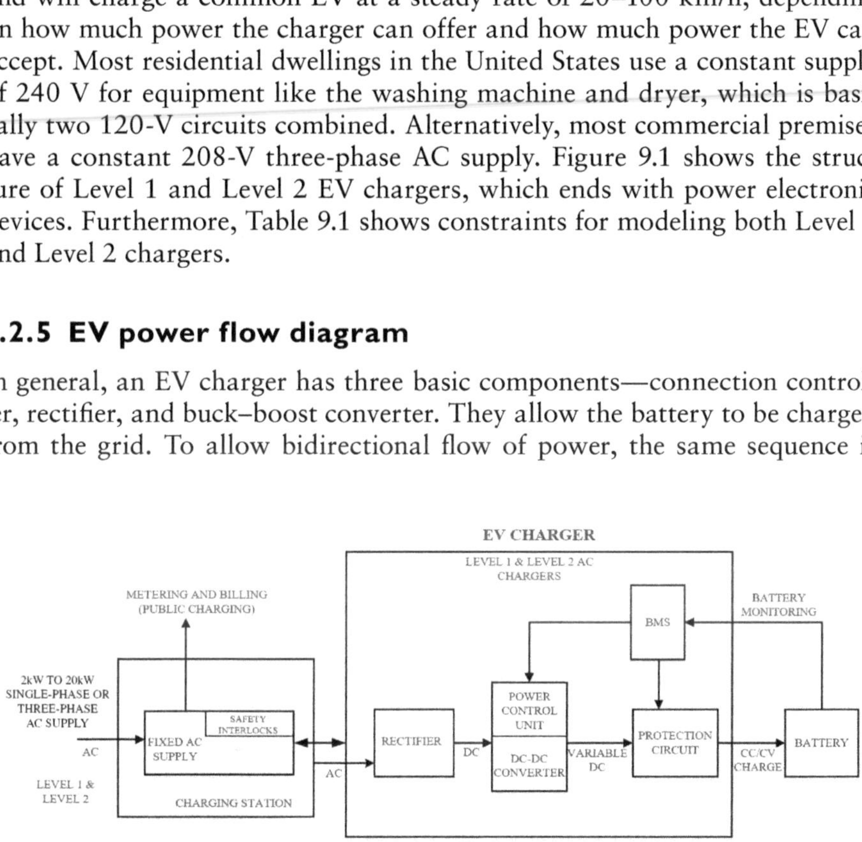

Figure 9.1 Levels 1 and 2 charging topology.

Table 9.1 Level 1 and Level 2 design factors

Design factor	*Level 1 charger DC–DC converter*	*Level 2 charger DC–DC converter*
AC input voltage	120 V	240 V
Topology	Nonisolated buck-boost converter	Full-bridge converter
Control Techniques	Basic pulse width modulator (PWM)	Advanced PWM, phase-shift modulation, digital control

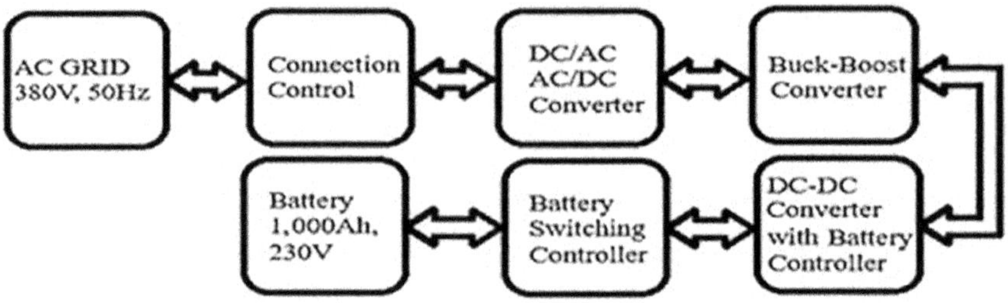

Figure 9.2 EV power flow diagram.

replicated from battery side, which helps converting the DC to AC and feeding back power toward the grid. Figure 9.2 gives an overview of the power flow diagram of an EV charger. AC power is fed to the connection control, which allows the electric power flow to the battery either from the grid side (during G2V mode) or from the battery side (during V2G mode). The signal is sent to a rectifier, which converts it from AC to DC, and then to a buck–boost converter, which will either step down or step up the signal during G2V or V2G operation, respectively.

An electrical grid is a connected network for electricity delivery from generators or the producers toward the load side or the consumers. Electrical grids vary in size and can span from a farm to the size of a small continent. An AC power source is used that provides 380 V and 50 Hz of power.

The connection control block comprises six switches, with three dedicated for charging and three for discharging. The operation mode is exclusively controlled by a gate input signal in parallel with a series RC snubber circuit. The circuit's switch has an inherent/internal resistance which is zero when on, and the inherent resistance is infinite when it is off. The switch model is in on state when the gate signal (g) is switched to 1 and the switch model is in off state when the gate signal is set to 0. An external pulse is provided which dictates the mode of operation, that is, charging or discharging of the battery.

A bidirectional converter is used for this architecture, and appropriate pulses are provided to it. To step up and step down, a buck–boost converter is used. During the charging mode, the buck–boost converter bucks the voltage and vice versa during discharge. Appropriate switching is done by processing the error through the PID controller. An ideal buck or boost converter has an active diode, which is a MOSFET and a passive diode which is any simple diode. MOSFET has been used for both ease of switching and smooth signal with minimal noise or distortion. PID controller is used for an accurate and stable controller.

An internal diode and four MOSFET switches are constructed in parallel with a series RC snubber circuit. The switch conducts in both directions and functions as a resistance (R_{on}) when a gate signal is applied. The antiparallel diode receives current if the gate signal drops to zero while the current is negative. To smooth the transmission or reception of the signal from either side, an additional DC–DC converter is utilized. PID is utilized for a precise and reliable controller.

A gate signal controls the switch, which is connected to a series RC snubber circuit. The switch model has an internal resistance when it is turned on (R_{on}). Internal resistance is infinite off state. When the gate signal (g) is set to 1, the switch model is in the on state. For charging or discharging, two ideal switches are used. The battery specification used in simulation is 230 V/100 Ah (23 kWh), which is equivalent to battery required by an all-terrain vehicle. A Level 1 wall charger box is a collection of bidirectional converters, inverters, and controllers to ensure that there is low distortion, minimal noise when battery is fed DC power during G2V mode, and grid is fed AC power during V2G mode. Figure 9.3 is the simulation of a grid to an EV-connected system operating under Level 1 architecture of charging. The simulation diagram for both Level and Level 2 chargers is modeled with MATLAB Simulink and the description and construction for the same is explained in the coming sections with block diagrams, design specifications.

9.2.6 Level I wall charger connection box

A Level 1 charging architecture of an electrical vehicle station follows a linear, symmetric charging topology consisting of basic switches for connection control, inverters, and bidirectional converters and the buck–boost converter. An external pulse is provided that controls the mode of operation under which the simulation is running, that is, either V2G or G2V. Figure 9.4 is the electric wall charger connection that consists of power electronic devices such as DC–AC/AC–DC, DC–DC converter, and buck–boost converter along with battery switching connection.

9.2.7 Connection controller

Figure 9.5 shows a basic six-switch ideal switch design, with three dedicated to charging and three to discharging. An input gate signal in parallel with a series RC snubber circuit controls the mode of operation exclusively. When the gate signal (g) is set to 1, the switch model is on, and when the gate signal is set to 0, the switch model is off. An external pulse is provided that dictates the mode of operation, that is, charging or discharging of the battery.

9.2.8 Bidirectional converter AC to DC and DC to AC

A bidirectional converter is a power electronic device where power can flow in either direction; this means that its power can be fed to the load and the load can also feed the power back to the source. A bidirectional converter is used to ensure the V2G and G2V operation is working successfully. An insulated-gate bipolar transistor (IGBT) power electronic device is used with a reference PWM generator. Separate universal bridges are used for charging

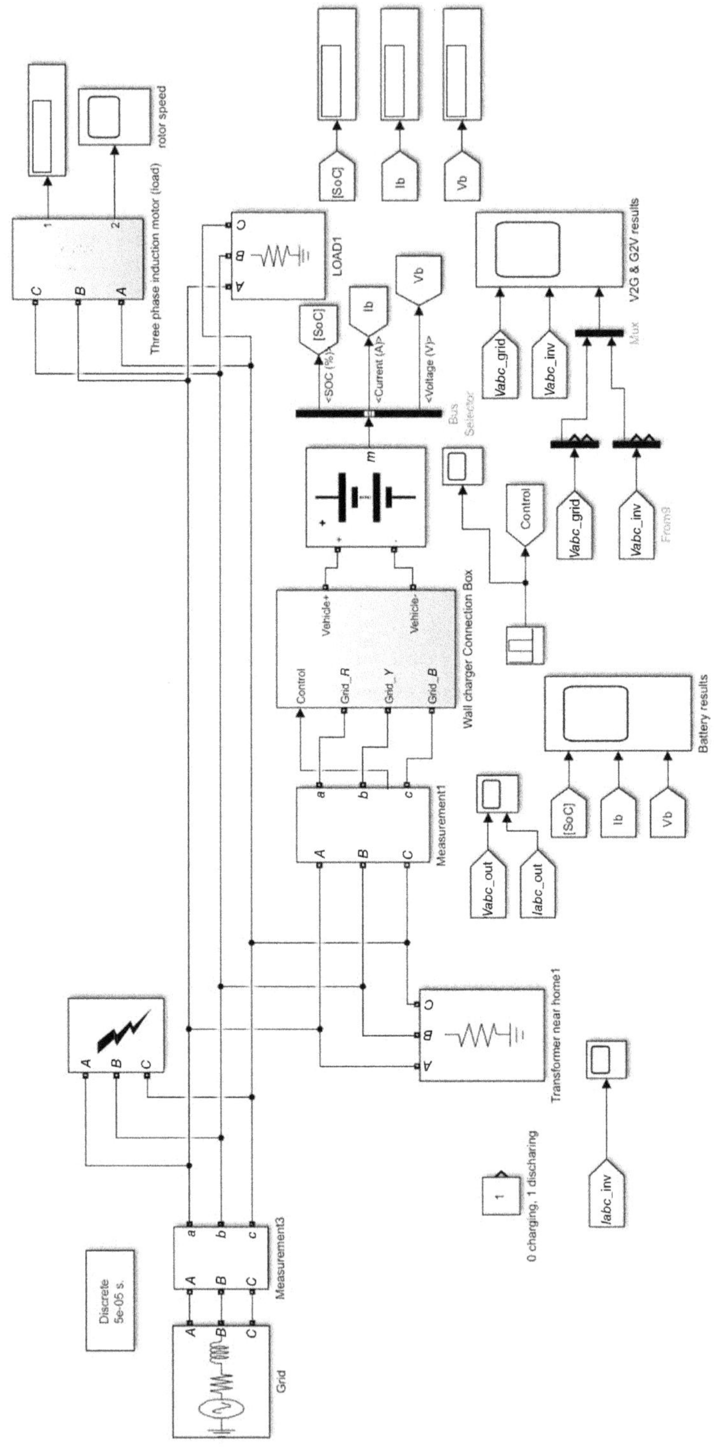

Figure 9.3 Level 1 simulation diagram.

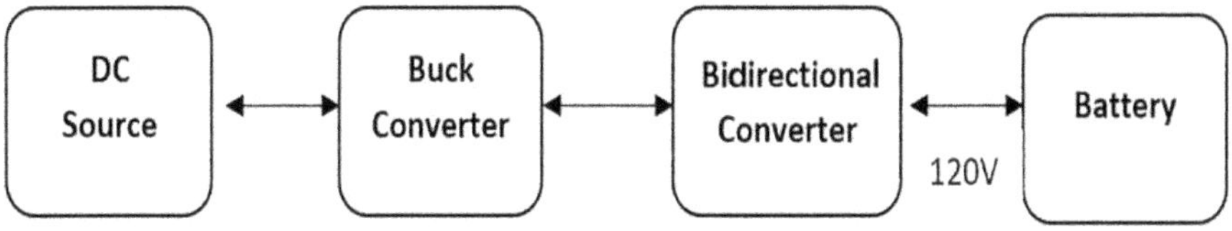

Figure 9.4 Level I wall charger block diagram.

and discharging. A bidirectional converter consists of a separate module for charging and discharging. It consists of a universal bridge connected to a PWM reference module, which provides six firing pulses that are required for suitable switching (Figure 9.6).

9.2.9 Buck–boost converter

To step up and step down, a buck–boost converter is used. The power required by the battery is low compared to the input provided by the grid, hence a buck converter is used, and when the power from battery is delivered back to the grid, boost converter is used. In Figure 9.7, depending on the mode of operation, that is, charging or discharging, a buck or boost converter uses a PID controller for their switching. An ideal buck or boost converter has an active diode, which is a MOSFET and a passive diode, which is any simple diode. The charger uses MOSFET switches for both ease of switching and smooth signal with minimal noise or distortion. PID is used for an accurate and stable controller.

Overall, MOSFETs are the recommended option for buck–boost converters due to their increased efficacy, quick switching speed, and better thermal performance. They provide for efficient power conversion with low losses, which is crucial for applications that need to run dependably and effectively such as electric cars. Some of the other main characteristics that make the MOSFET put in upper hand are safe and better operating switch and efficient energy management while offering battery packs flexibility and lifetime.

9.2.10 DC–DC converter with battery controller

Figure 9.8a–d depicts the DC–DC converter modeled using the MATLAB Simulink software. A series RC snubber circuit is used in parallel with four switches, and an internal diode, as shown in Figure 9.8a. When a gate signal is supplied, the MOSFET conducts in both directions and behaves as a resistance (R_{on}). When current is negative, the gate signal drops to zero, and current is transmitted to the antiparallel diode, which is depicted in Figure 9.8b and c. PID is a type of controller that is both accurate and stable, picturized in Figure 9.8c and d. PID controllers are a common option for controlling DC–DC converters in battery applications due to its simplicity, durability, stability, robustness, and efficacy in obtaining desired performance.

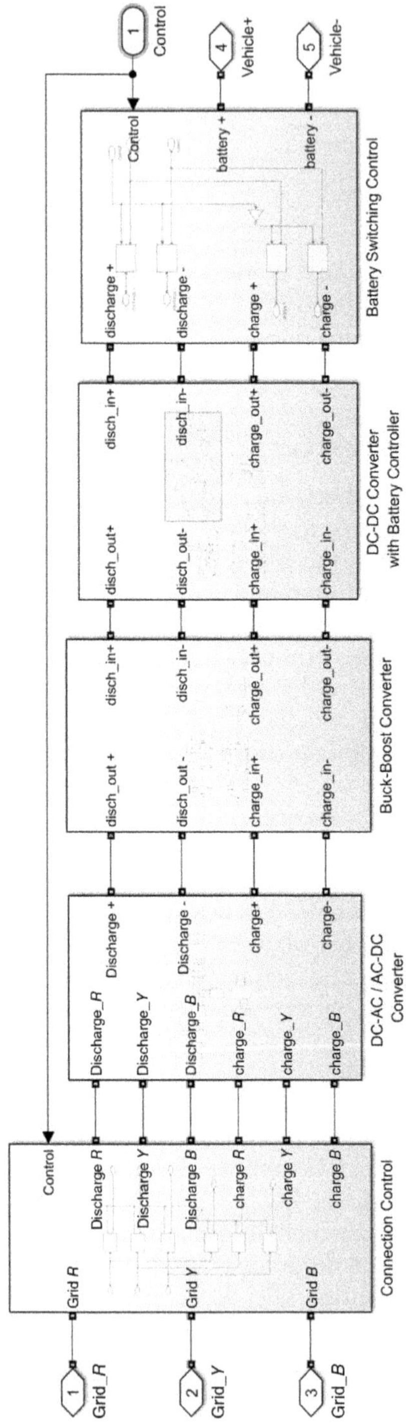

Figure 9.5 Level 1 wall charger connection box Simulink diagram.

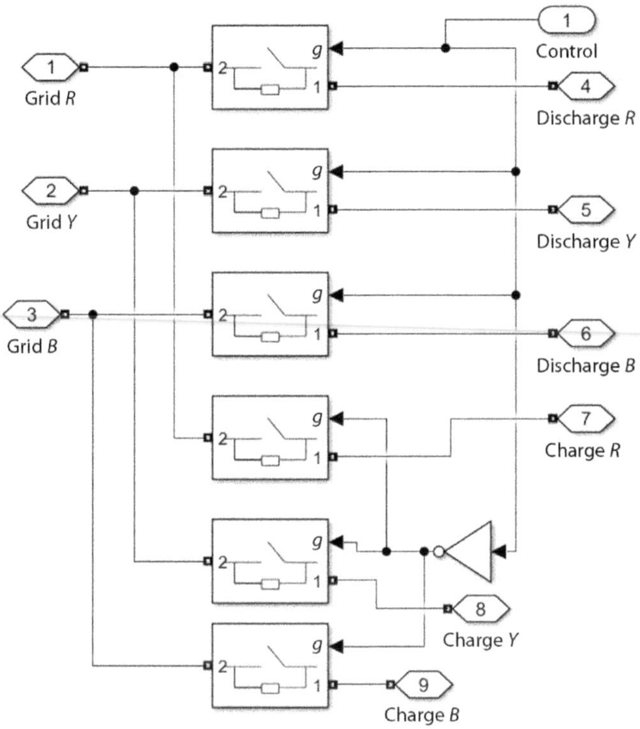

Figure 9.6 Connection controller using simulation.

9.2.11 Battery-switching controller

To maximize charging, discharging, and overall battery management in EVs, a battery-switching controller in an EV charger controls the activity of several battery packs or modules. With the main characteristics such as safe operation and efficacy of energy management along with battery packs flexibility, improved lifespan is ensured by the battery-switching controller system.

To maintain a reliable operation of the circuit and to have an improved lifespan, electronics components subjected to transient voltage along with mitigating the electromagnetic interference in an efficient manner snubber circuits are used. Also, an augmented switching performance is accomplished when this circuit is used.

Figure 9.9 depicts a switch controlled by a gate signal and a series RC snubber circuit. When the switch model is turned on, it has an internal resistance (R_{on}). In the off state, internal resistance is infinite. The switch model is in the on state when the gate signal (g) is set to 1. Two ideal switches are used for charging and two separate ideal switches are used for discharging; since power fed and given is DC signal only, two dedicated switches for each mode are needed.

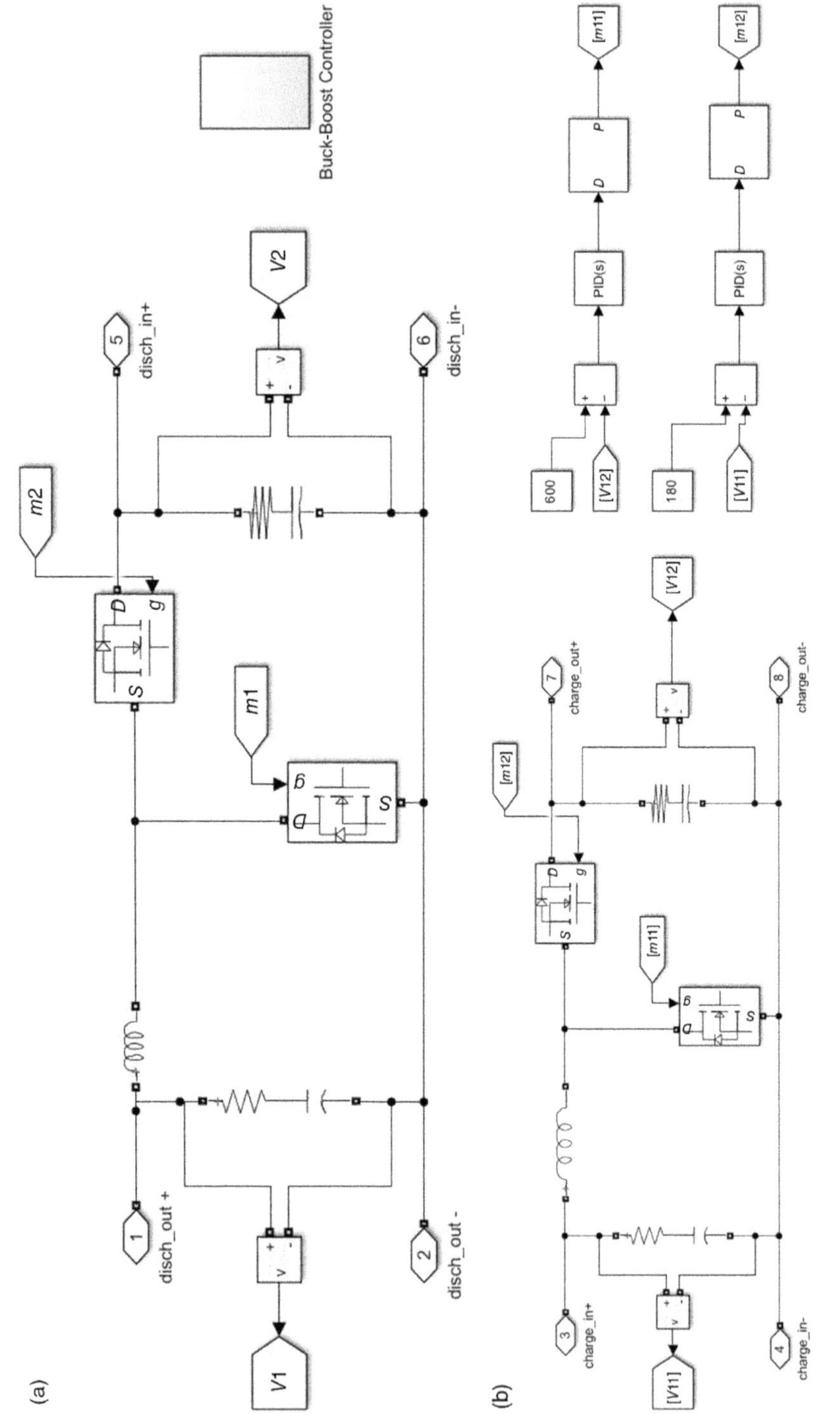

Figure 9.7 (a) Buck–boost converter in MATLAB Simulink. (b) Converter with PID controller.

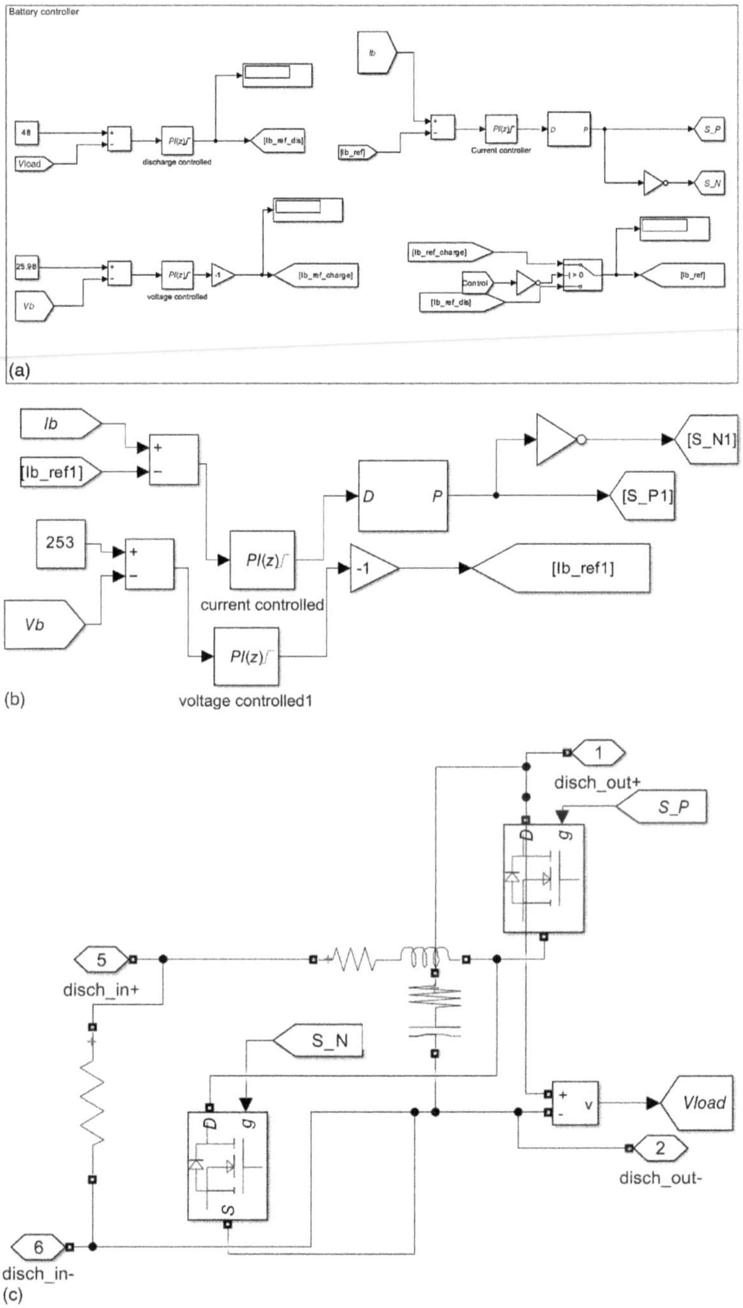

Figure 9.8 DC–DC converter with battery controller stage using MATLAB Simulink. (a) Controller circuit for DC–DC Converter with battery in discharge mode. (b) Controller circuit for DC–DC Converter with battery in charging mode. (c) DC–DC Converter configuration during battery discharging mode. (d) DC–DC Converter configuration during battery charging mode. *(Continued)*

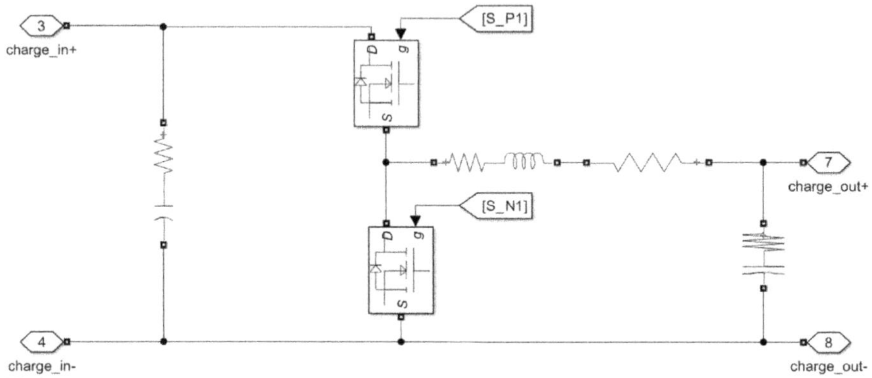

(d)

Figure 9.8 (Continued)

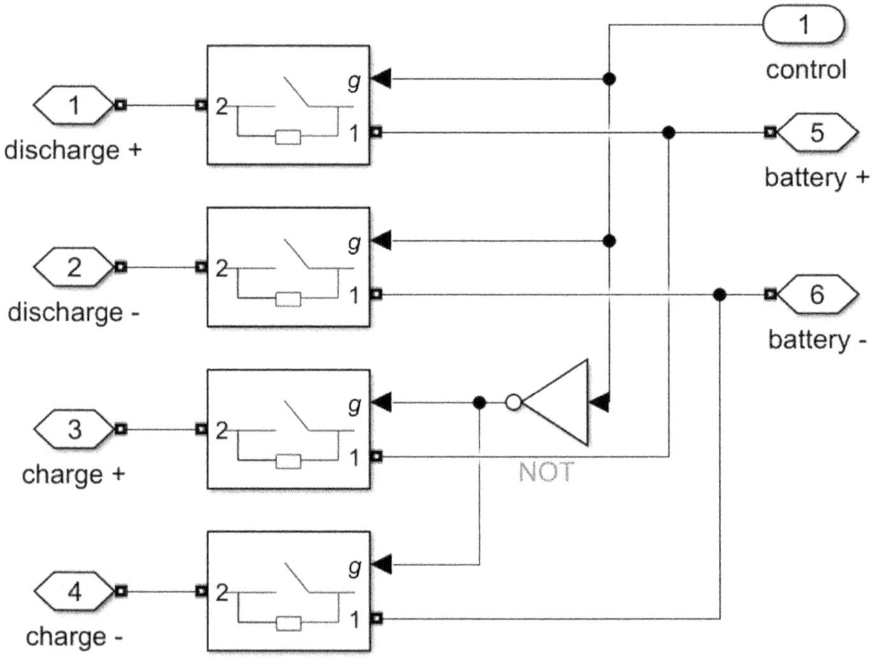

Figure 9.9 Battery-switching controller.

9.3 BATTERY SPECIFICATION

Ideal electric scooter battery has a specification of 48 V/24 Ah (1,152 Wh). Ideal electric car battery has a specification of 100 kWh. The specification of battery in simulation is 230 V/100 Ah (23 kWh). Below is the MATLAB simulated block of the Li-ion battery used in this work for the specified rating. Li-ion batteries are preferred because of their high energy density, extended life cycle, fast charging speed, and general efficacy. Because of these qualities, they are especially well-suited for contemporary applications, particularly given the rising demand for renewable energy storage and EVs. It is anticipated that Li-ion battery will continue to play a significant role in the energy storage market as technology advances.

9.3.1 Level 2 charging/discharging architecture in MATLAB

To accommodate varying customer needs, charging scenarios, and operational requirements, it is advantageous to have both Level 1 and Level 2 charging alternatives for EVs. EV infrastructure can better serve a broader range of user needs, increase convenience, better manage grid loads, and foster the expansion of EV adoption by providing both Level 1 and Level 2 charging choices. The general usability and practicality of EVs are improved for a range of drivers and usage situations by this combination.

A Level 2 wall charger box is a collection of bidirectional converters, inverters, and controllers to ensure that there is low distortion, minimal noise when battery is fed DC power during G2V mode as well as when the grid is fed with AC power during V2G mode. Figure 9.10 depicts the block diagram of the Level 2 charger proposed.

Figure 9.11 is a combination of power electronics devices that are used along with PID controllers which dictate switching of the circuit. A simple six-switches configuration is used with six firing pulses, driven by individual PWM input. A buck–boost converter is used to step up and step down the bidirectional voltage. This ideal six-switches configuration receives the input signal from six PWMs and gives the desired output required by Level 2 EV charger. Also, the simulation for the same is shown in Figure 9.11.

An ideal six-switching configuration is a combination of power electronics devices that are used along with PID controllers which dictate the switching of the circuit. A simple six-switches configuration is used with six firing pulses, driven by individual PWM input. A buck–boost converter is used to

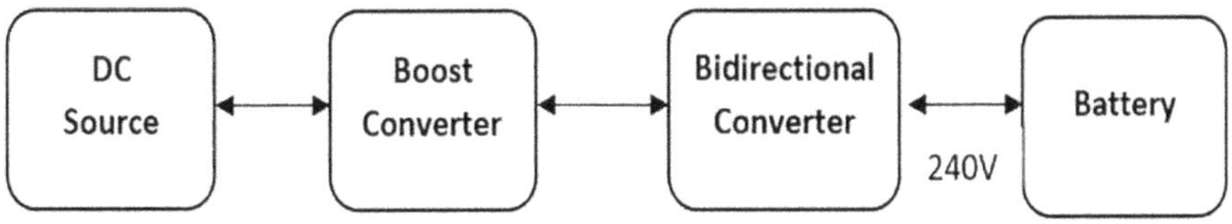

Figure 9.10 Level 2 block diagram.

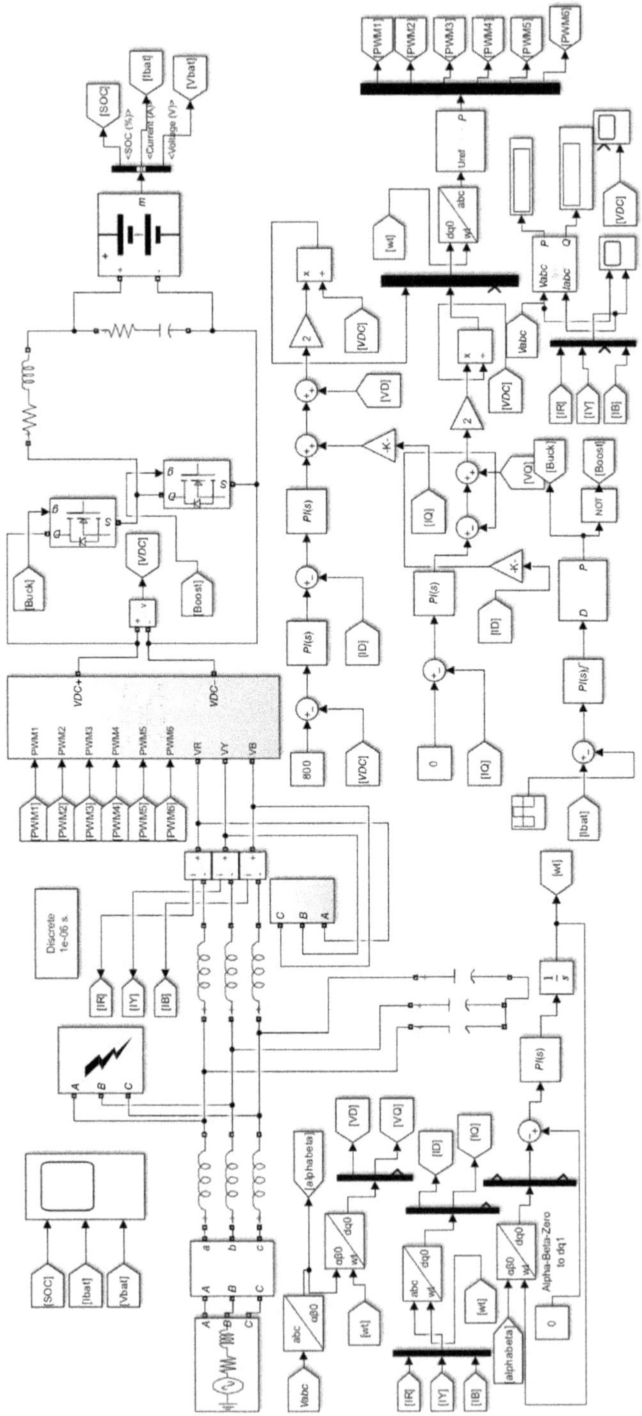

Figure 9.11 Level 2 simulation diagram.

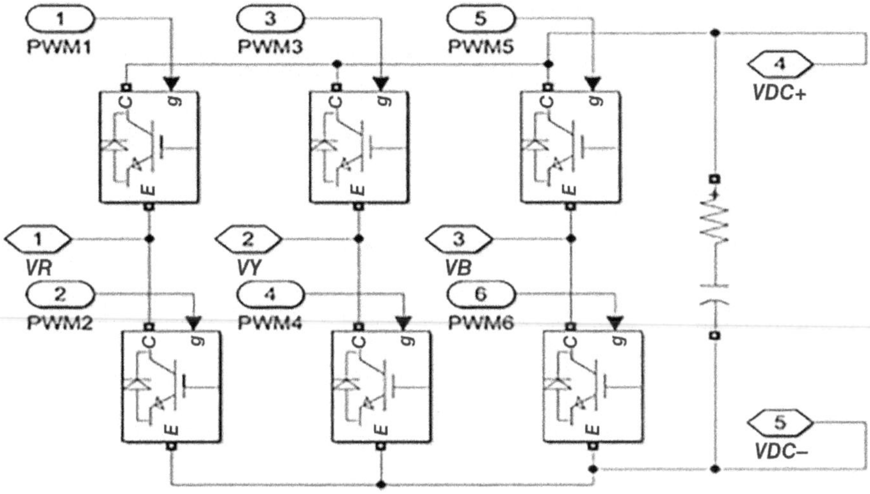

Figure 9.12 Ideal six-switches configuration.

step up and step down the bidirectional voltage. Figure 9.12 is an ideal six-switches configuration that takes an input signal from the six PWMs and gives the desired output required by Level 2 EV charger.

9.4 RESULT AND DISCUSSION

The system primarily employs the use of a grid that is connected to a load. This grid-to-load coupling, where an electric car (load) wall charger plugs, links the system to the EV battery between the source side and the load side. A wall charger is the electric converter that can convert grid power into DC power needed for charging a battery or it can make connection between the grid and the EV in vehicle mode to power the grid. In the V2G mode, power transfer from EV, i.e., the DC power from the battery, is fed back into the AC grid after the wall charger has converted the voltage from DC to AC. EV charging architecture at Level 1 and Level 2 has been designed for this purpose. Table 9.2 shows the specifications used for both the levels considered in this work.

There are two primary testing cases used for the simulation using MATLAB: Level 1 and Level 2 charging architecture of EV. Simulation of both Level 1 and Level 2 charging architecture is built, which is connected to grid as well as a residential load.

Under both conditions, three phase balanced faults are introduced and an additional circuit breaker is installed, which ensures that there is no additional voltage or current noise delivered from the grid side. During fault intervals, the battery from both Level 1 and Level 2 configuration switches from

Table 9.2 Level 1 and Level 2 parameters

S. No.	*Charging architecture*	*Grid input*	*Charging system*	*Residential load specification*
1	Level 1	380 V, 50 Hz AC	120 V	5.4 HP (4 kW), 400 V 50 Hz 1,430 RPM
2	Level 2	380 V, 50 Hz AC	240 V	5.4 HP (4 kW), 400 V 50 Hz 1,430 RPM

charging to discharging and delivers power back to the grid and load side, thereby maintaining a better operation of the system during the critical load conditions as well.

9.4.1 Level 1 architecture: A test case 1

The grid voltage considered in the work is 380 V and the three-phase fault is introduced into the system from 0.2 second to 0.4 second. During the fault condition, the drop in grid voltage and a surge in grid current is observed. This is because power is proportional to voltage and current; the Level 1 architecture under fault condition is depicted in Figures 9.12–9.14.

During no fault condition, the grid voltage of 380 V is observed. When the grid fault occurs, the inverter will serve the load; the simulation result for the same is depicted in Figure 9.15.

The significance of this work is achieving the objective of uninterrupted power supply during the fault in the system considered. As per the above objective, an automatic discharging should take place to make sure that the connected residential load is receiving uninterrupted power supply during the fault condition. During the period of fault subjected at 0.2–0.4 second, the switch gets connected to discharging mode of inverter and hence the critical load is powered from inverter instead of the grid. Figure 9.16 shows the inverter voltage during the Level 1 mode of operation. During a fault or discharging period, the bidirectional converter in the battery converts 120 V DC

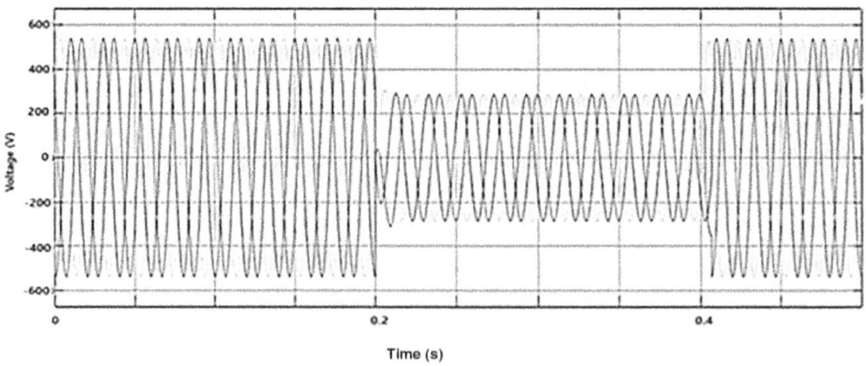

Figure 9.13 Grid voltage during fault.

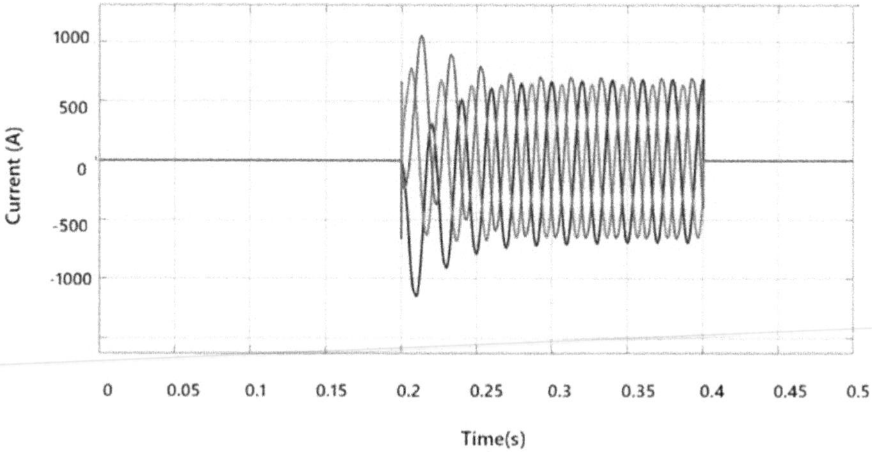

Figure 9.14 Grid current during fault.

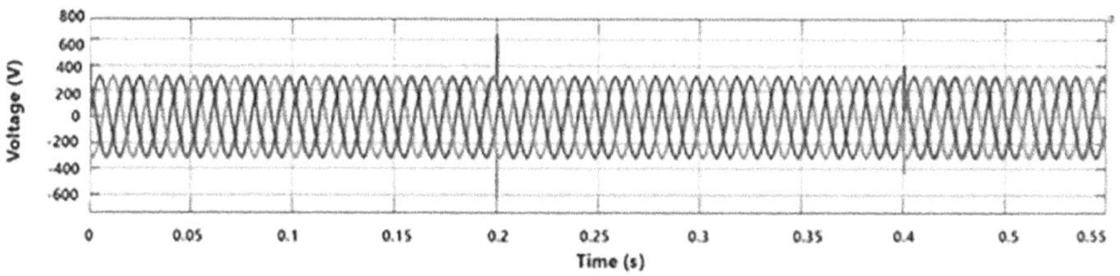

Figure 9.15 Grid voltage without fault.

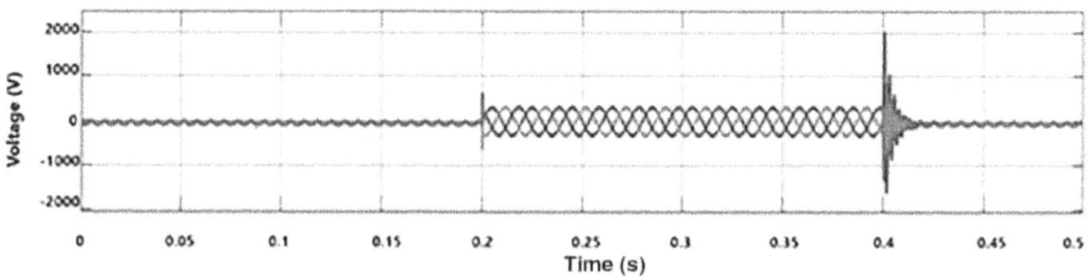

Figure 9.16 Inverter voltage during fault.

back to 380 V AC that is suitable for the grid and for the residential loads. Figure 9.17 shows the grid and battery voltage synchronization condition during V2G operation.

Figure 9.18 depicts the state of charge of the battery under Level 1 charging at 120 V, this being switched from charging to discharging from 0.2 second and vice versa from 0.4 second onward during fault and unfaulty condition.

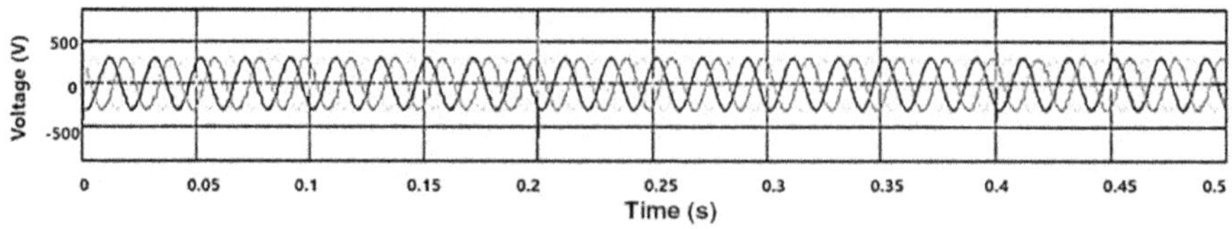

Figure 9.17 Grid and battery voltage synchronized during fault.

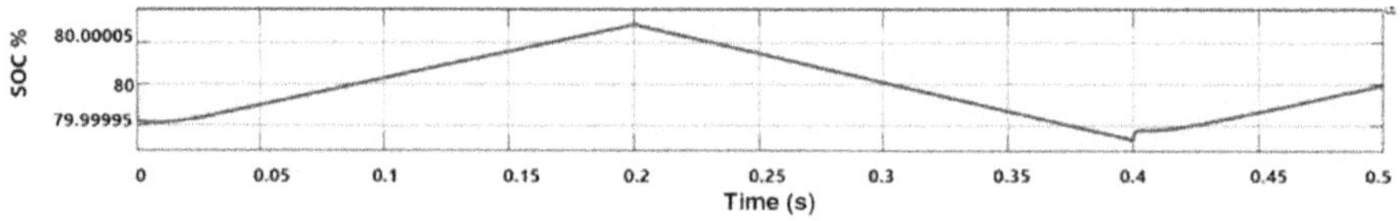

Figure 9.18 State of charge battery.

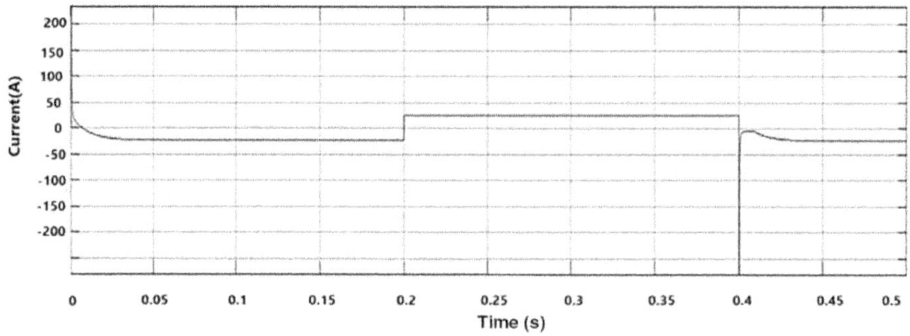

Figure 9.19 Battery current.

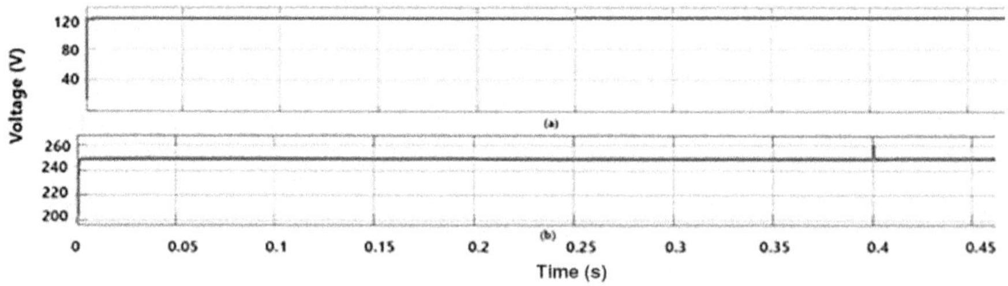

Figure 9.20 Battery voltage at Level 1 and Level 2.

Figure 9.19 shows the battery current. When the battery is charged, a stable current of 15–20 A is delivered to the battery and when it is discharged (the flow of electricity is reversed to load), the current value of 15–20 A is obtained as in simulation. Figure 9.20 shows a stable voltage of about 120 V being fed and released to the battery in Level 1 and Level 2 battery at voltage 240 V.

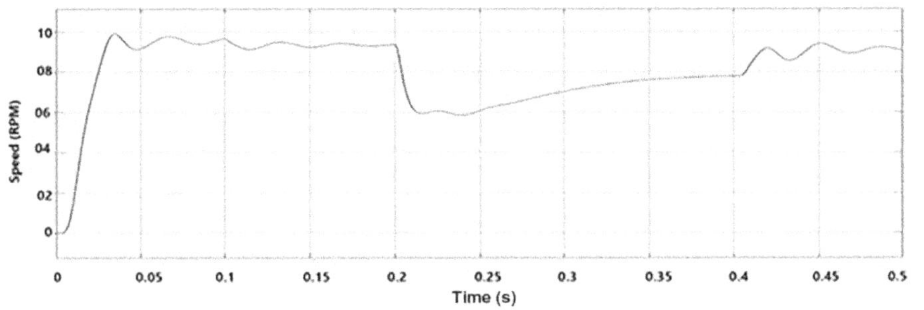

Figure 9.21 Asynchronous motor speed (RPM) (residential load).

Figure 9.21 shows that with Level 1 charging architecture, when fault occurs from 0.2 second to 0.4 second, power is cut from the grid and power from EV battery alone is provided to the asynchronous machine. There is a rise in motor speed after which speed stabilizes and hence operation of critical load is carried out under fault conditions. Figure 9.22a shows the state of charge of the battery under Level 2 charging, it is being switched from charging to discharging from 0.5 second onward. Figure 9.22b depicts the current in and out of the battery. When battery is being charged, a stable current of 40 A is delivered to the battery. Also, Figure 9.22c shows a stable voltage of about 220 V being fed to the battery.

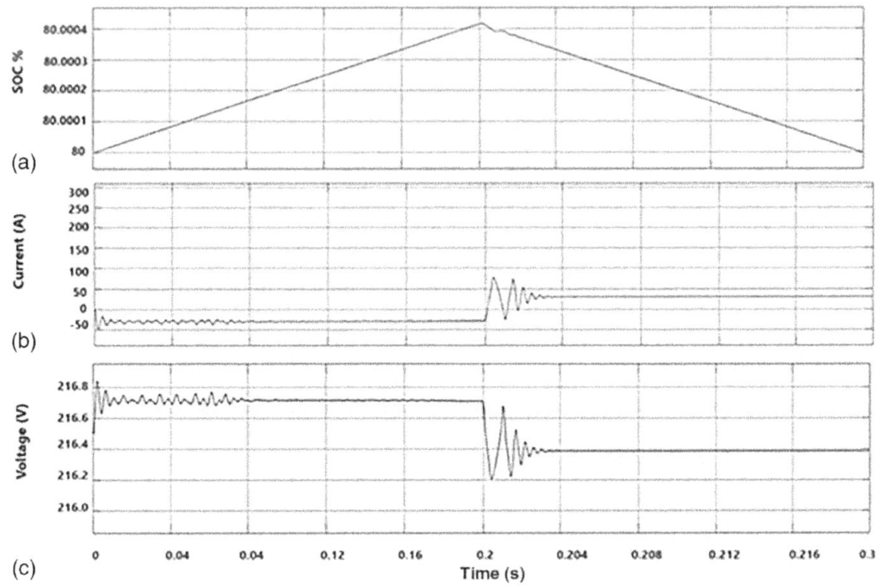

Figure 9.22 Level 2 (a) SOC, (b) current, and (c) voltage graph.

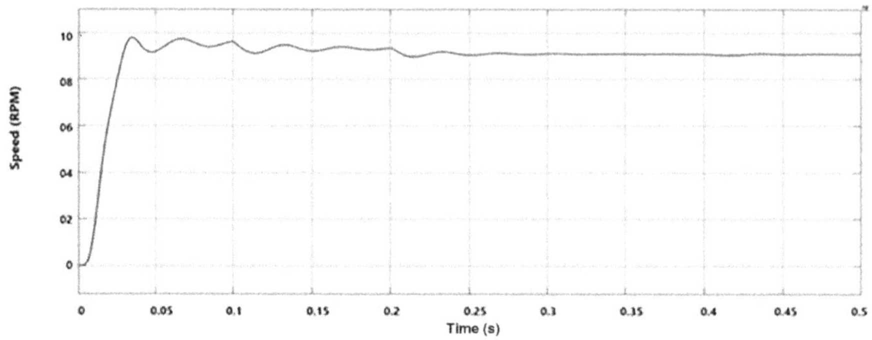

Figure 9.23 Asynchronous motor speed Level 2.

Figure 9.23 shows the RPM at which the load (asynchronous motor) considered is connected to Level 2 system (residential load). With Level 2 charging architecture, a fault is introduced from 0.2 second to 0.4 second, that makes the load connection from the grid to be disconnected (due to fault condition) and gets connected to battery such that it provides power to the asynchronous machine. There is a rise in motor speed after which the speed stabilizes over few seconds and hence the operation of critical load is carried out under fault conditions.

Further a comparison is made between the Level 1 and Level 2 architecture, and its observations are summarized. It is observed that Level 1 charger (120 V, 15 A) discharges the battery to approximately 79.9996% SOC in a time frame of 1 second. Also, Level 2 charger (240 V, 40 A) charges the battery to approximately 79.9992% SOC in a time frame of 1 second.

Also, from the result of Level 1 charger (120 V, 15 A), it charges the battery to approximately 80.006% SOC in a time frame of 1 second and the Level 2 charger (240 V, 40 A) charges the battery to approximately 80.008% SOC in a time frame of 1 second, as mentioned in Table 9.3.

When an asynchronous motor load is connected to the proposed Level 1, Level 2 system and the simulation result of speed versus time graphs are as depicted in Figure 9.24. It is observed that the motor speed achieved by Level 1 EV architecture peaks at around 9.6 RPM, after which the speed stabilizes and gets maintained at an operational speed of 7.3 RPM. The highest motor speed achieved by Level 2 EV architecture is around 9.8 RPM after which the speed stabilizes and is maintained at an operational speed of 9 RPM, thereby making Level 2 with higher efficacy compared to Level 1 architecture.

Table 9.3 Comparison between Level 1 and Level 2 performance during load condition

Name	*Rating of the system*	*SOC %*	*Time*
Level 1	120 V, 15 A	79.9996%	1
Level 2	240 V, 40 A	79.9992%	1

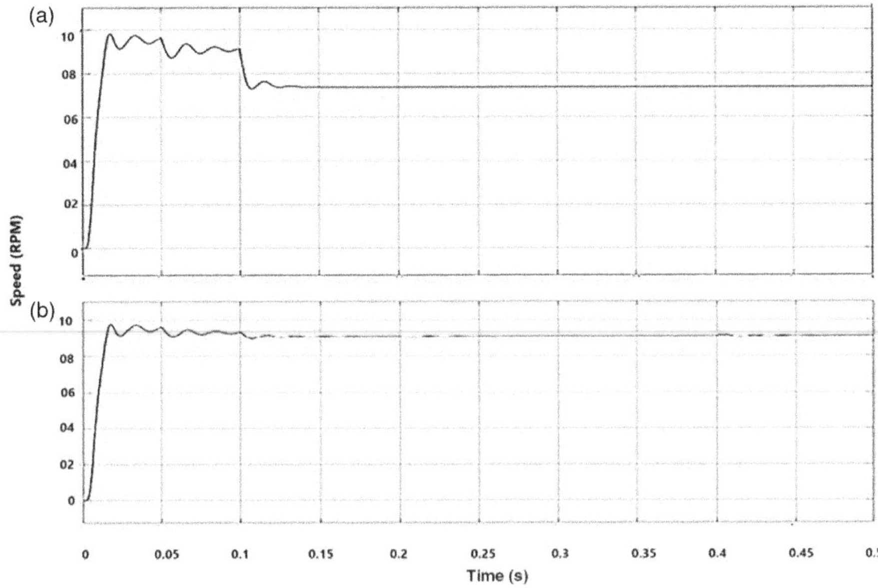

Figure 9.24 Levels 1 and 2 asynchronous motor speed.

9.5 CONCLUSION

This chapter proposes the Level 1 and Level 2 charging systems of EV which are designed for charging with a standard voltage of 120 V with a portable charger and an enhanced version of it comes with a supply voltage of 240 V. The majority of the plug-in hybrid vehicles in today's world are designed to draw a minimum current of 15 A of power and the charging stations built are being mass produced with basic Level 1 architecture of charging. Since this level of architecture employs the use of slow charging, trickle charging will maintain a battery charge at the same rate the self-discharging is taking place; this state happens nearly exclusively when the battery is not loaded. Level 1 and Level 2 charging architecture can support critical loads that require voltage support to run during grid faults or during power surges that happen during peak hours. The MATLAB simulation results provide proof that during fault, the residential load (asynchronous motor) can operate smoothly at a stable 8–9 RPM. While Level 1 charger is widely adopted in the automobile industry, there is not enough awareness about the potential advantages of Level 2 EV charger.

Through MATLAB simulation, both Level 1 and Level 2 chargers are developed in this work and the results are compared and tabulated. The charging speed for a Level 2 charger is higher than that of a Level 1 charger, under standard condition when the SOC of the battery is 80%. The discharging speed for a Level 2 charger is faster than that of a Level 1 charger, under standard conditions when the SOC of the battery is 80%. The critical load operating under fault condition or during peak hour requires stable input power at all times, and this is provided by the Level 2 charging architecture, which is able to respond faster and deliver power backup to the load. The response time for the asynchronous motor as a critical load under Level 2 charging is shorter; also, the load connected is able to operate steadily, quickly, and smoothly compared to Level 1 charger. Also, the Level 2 architecture gives a shorter trickle phase with faster charging, smart cutoff and high power output.

ABBREVIATIONS

BEV	Battery-operated electric vehicles
DAB	Dual active bridge
DTF	Distribution transformer
EV	Electric vehicles
EVCS	Electric vehicle charging stations
G2V	Grid-to-vehicle
HEV	Hybrid electric vehicles
Li-IB	Lithium-ion batteries
LL	Load leveling
LVRT	Low-voltage ride-through capability
MOSFET	Metal-oxide-semiconductor field-effect transistor
MPC	Model predictive control
PHEV	Plug-in hybrid electric vehicles
PLS	Peak load shaving
TCOPF	Time-coordinated optimal power flow formulation
V2G	Vehicle-to-grid

REFERENCES

1. S. Acha, T. C. Green, N. Shah, "Effects of optimised plug-in hybrid vehicle charging strategies on electric distribution network losses," IEEE PES T&D 2010, New Orleans, LA, USA, 2010. doi: 10.1109/TDC.2010.5484397.
2. M. J. Rutherford, V. Yousefzadeh, "The impact of electric vehicle battery charging on distribution transformers," 2011 Twenty-Sixth Annual IEEE Applied Power Electronics Conference and Exposition (APEC), Fort Worth, TX, USA, 2011. doi: 10.1109/APEC.2011.5744627.
3. S. Shahidinejad, S. Filizadeh, E. Bibeau, "Profile of charging load on the grid due to plug-in vehicles," IEEE Transactions on Smart Grid, vol. 3, no. 1, pp. 135–141, 2012. doi: 10.1109/TSG.2011.2165227.
4. Jira Jia, I. Livuan Chen-Zen, "Design of a bidirectional power interface for V2G technology with smaller DC-link capacitance," International Journal of Smart Grid and Clean Energy, vol. 3, no. 1, pp. 104–110, 2014. doi: 10.12720/sgce.3.1.104-110.
5. T.U. Daim, X. Wang, K. Cowan, "Technology roadmap for smart electric vehicle-to-grid (V2G) of residential chargers," Journal of Innovation and Entrepreneurship, vol. 5, no. 15, 2016. https://doi.org/10.1186/s13731-016-0043-y.

6. S. Deb, K. Kalita, P. Mahanta, "Review of impact of electric vehicle charging station on the power grid," 2017 International Conference on Technological Advancements in Power and Energy (TAP Energy), 2017. doi: 10.1109/TAPENERGY.2017.8397215.
7. K. Mahmud, S. Morsalin, M. J. Hossain, G. E. Town, "Domestic peak-load management including vehicle-to-grid and battery storage unit using an artificial neural network," 2017 IEEE International Conference on Industrial Technology (ICIT), 2017. doi: 10.1109/ICIT.2017.7915424.
8. F. M. Shakeel, O. P. Malik, "Vehicle-to-grid technology in a micro-grid using DC fast charging architecture," 2019 IEEE Canadian Conference of Electrical and Computer Engineering (CCECE), 2019. doi: 10.1109/CCECE.2019.8861592.
9. Femmina Mohammed Shaqeel, Om P. Malik, "V2G simulation for frequency regulation in micro grid with solar, wind and diesel power generation environment," International Research Journal of Engineering and Technology (IRJET), 2020. doi:10.1109/SeFet48154.2021.9375722.
10. N. Mamatha, Arpita Banik, "A study on vehicle to grid technology and its scope," International Journal of Applied Engineering Research, vol. 14, no. 1, pp. 8–13, 2019.
11. Bijhan Bibaka, Hattice Teckiner-Mogulkocb, "A comprehensive analysis of vehicle to grid (V2G) systems and scholarly literature on the application of such systems," Renewable Energy Focus, vol. 36, pp. 1–20, 2021. doi. 10.1016/j.ref.2020.10.001.
12. Morris Brenna, Federica Foiadelli, Carola Leone, Michela Longo, "Electric vehicles charging technology review and optimal size estimation," Journal of Electrical Engineering & Technology, vol. 15, pp. 2539–2552, 2020. doi: 10.1007/s42835-020-00547-x
13. Amit Sachdeva, Dadaso D. Mohite, "Recent development in green fuel vehicles and their future advancements," International Research Journal of Engineering and Technology, vol. 7, no. 10, pp. 1175–1192, 2020.
14. Barthłomiej Moroczek, Amanduh Cołodyńska, "The V2G process with the predictive model," IEEE Access, vol. 8, pp. 86947–86956, 2020. doi: 10.1109/ACCESS.2020.2991329.
15. Michael O. Dioha, Lei Duan, Tyler H. Ruggles, Sara Bellocchi and Ken Caldeira, "Exploring the role of electric vehicles in Africa's energy transition: A Nigerian case study," I Science, vol. 25, no. 3, pp. 103926, 2022. doi: 10.1016/j.isci.2022.103926.
16. Sonali Goel, Renu Sharma, Akshay Kumar Rathore, "A review on barrier and challenges of electric vehicle in India and vehicle to grid optimisation," Transportation Engineering, vol. 4, pp. 1–14, 2021.
17. M. Noori, Yang Zhao, N. Onat, Stephanie Gardner, O. Tatari, "Light-duty electric vehicles to improve the integrity of the electricity grid through vehicle-to-grid technology: Analysis of regional net revenue and emissions savings," Applied Energy, vol. 168, no. 15, pp. 146–158, 2016.
18. Kang Miao Tan, Vigna K. Ramachandaramurthy, Jia Ying Yong, "Integration of electric vehicles in smart grid: A review on vehicle to grid technologies and optimization techniques," Renewable and Sustainable Energy Reviews, vol. 53, pp. 720–732, 2016.
19. S. S. Ravi, M. Aziz, "Utilization of electric vehicles for vehicle-to-grid services: Progress and perspectives," Energies, vol. 15, pp. 589–616, 2022.
20. A. Mohammad, R. Zamora, T.T. Lie, "Integration of electric vehicles in the distribution network: A review of PV based electric vehicle modelling," Energies, vol. 13, no. 17, pp. 4541–4561, 2020.
21. F. Shaaban, A.A. Eajal, E.F. El-Saadany, "Coordinated charging of plug-in hybrid electric vehicles in smart hybrid AC/DC distribution systems," Renewable Energy, vol. 82, pp. 92–99, 2014.

22. Y. Ota, H. Taniguchi, T. Nakajima, K. M. Liyanage, J. Baba, A. Member, Yokoyama, "Autonomous distributed V2G (vehicle-to-grid) satisfying scheduled charging," IEEE Transactions on Smart Grid, vol. 3, no. 1, pp. 559–564, 2012.
23. M. Brenna, F. Foiadelli, C. Leone, M. Longo, "Electric vehicles charging technology review and optimal size estimation," Journal of Electrical Engineering & Technology, vol. 15, pp. 2539–2552, 2020.
24. R. Vijayakumar, K. Malarvizhi, S. Divyapriya., "Design of public plug-in electric vehicle charging station for improving LVRT capability of grid connected wind power generation," International Conference Proceedings on Soft-Computing and Network Security, pp. 1–6, 2018.
25. Rahulkumar Jayaraman, R Narayanamoorthi, Pradeep Vishnuram, Mohit Bajaj, Vojtech Blazek, Lukas Prokop, Stanislav Misak, "An empirical survey on wireless inductive power pad and resonant magnetic field coupling for in-motion EV charging system," IEEE Access, vol. 11, pp. 4660–4693, 2022. doi: 10.1109/ACCESS.2022.3232852.
26. Yuvaraja Shanmugam et al, "Solar-powered five-leg inverter-driven quasi-dynamic charging for a slow-moving vehicle," Frontiers in Energy Research, vol. 11, p. 1115262, 2023.
27. Dominic Savio Abraham et al, "Fuzzy-based efficient control of DC microgrid configuration for PV-energized EV charging station," Energies, vol. 16, no. 6, p. 2753, 2023.
28. B. Anil Kumar et al, "A novel framework for enhancing the power quality of electrical vehicle battery charging based on a modified Ferdowsi converter," Energy Reports, vol. 10, pp. 2394–2416, 2023.
29. Mohammed Abdullah Ravindran et al, "A novel technological review on fast charging infrastructure for electrical vehicles: Challenges, solutions, and future research directions," Alexandria Engineering Journal, vol. 82, pp. 260–290, 2023.
30. B. Anil Kumar et al, "A novel strategy towards efficient and reliable electric vehicle charging for the realisation of a true sustainable transportation landscape," Scientific Reports, vol. 14 no. 1, p. 3261, 2024.
31. Boya Anil Kumar et al, "Hybrid genetic algorithm-simulated annealing based electric vehicle charging station placement for optimizing distribution network resilience," Scientific Reports, vol. 14, no. 1, p. 7637, 2024.
32. Sidharth Sabyasachi et al, "Reimagining E-mobility: A holistic business model for the electric vehicle charging ecosystem," Alexandria Engineering Journal, vol. 93, pp. 236–258, 2024.
33. Ch SV Prasada Rao et al, "Unified power quality conditioner-based solar EV charging station using the GBDT–JS technique," Frontiers in Energy Research, vol. 12, p. 1343635, 2024.

Chapter 10

Open Charge Point Protocol (OCPP)

A survey of enabling interoperability and innovation in electric vehicle charging infrastructure

Veerendra Dakulagi, Shashidhara K.S., Raghu J., Hemapriya B.C., Ravi Raushan, and Rohini Dakulagi

10.1 INTRODUCTION

The Open Charge Point Protocol (OCPP) stands as a pivotal advancement in the realm of electric vehicle (EV) charging infrastructure, heralding a new era of standardization and interoperability. Created to facilitate seamless communication between EV charging stations (CS) and central management systems (CMSs), OCPP addresses fundamental challenges that have historically hindered the widespread adoption and efficient operation of EV charging networks. Many commercial charging points (CPs) utilize OCPP for bidirectional publish-and-subscribe communication with a charge point operator (CPO). When setting up a CPO infrastructure on Amazon Web Services (AWS), one must establish an OCPP WebSocket endpoint. This endpoint, referred to as the OCPP Gateway, serves as an intermediary between OCPP and Message Queuing Telemetry Transport (MQTT) protocols, facilitating integration with AWS Internet of Things (IoT) Core and other downstream CPO services hosted on AWS.

The upcoming architecture diagram provides a detailed overview of the comprehensive end-to-end solution that will be developed and discussed in this chapter.

Figure 10.1 illustrates the communication flow where OCPP messages from a charge point are proxied to a CPO service. This communication occurs through a one-to-one relationship facilitated by a WebSocket connection and MQTT topic. The WebSocket connection allows real-time bidirectional communication, while MQTT provides a lightweight messaging protocol for publishing OCPP messages over the designated topic. This setup ensures efficient and reliable data exchange between the charge point and the CPO service, enabling seamless operation and management of the EV charging network.

The OCPP has been extensively studied for its role in enabling seamless communication in EV charging systems. Early research by Han and Xiao focused on privacy issues in vehicle-to-grid (V2G) networks, highlighting gaps in location and identity protection but neglecting protocol vulnerabilities.

DOI: 10.1201/9781003515807-10

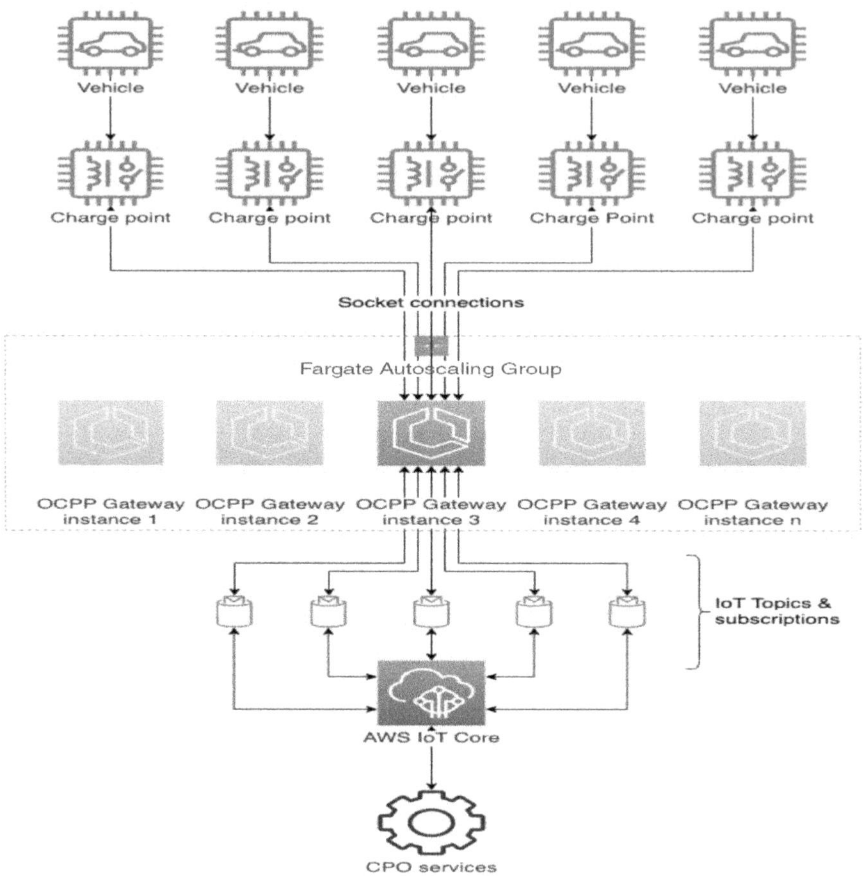

Figure 10.1 Charge point OCPP message proxied to CPO service.

Despite progress, gaps remain in integrating blockchain, AI, and PKI technologies, and linking security vulnerabilities to specific protocol versions. The literature underscores the critical need for adaptive security measures and deeper exploration of OCPP 2.0's features to address emerging threats in EV charging networks.

The integration of EVs with smart grids and charging infrastructure is an evolving area of research aimed at enhancing operational efficiency, cybersecurity, and interoperability. A key development in this field is the use of open standards, particularly the OCPP, which promotes interoperability and reduces vendor lock-in. Greenlots' white paper underscores the advantages of OCPP, emphasizing its ability to support diverse and scalable charging networks, which is crucial for the widespread adoption of EVs and their seamless integration into existing energy systems [1].

Smart charging technologies play a vital role in optimizing the energy use of EVs and enhancing the efficiency of smart grids. Bhargavi et al. highlight how these technologies can balance grid loads by adjusting charging times and power demands based on grid conditions. This dynamic interaction between EVs and the grid helps optimize energy distribution, improving grid stability while supporting the transition to cleaner, more sustainable energy sources [2]. As the adoption of EVs grows, smart charging becomes increasingly essential for managing energy consumption and ensuring that the grid can handle the additional demand without compromising efficiency.

However, the rapid expansion of EV charging infrastructure introduces significant cybersecurity risks. El Hussini et al. address these concerns, focusing on vulnerabilities such as data breaches, hacking, and operational disruptions that could undermine the security and reliability of EV CSs. They propose a comprehensive cybersecurity framework to mitigate these risks and protect the integrity of charging networks [3]. Furthermore, they discuss potential cyberattacks on autonomous vehicles, including threats to the communication systems between EVs and CSs, and suggest defense mechanisms to enhance system security [4–8]. The chapter provides a detailed analysis of security risks in EV charging station management systems (CSMS), identifying critical vulnerabilities such as power jacking and unauthorized data access, and recommending strategies to safeguard against these threats [9–11].

The Open Charge Alliance (OCA) continues to promote the adoption of OCPP, particularly its latest version, OCPP 2.0.1. This updated version includes improvements in secure firmware updates and mutual authentication mechanisms, which enhance the security of CSs and the integrity of data exchange between EVs and the grid. The chapter further emphasizes the importance of OCPP in enabling secure and efficient smart grid networks, positioning it as a crucial component for the successful integration of EVs into the broader energy ecosystem [12].

The rapid rise of cybersecurity threats targeting EVs and their associated infrastructure has become a critical concern. A staggering 225% increase in cyberattacks on EVs within just three years has been shown in Ref. [13], emphasizing the urgent need for enhanced security frameworks to protect these systems from evolving threats. Reference [14] raises concerns that public EV charging data, being highly accessible, could become a prime target for cyberattacks. Implementing robust privacy measures to safeguard sensitive information and ensure that EV users' data remains secure has been advocated [15]. These observations underscore the central role of cybersecurity in advancing the EV ecosystem, particularly in terms of protecting the privacy and integrity of data shared between EVs, CSs, and grid operators [16].

Additionally, van Amstel et al. emphasize the advantages of OCPP in overcoming the constraints of proprietary protocols. OCPP's flexibility allows it to adapt to the diverse needs of the EV industry, offering an essential framework for creating charging solutions that work across various station types and manufacturers [17].

The most recent version of the protocol, OCPP 2.0.1, offers notable improvements to its topology and design [18]. For EV charging networks to function dependably, these changes are intended to facilitate safer and more effective communication between EVs and CSs. By including a smart charging extension with sophisticated reservation features, Orcioni and Conti build upon the OCPP standard. This modification makes it possible to schedule charging sessions more effectively, which maximizes energy use and lessens the load on the grid during peak hours [19]. Similar to this, Schlund et al. investigate the possibility of bidirectional flexibility in EV charging, emphasizing the ways in which sizable EV fleets might be used to support grid balancing and offer other services like energy distribution and storage [20].

In order to preserve grid stability, smart grid technologies must have low-latency connectivity [21]. The optimal operation of EV charging systems and grid stability depend on minimizing communication delays, as highlighted by Heron et al.'s investigation of demand–response round-trip latency in IoT topologies [22]. In a different study, Heron and Sun assess the effects of ideal and realistic communication models on system performance and charging efficiency for EV charging in smart grids. Their results show how crucial a strong communication infrastructure is to improving charging networks' overall performance [23].

The concept of V2G is gaining attention as a transformative solution for improving grid resilience and sustainability. Noel et al. examine V2G as a sociotechnical transition that not only supports electric mobility but also strengthens the grid by enabling bidirectional energy flow between EVs and the grid [24]. However, with the increased complexity of such systems, cybersecurity remains a concern. Van Aubel and Poll assess the security risks of EV charging protocols and propose strategies to mitigate vulnerabilities such as unauthorized access and data breaches [25]. In line with this, a review opens protocols for e-mobility and argues that standardized communication frameworks are essential for ensuring secure and interoperable charging networks [26–28].

To accommodate the growing demand for EVs, Wellisch et al. propose a modular software implementation for smart CSs, which can be adapted and scaled to meet the needs of expanding charging networks [29]. This aligns with the work of Emre et al., who review existing power transfer solutions for EVs, focusing on the technological advancements made under the EU's Seventh Framework Programme [28]. Finally, the OCA's background paper outlines the organization's efforts to promote open standards and facilitate the deployment of interoperable EV CSs, which are vital for the widespread adoption of electric mobility [30]. These studies collectively highlight the ongoing efforts to ensure that EV charging systems are secure, efficient, and capable of meeting the demands of a growing global requirement. The development of V2G communication standards plays a pivotal role in the successful integration of

EVs with smart grids, enabling secure and efficient energy exchanges between vehicles and the grid. A key standard in this context is ISO 15118-2, which specifies the network and application protocol requirements for V2G communication. This standard ensures that the system supports bidirectional energy flow and is compatible with a wide variety of CSs and grid configurations, thus enhancing the interoperability and scalability of V2G systems [31]. Complementing ISO 15118-2 is ISO 15118-1, which provides foundational information and use-case definitions for V2G communication. This helps create a common understanding of how EVs and the grid interact, particularly in areas such as energy management, charging optimization, and ensuring grid stability during peak and off-peak hours [32].

ISO 15118-3 further supports the V2G framework by detailing the physical and data link layer requirements for communication [33]. This standard outlines the technical specifications for real-time data exchange between EVs and CSs, ensuring that both hardware and communication protocols meet stringent performance and security standards. Such specifications are crucial for maintaining a secure and reliable communication channel, which is the backbone of any successful V2G deployment, allowing for seamless energy transfers without compromising safety or efficiency [34].

The OCPP2.0.1 is a notable advancement in the continuing development of EV charging infrastructure. Buve, Klapwijk, and de Leeuw stress how crucial OCPP 2.0.1 is to improving compatibility among various CS suppliers. Through the provision of an open communication protocol, OCPP facilitates the efficient communication between various CSs, allowing for smooth user experiences and optimal energy management throughout a network of disparate infrastructure and equipment.

The construction of fast-charging infrastructure becomes increasingly important as EV usage keeps growing. Parchomiuk et al. offer a thorough analysis of the prospects and difficulties related to EV rapid charging [35]. The ecosystem surrounding EV charging is still quite concerned about cybersecurity, particularly when the integration of V2G systems creates new threats. In order to protect EVs and CSs from potential intrusions, Hodge et al. discuss the cybersecurity risks unique to V2G communication interfaces and offer strong mitigating techniques [36]. Similarly, Buschlinger et al. provide the "Plug-and-Patch" approach, a safe way to add value-added services to EV CSs without jeopardizing the infrastructure's fundamental security [37].

For better data management and business operations, Jakó and Knapp present a business scenario and data flow model within the NeMo hypernetwork, which supports EV charging infrastructure and its integration with smart grids. This model facilitates improved data interoperability, allowing for more efficient resource management and enhanced communication across different parts of the EV charging ecosystem [38].

Antoun et al. provide a comprehensive security assessment of the entire EV charging ecosystem, focusing on the vulnerabilities introduced by V2G integration. Their work offers practical solutions to mitigate cybersecurity risks, ensuring that the transition to V2G technology does not expose the system to malicious actors [39]. Meanwhile, Bilolikar et al. examine best practices for expanding EV charging infrastructure. Their recommendations serve as a guide for policymakers and industry stakeholders to address the growing demands of EVs, providing strategies to scale infrastructure and improve its efficiency to accommodate the increasing number of EVs on the road [40].

These studies collectively underline the importance of comprehensive communication standards, robust security protocols, and scalable infrastructure for the successful integration of EVs into smart grids. As EV technology continues to evolve, ensuring the interoperability, security, and efficiency of charging systems remains paramount to realizing the full potential of electric mobility.

The integration of EVs with smart grids through V2G technology is a rapidly evolving field, with significant advancements being made in communication standards and infrastructure to support efficient energy exchanges. A prominent example of V2G implementation is the Los Angeles Air Force Base Vehicle-to-Grid demonstration project, as reported by Douglas et al. in their final project report for the California Energy Commission [41]. This demonstration project highlights the real-world application of V2G technology, showcasing how EVs can support grid stability by providing bidirectional energy flow. The project not only demonstrated the potential of V2G systems but also revealed the challenges related to grid management, communication, and interoperability among different stakeholders in the energy ecosystem.

Communication standards are fundamental to the seamless operation of V2G systems. The ECHONET Lite specification, as outlined by the ECHONET Consortium, is one such communication standard that plays a key role in ensuring interoperability in smart grid environments [42]. ECHONET Lite is a lightweight communication protocol designed to connect various smart devices, including EVs, to the grid. Its specification is continually updated, with the latest version providing enhanced features to support the growing demand for energy-efficient solutions in homes and buildings. This protocol is part of a larger trend toward standardization in the smart grid space, where efficient data exchange and secure communication are paramount.

In the context of grid integration, Willrett's work on the development of e-mobility and the associated challenges offers valuable insights into the evolving landscape of V2G systems [43]. The integration of electric mobility into the grid introduces complexities such as managing fluctuating energy demands, ensuring fast charging capabilities, and optimizing grid stability. Willrett discusses the technological advancements and regulatory frameworks that are crucial for overcoming these challenges, such as the role of V2G in balancing supply and demand in smart grids.

The IoT is another key enabler of V2G technology, with network simulators such as the one described by Pham et al. providing valuable tools for generating datasets to test and optimize communication systems in real-world scenarios [44]. IoT-based simulations are essential for evaluating the performance of V2G systems under different conditions and ensuring that the communication networks between EVs, CSs, and the grid are robust and reliable.

The development of smart charging standards has also been a focus in the literature, with Willrett further discussing the importance of implementing standardized charging systems to support the widespread adoption of EVs [45]. Smart charging protocols ensure that the charging process is efficient and adaptive to grid requirements, allowing EVs to be charged at optimal times based on electricity demand and grid conditions. These standards are crucial for ensuring that V2G systems can be integrated effectively into the broader energy ecosystem.

Energy management within the smart grid, particularly in the context of a low-carbon economy, is another important aspect of V2G systems. Erol-Kantarci and Mouftah explore how pervasive energy management solutions can be used to optimize energy consumption and distribution, with a focus on reducing carbon emissions and supporting the transition to sustainable energy [46]. Their work emphasizes the need for smart grid systems to be both energy-efficient and capable of handling the dynamic nature of renewable energy sources, which are integral to many V2G implementations.

Open charging protocols, such as the Open Intercharge Protocol (OICP), are also central to ensuring the interoperability of EV charging infrastructure. The OICP, developed by Hubject, enables the exchange of data between different EV CSs, ensuring that users can access CPs regardless of the station vendor [47]. This open protocol supports the expansion of EV charging networks and facilitates the adoption of V2G technology by promoting standardization in communication and transaction handling.

The ISO 15118 standard has been instrumental in shaping V2G communication requirements. Specifically, Parts 4, 5, and 8 of ISO 15118 define the protocols for network and application conformance, as well as the physical and data link layer requirements for wired and wireless communication [48–50]. These standards establish the technical foundations for reliable and secure communication between EVs, CSs, and the grid, ensuring that V2G systems operate seamlessly and safely.

Together, these studies and standards underscore the importance of establishing robust communication frameworks, security protocols, and energy management strategies to enable the successful integration of EVs with smart grids. As V2G technology continues to evolve, the ongoing development of these standards will be critical in addressing the challenges associated with grid stability, energy efficiency, and the widespread adoption of EVs. Comparison of EV charging protocols, standards, and technologies is mentioned in Table 10.1.

Table 10.1 Comparison of EV charging protocols, standards, and technologies

Feature/aspect	*OCPP (47)*	*V2G (48)*	*Grid integration e-mobility (43)*	*Security of EV charging protocols (25)*	*Smart charging implementation (45)*	*EV fast charging infrastructure (35)*	*Vehicle cybersecurity (36)*
Focus area	EV charging interoperability	V2G communication	E-mobility grid integration	Security in charging protocols	Standards for smart charging	Fast charging solutions	Cybersecurity in EV infrastructure
Standard/ Protocol	OCPP	ISO 15118-4, 15118-5	IEC standards	OCPP security evaluation	OCPP 2.0.1	EV fast charging	Cyber defense protocols for EVs
Communication technology	Open and interoperable for chargers	Wireless and wired communication	IoT integration	OCPP evaluation for security	OCPP standards for integration	Fast-charging protocols	Threat mitigation for EV charging
Interoperability	Very high	Medium	Medium	Low	High	Medium	High
Implementation year	2018	2018	2019	2021	2020	2019	2019
Security focus	High	High	Low	Very high	Medium	Low	Very high
Key contribution	Interoperability across networks	V2G standards	Smart charging and energy integration	Evaluation of security protocols	OCPP role in global charging	EV fast-charging solutions	Mitigation of cybersecurity threats

10.2 KEY COMPONENTS OF OCPP

The OCPP is a pivotal standard in EV charging, enabling interoperability across diverse systems.

The operation and management of EV charging infrastructure involves multiple components working together to ensure efficient, secure, and reliable service delivery. The OCPP plays a central role in enabling communication between these components, supporting the secure exchange of data and the enforcement of smart charging policies. Here's a detailed explanation of the core components involved in EV charging systems and how OCPP facilitates their interaction:

Charging Station: CS is a critical hardware component that interfaces directly with the electric vehicle supply equipment (EVSE) and manages the local operations related to EV charging. CS serves as the physical location where EVs are plugged in for charging, and it is responsible for providing power to the vehicle, monitoring its charging status, and communicating with the CSMS. OCPP facilitates secure communication between CSs and CSMS, ensuring that various functions such as power management, transaction handling, and operational policies are effectively implemented. This communication includes reporting charging progress, sending fault notifications, and enforcing power limits to ensure safe charging and prevent overloading.

Charging Station Management System: CSMS acts as the control center for the entire network of CSs. It oversees the operation and monitoring of the CSs, manages user interfaces, and processes transactions related to EV charging. CSMS uses OCPP to interact with networked CSs, enabling it to implement smart charging policies. These policies can include prioritizing the charging of certain vehicles, limiting charging rates based on grid demand, and ensuring energy efficiency. Additionally, the CSMS can manage user authentication, handle payments, and provide real-time usage data for analysis. By implementing smart charging policies, the CSMS ensures that the network is used optimally, balancing demand, energy costs, and grid stability while providing a seamless user experience.

Electric Vehicle Supply Equipment: EVSE is the edge subsystem that connects the EV to the CS and ensures both physical and data connectivity during the charging process. EVSE typically includes the charging cable, connectors, and other hardware necessary for the safe and efficient transfer of electricity to the EV. OCPP facilitates communication between EVSE and CSMS, ensuring that the data exchanged—such as vehicle identification, charging status, and energy usage—is secure and accurate. This interaction ensures transaction integrity, helping to maintain reliable service management and enabling features such as remote monitoring and control of the charging process.

Local Controller (LC) and Proxy (LP): The LC and LP are supplementary components that enhance the functionality and reliability of the charging infrastructure. The LC provides redundancy and backup communication in the event of a failure or interruption in the CSMS, ensuring that CSs can continue to operate even if the central system is temporarily unavailable. The LP, on the other hand, acts as a message router, enabling CSs without direct network access to still communicate with the CSMS through an intermediary device. OCPP's flexible structure allows these components to integrate seamlessly into the overall charging system, ensuring continuity of operations and improving fault tolerance.

Distribution System Operator (DSO): While the DSO is not directly managed by OCPP, it plays an essential role in maintaining grid stability by regulating the flow of electricity based on real-time data. The DSO's role is crucial for ensuring that the grid can handle the additional load caused by EV charging, particularly during peak demand periods. OCPP indirectly supports the DSO's operations by providing communication between the CSMS and the grid, which allows for better coordination and the implementation of demand response strategies. For instance, smart charging policies implemented through OCPP can enable the CSMS to adjust charging rates based on grid conditions, ensuring that the DSO can maintain grid balance.

Security Mechanisms: Security is a fundamental concern in any network system, and the EV charging ecosystem is no exception. OCPP 2.0 addresses security through various mechanisms, with transport layer security (TLS) being a key feature. TLS ensures that data exchanged between CS, CSMS, and other components is encrypted, preventing unauthorized access and ensuring the integrity of communication. Additionally, OCPP 2.0 includes features such as secure firmware updates, event logging, and notifications to further strengthen the security of the charging infrastructure. These security measures protect against cyberattacks, prevent unauthorized access, and ensure the safe and reliable operation of EV charging networks.

10.3 ADVANTAGES OF OCPP

- By standardizing communication protocols and message formats, OCPP promotes interoperability among different EV charging equipment and management systems. This interoperability allows EV drivers to use various charging networks with the assurance that their vehicles can communicate effectively with different CPs.
- OCPP's modular architecture and standardized protocols support the scalability of EV charging networks. As the demand for EVs and charging infrastructure grows, OCPP facilitates the integration of new CPs

and the expansion of existing networks without significant compatibility issues.

- The protocol's flexibility allows for continuous evolution and adaptation to new technological advancements and industry requirements. It supports innovation in areas such as smart grid integration, energy management, billing systems, and V2G communication, thereby fostering a dynamic and competitive market environment.
- OCPP includes provisions for secure communication between CPs and CPOs, including authentication mechanisms, encryption, and data integrity checks. These security measures protect against unauthorized access, tampering, and data breaches, ensuring the confidentiality and integrity of EV charging transactions.
- OCPP's widespread adoption is driven by its open-source nature, community support, and endorsement by industry stakeholders. Many manufacturers of EV charging equipment, CPOs, software developers, and service providers adhere to OCPP standards to leverage its benefits and contribute to the growth of the EV charging infrastructure.

10.4 ROLE OF THE OCA IN DEVELOPING OCPP AS AN OPEN-SOURCE STANDARD

The development and uptake of the OCPP as a standardized open-source framework in the EV charging sector is greatly aided by the OCA. The OCA was established with the goal of developing an accessible and interoperable charging infrastructure. Its main objective is to encourage cooperation between stakeholders in order to remove obstacles to the adoption of EVs. Fundamentally, OCPP facilitates communication between CMSs and CSs, guaranteeing interoperability and smooth functioning among various service providers and manufacturers. The OCA promotes innovation and lowers the price of proprietary solutions by making OCPP an open-source standard, which speeds up the expansion of the EV charging network around the world.

The OCA's role in developing OCPP involves several key functions. First, it coordinates efforts among member organizations to define and refine the protocol's specifications, ensuring it evolves to meet industry needs and technological advancements. Second, the alliance promotes OCPP through educational initiatives, workshops, and conferences, raising awareness and facilitating its implementation across diverse markets and applications.

Moreover, the OCA provides a platform for collaboration and knowledge-sharing, enabling stakeholders to exchange best practices, address technical challenges, and advocate for supportive policies. This collaborative approach not only enhances the robustness and reliability of the protocol but also cultivates a community-driven ecosystem that supports sustainable and scalable EV infrastructure deployment. Standardization and interoperability are crucial factors in driving the widespread adoption of EVs by ensuring seamless integration, reliability, and accessibility of charging infrastructure. Here's why they are important:

- Standardized protocols, such as the OCPP, ensure that EVs from different manufacturers can reliably connect and charge at various CSs. This interoperability reduces barriers for EV drivers, allowing them to use any public or private charger without compatibility concerns.
- Standardization fosters a competitive marketplace for EV infrastructure by lowering development costs and encouraging investment. When CS operators and manufacturers adhere to a common standard, they can scale their operations more efficiently, attract more users, and contribute to the expansion of charging networks.
- Consistent user experiences across different charging networks build trust among consumers and businesses alike. Drivers are more likely to choose EVs when they have confidence in the availability and reliability of charging infrastructure, supported by standardized protocols.
- Governments and regulatory bodies often favor standards-based approaches because they simplify compliance, monitoring, and enforcement of regulations related to EV charging. Clear standards also facilitate the implementation of incentives and policies aimed at promoting EV adoption.
- Standardization does not stifle innovation; instead, it encourages it by providing a stable foundation upon which new technologies and services can be built. It enables interoperability between emerging technologies like smart grids, renewable energy integration, and V2G systems, thereby future-proofing the EV ecosystem.

10.5 PRIVACY ISSUES IN EV CHARGING INFRASTRUCTURE

The enormous volume of data that is transferred between CSs, cars, and the CMS raises privacy concerns in EV charging infrastructure, especially in the context of OCPP. This data contains the EV users' personal information, charging habits, and maybe private geographical information. This data may be subject to malicious assaults, misuse, or illegal access if improperly handled, which would violate user privacy.

To get around these problems, a variety of detection and deflection techniques are employed. Finding and preventing potential privacy violations, such as unauthorized data access or cyberattacks that target user data, is the aim of detection techniques. Methods including data encryption, intrusion detection systems (IDS), and anomaly detection are frequently employed to keep an eye on the network and notify system administrators of any privacy threats.

By taking proactive steps, deflection mechanisms, on the other hand, aim to stop or lessen the effects of privacy threats. These could include data anonymization to safeguard user identities, access control systems to guarantee that only authorized parties have access to sensitive information, and secure communication protocols (e.g., TLS).

Together, these mechanisms ensure that privacy concerns are adequately addressed, safeguarding users' personal information and enhancing the trustworthiness of OCPP-enabled EV charging networks. Privacy issues and detection/deflection mechanisms for OCPP are mentioned in Table 10.2.

Table 10.2 Privacy issues and detection/deflection mechanisms for OCPP

Privacy issue	*Description*	*Detection mechanism*	*Deflection mechanism*	*References*
Data confidentiality	Unauthorized access to sensitive data (e.g., user details, payment info) during communication	Encryption of data transmitted between the CS and EV	Implement end-to-end encryption and secure communication protocols (TLS)	[1–3]
Location privacy	Risk of exposing user's geographic location when charging in public areas	Geofencing detection using GPS data and charging logs	Anonymizing user location through proxy servers	[4–6]
Data integrity	Tampering with the data sent from the EV or the CS	Integrity check algorithms like checksums or cryptographic signatures	Use of blockchain technology for immutable transaction records	[7, 8]
Authentication of users and devices	Unauthorized access by malicious third parties to CPs or user data	Multifactor authentication (MFA), public-key infrastructure (PKI)	Smart card or RFID authentication, device fingerprinting	[9–11]
Usage anonymity	Risk of revealing user's charging behavior patterns and habits	Traffic analysis to monitor charging patterns and user interactions	Implementing pseudonymous identifiers and anonymized transaction records	[12–14]
Interoperability and compatibility	Difficulties in ensuring secure and private data exchange between different charging networks	Cross-network monitoring tools for detecting incompatible protocols	Standardization of data protocols across different EV networks	[15–17]
Payment system privacy	Exposure of financial transaction details during payments for charging services	Secure payment gateways (e.g., tokenization, encryption)	Payment deflection through third-party secure payment services	[18–20]

10.6 IMPLEMENTATION OF OCPP IN EV CHARGING INFRASTRUCTURE

Implementing the OCPP within EV charging infrastructure is a critical step toward enhancing interoperability and fostering innovation. This section delves into the practical aspects of deploying OCPP-compliant systems, highlighting key considerations and strategies for successful implementation.

System Design and Integration

Designing an OCPP-compliant system begins with selecting compatible hardware and software components that adhere to OCPP specifications. Manufacturers and operators must ensure that CSs, CMSs, and backend servers support OCPP protocols for seamless communication.

- *Hardware Selection:* Choose CSs equipped with OCPP-compatible controllers and interfaces. Ensure compatibility with both AC and DC charging standards to cater to diverse EV charging needs.
- *Software Integration:* Integrate OCPP libraries or software development kits (SDKs) into CMS and backend servers. Implement necessary interfaces for handling OCPP messages, including real-time data exchange, session management, and remote control functionalities.

Protocol Implementation and Compliance

Achieving OCPP compliance involves rigorous testing and validation to ensure adherence to protocol standards. This step is crucial for interoperability across different EV charging networks and management systems.

- *Testing Procedures:* Conduct thorough interoperability testing with OCPP-certified devices and platforms. Verify compliance with OCPP versions (e.g., OCPP 1.6, OCPP 2.0) to guarantee seamless integration and functionality.
- *Certification Requirements:* Obtain OCPP certification from recognized testing laboratories or alliances. Compliance with OCPP standards ensures reliability, security, and compatibility in EV charging operations.

Network Deployment and Scalability

Deploying OCPP networks requires careful planning to accommodate scalability and future growth in EV adoption. Considerations include network architecture, scalability options, and infrastructure readiness.

- *Network Architecture:* Design-scalable network architectures capable of supporting increasing numbers of CSs and users. Implement load management strategies and dynamic pricing mechanisms supported by OCPP for efficient resource allocation.

- *Scalability Planning:* Prepare for network expansion by deploying modular and scalable infrastructure. Plan for upgrades and additions to accommodate new technologies and evolving market demands.

Operational Management and Maintenance

Effective operational management ensures optimal performance and reliability of OCPP-enabled EV charging infrastructure. Implement proactive maintenance practices and monitoring systems to address potential issues promptly.

- *Monitoring and Control:* Utilize OCPP functionalities for real-time monitoring of charging sessions, station statuses, and energy consumption. Implement remote control capabilities for troubleshooting and maintenance tasks.
- *Data Security and Privacy:* Implement robust security measures to protect OCPP communication and user data. Ensure compliance with data privacy regulations (e.g., GDPR) to safeguard user information and maintain trust.

Continuous Improvement and Innovation

Embrace a culture of continuous improvement and innovation to leverage OCPP's open-source framework. Foster collaboration among stakeholders to drive advancements in EV charging technologies and services.

- *Innovation Initiatives:* Engage in collaborative projects and research to enhance OCPP functionalities and capabilities. Explore new applications such as V2G integration and smart grid interoperability.
- *Feedback and Adaptation:* Gather feedback from users, operators, and manufacturers to identify areas for enhancement. Adapt OCPP implementations to meet evolving market trends and technological advancements.

Comparison of OCPP with other methods in EV charging infrastructure is presented in Table 10.3.

10.7 A NEW ERA IN EV SMART CHARGING: OCPP 2.0

The OCPP is a global, open-standard facilitating communication between EV CSs and CMSs. First introduced in 2009, OCPP is free to use and license-free, enabling seamless interoperability across manufacturers, utilities, CPOs, and software providers. By being vendor-independent, OCPP empowers infrastructure operators to choose CSs flexibly and enables manufacturers to cater to diverse markets. The Open Charging Alliance (OCA) oversees its evolution, ensuring adaptability and relevance in a dynamic EV landscape.

Table 10.3 Comparison of OCPP with other methods in EV charging infrastructure

Parameter	*OCPP*	*Other methods*
Flexibility	Interoperability across manufacturers, scalable	Proprietary, limited interoperability [3–5]
Standardization	Open standard, widespread adoption	Lack of standardization, compatibility issues [6–8]
Scalability	Supports from single points to large networks	Limited scalability, additional infrastructure [9–11]
Security	Robust security measures, encryption, authentication	Varying security levels, customization needs [12–14]
Ease of integration	Well-defined APIs, documentation	Limited APIs, documentation [15–17]
Remote management	Remote monitoring, diagnostics, updates	Limited remote capabilities, proprietary [18]
Vendor neutrality	Promotes neutrality, hardware/software choices	Proprietary solutions, vendor lock-in [19]

OCPP supports critical IoT functionalities, including user authentication, energy tracking, and charging commands, promoting efficient communication between CSs and backend systems.

10.8 COMPARING OCPP 2.0 AND OCPP 1.6

OCPP 2.0 introduces advanced features over its predecessor, OCPP 1.6, addressing new trends in EVs and the energy market as shown in Figure 10.2. It enhances data communication, security, and interoperability. Key innovations include ISO 15118, which supports secure authentication (similar to PKI) and efficient data exchanges between vehicles and charge points.

Notable features of OCPP 2.0 include the following:

- *Enhanced Smart Charging:* Utilities and CPOs can use sophisticated control functionalities, optimizing energy delivery based on real-time data.
- *Improved Communication:* Detailed data fields enable up-to-date reporting on vehicles, energy usage, and system performance.

Despite its 2018 release, many systems still use OCPP 1.6, but the shift to OCPP 2.0 is underway, with early adopters like Ebee leading the transition.

10.9 ADVANCEMENTS IN IOT COMMUNICATION WITH OCPP 2.0

OCPP 2.0 improves interactions between EVs, charge points, and management systems through advanced protocols. For example, when an EV user

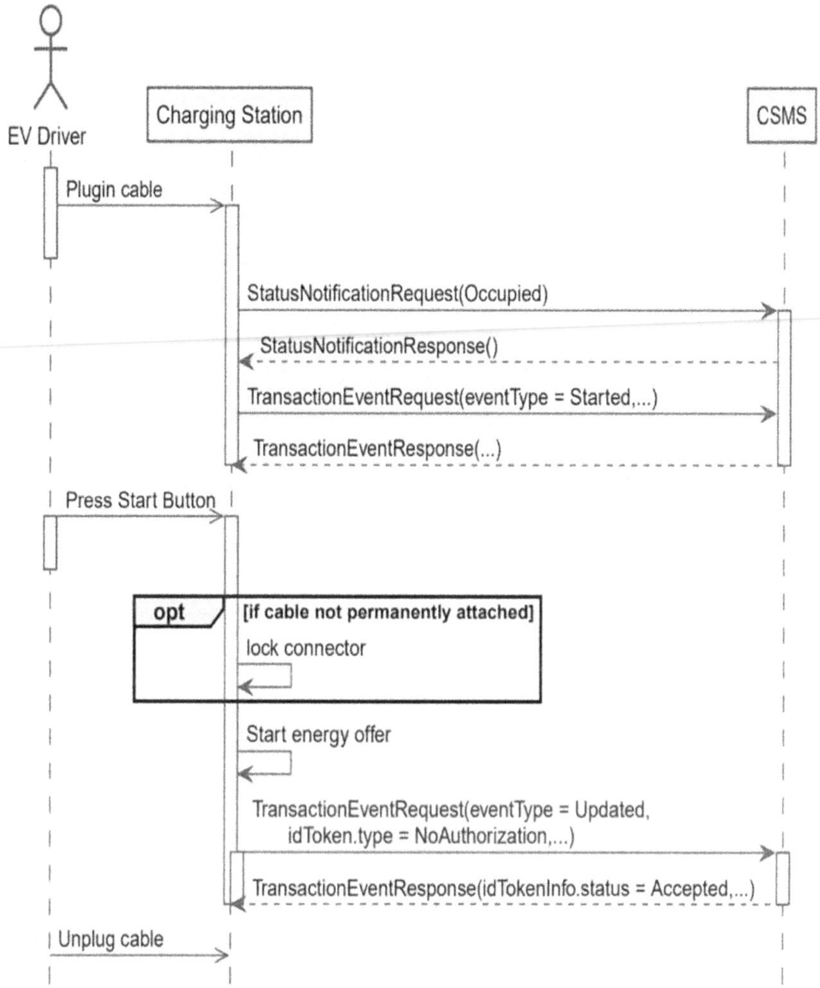

Figure 10.2 Comparison of OCPP 2.0 and OCPP 1.6 in EV charging networks.

connects to a charge point, the vehicle communicates its required energy via ISO 15118, as shown in Figure 10.3. This information is relayed to the central charging system, enabling precise load planning.

The protocol supports *High-Level Communication for Load Leveling*, ensuring efficient energy distribution and optimal usage of the CSMS. The process includes the following:

1. The EV sends a *ChargeParameterDiscoveryReq* to the CS.
2. The CS communicates needs to the CSMS via a *NotifyEVCharging NeedsRequest*.
3. The CSMS responds with a *NotifyEVChargingNeedsResponse*.

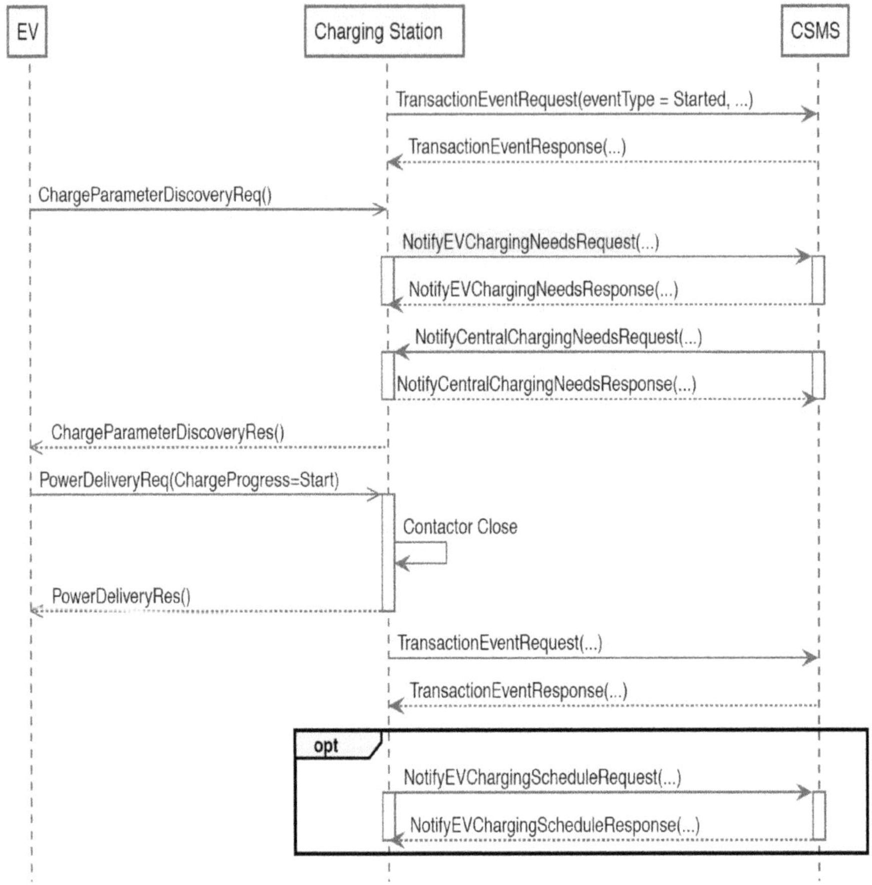

Figure 10.3 Enhanced IoT communication among EVs, CPs, and OCPP backend using OCPP 2.0.

These enhancements make OCPP 2.0 pivotal for building scalable, intelligent EV charging infrastructure, aligning with future energy demands and fostering sustainable growth.

10.10 CONCLUSION

The OCPP stands at the forefront of revolutionizing EV charging infrastructure by establishing a standardized communication framework. This chapter has explored how OCPP addresses critical challenges such as compatibility, scalability, and fostering innovation within the EV ecosystem. OCPP's ability to define a common language through structured data and messages ensures seamless interoperability across diverse CSs and management systems, irrespective of manufacturer differences. This interoperability not only simplifies deployment but also enhances user accessibility and reliability of EV charging services. Moreover, OCPP's standardized framework supports scalable deployment, accommodating varied charging scenarios, power levels, and advanced functionalities like load management and dynamic pricing. This scalability is crucial for building robust and future-ready EV charging networks that can meet the growing demand efficiently. The open-source nature of OCPP fosters collaboration among stakeholders, driving continuous innovation in EV charging operations. This collaborative environment promotes the development of new functionalities and efficiencies, thereby advancing the overall sustainability and effectiveness of electric mobility solutions. The adoption of OCPP not only enhances market expansion and grid integration but also contributes to improved user experiences and environmental sustainability through reduced greenhouse gas emissions. By embracing OCPP, stakeholders across the EV charging ecosystem can leverage its benefits to propel the industry toward a more interconnected, efficient, and sustainable future.

10.11 FUTURE WORK

Looking ahead, several avenues for future work in OCPP implementation and EV charging infrastructure development emerge:

- *Enhanced Security Measures:* Further strengthen OCPP protocols to address evolving cybersecurity threats and enhance data protection measures across EV charging networks.
- *Integration with Smart Grid Technologies:* Explore deeper integration of OCPP with smart grid technologies to optimize energy management, support bidirectional energy flows (V2G), and improve grid stability.
- *Standardization Updates:* Continuously update OCPP standards to incorporate emerging technologies and industry requirements, ensuring compatibility with future EV innovations.
- *User-Centric Innovations:* Focus on developing user-centric features such as mobile app integrations, seamless payment solutions, and personalized charging experiences to enhance user satisfaction.
- *Global Adoption and Regulatory Alignment:* Promote global adoption of OCPP standards while aligning with regional regulatory frameworks to facilitate interoperability and regulatory compliance.
- *Sustainability Initiatives:* Expand efforts in integrating renewable energy sources and promoting sustainable practices within EV charging infrastructure, reducing carbon footprints and promoting environmental stewardship.
- *Data Analytics and Predictive Maintenance:* Implement advanced analytics and machine learning algorithms to optimize CS operations, predict maintenance needs, and enhance overall system efficiency.

REFERENCES

1. Greenlots, "Open vs. closed charging stations: Advantages and disadvantages," White Paper, Los Angeles, CA, 2018. Available at: https://greenlots.com/wp-content/uploads/2018/10/Open-Standards-White_Paper-compressed.pdf
2. K. Bhargavi, N. Jayalaksmi, S. Malagi, and V. K. Jadoun, "Integration of plug-in electric vehicles in smart grid: A review," 2020 International Conference on Power Electronics & IoT Applications in Renewable Energy and *I*ts Control *(PARC)*, IEEE, pp. 214–219, 2020.
3. H. El Hussini, C. Assi, B. Moussa, R. Atallah, and A. Ghrayeb, "A tale of two entities: Contextualizing the security of electric vehicle charging stations on the power grid," in *ACM Trans. Internet Things*, vol. 2, no. 2, pp. 1–21, 2021.
4. K. Kim, J. S. Kim, S. Jeong, J. H. Park, and H. K. Kim, "Cybersecurity for autonomous vehicles: Review of attacks and defense," in *Comput. Security*, vol. 103. 2021, 102150.
5. Open Charge Alliance, 2020. Available at: https://www.openchargealliance.org/
6. Ampeco, "Enable innovation and cost efficiency with OCPP," 2022. Available at: https://www.ampeco.com/ocpp-open-charge-point-protocol/#cpo-benefits
7. Current, "Innovation and cost-efficiency in four letters: OCPP," 2021. Available at: http://www.current.eco/platform/ocpp
8. Open Charge Alliance, "Open charge alliance: Our mission," 2022. Available at: https://www.openchargealliance.org/about-us/
9. H. T. Mouftah and M. Erol-Kantarci, *Smart Grid: Networking, Data Management, and Business Models*, Hoboken, NJ: CRC Press, 2017.
10. B. Blum, "Cyberattacks on cars increased 2% in last three years," ISRAEL21c, 2022. Available at: https://www.israel21c.org/cyberattacks-on-cars-increased-225-in-last-three-years/
11. T. Nasr, S. Torabi, E. Bou-Harb, C. Fachkha, and C. Assi, "Power jacking your station: In-depth security analysis of electric vehicle charging station management systems," in *Comput. Security*, vol. 112, 2022, 102511.
12. S. Acharya, Y. Dvorkin, and R. Karri, "Public plug-in electric vehicles+grid data: Is a new cyberattack vector viable," in *IEEE Trans. Smart Grid*, vol. 11, no. 6, pp. 5099–5113, 2020.
13. Open Charge Alliance, "Open Charge Point Protocol 2.0.1," 2018. Available at: https://www.openchargealliance.org/protocols/ocpp-201/
14. Open Charge Alliance, "Open Charge Point Protocol 1.6," 2015. Available at: https://www.openchargealliance.org/protocols/ocpp-16/
15. F. Buve, M. Jansen, P. Klapwijk, and R. Leeuw, "OCPP 2.0.1, Part 2: Specification," Open Charge Alliance, Arnhem, The Netherlands, Rep. 2020-03-31, 2020. Available at: https://www.openchargealliance.org/protocols/ocpp-201/
16. A. Wargers, and D. Frenkel, "The world's first large-scale migration of OCPP based PEV charging infrastructure," Open Charge Alliance, Arnhem, The Netherlands, White Paper, 2023. Available at: https://www.openchargealliance.org/uploads/files/OCA-White_paper_on_OCPP_based_migration_version_5.0.pdf
17. M. van Amstel, R. Ghatikar, and A. Wargers, "Importance of Open Charge Point Protocol for the electric vehicle industry," Open Charge Alliance, Arnhem, The Netherlands, White Paper, 2016. Available at: https://www.openchargealliance.org/uploads/files/OCA-EN_whitepaper_OCPP_vs_proprietary_protocols_v1.0.pdf
18. F. Buve, M. Jansen, and P. Klapwijk, "OCPP 2.0.1, Part 1: Architecture & topology," Open Charge Alliance, Arnhem, The Netherlands, Rep. 2020-03-31, 2020. Available at: https://www.openchargealliance.org/protocols/ocpp-201/

19. S. Orcioni and M. Conti, "EV smart charging with advance reservation extension to the OCPP standard," in *Energies*, vol. 13, no. 12, pp. 3263–3284, 2020.
20. J. Schlund, M. Pruckner, and R. German, "Flexibility-modeling and maximizing the bidirectional flexibility availability of unidirectional charging of large pools of electric vehicles," *Proceedings of the Eleventh ACM International Conference on Future Energy Systems*, Association for Computing Machinery, 2020, pp. 121–132.
21. E. Ancillotti, R. Bruno, S. Palumbo, C. Capasso, and O. Veneri, "Experimental set-up of DC PEV charging station supported by open and interoperable communication technologies," 2016 International Symposium on Power Electronics, Electrical Drives, Automation and Motion *(SPEEDAM)*, 2016, IEEE, pp. 677–682.
22. J. W. Heron, J. Jiang, H. Sun, V. Gezerlis, and T. Doukoglou, "Demand-response round-trip latency of IoT smartgrid network topologies," in *IEEE Access*, vol. 6, pp. 22930–22937. 2018.
23. J. W. Heron and H. Sun, "Smart electric vehicle charging with ideal and practical communications in smart grids," 2019 IEEE Global Communications Conference *(GLOBECOM)*, IEEE, 2019, pp. 1–6.
24. L. Noel, G. Z. de Rubens, J. Kester, and B. K. Sovacool, *Vehicle-to-grid: A Sociotechnical Transition beyond Electric Mobility*, Cham, Switzerland: Springer Nature, 2019.
25. P. Van Aubel and E. Poll, "Security of EV-charging protocols," 2021. Available at: https://www.polvanaubel.com/research/chargego/protocol-security-evaluation/protocol-security-evaluation-draft-2020-03-10.pdf
26. I. Buamod, E. Abdelmoghith, and H. T. Mouftah, "A review of OSI-based charging standards and eMobility open protocols," 2015 6th International Conference on the Network of the *Future (NOF)*, IEEE, 2015, pp. 1–9.
27. D. Wellisch, J. Lenz, A. Faschingbauer, R. Pöschl, and S. Kunze, "Vehicle-to-grid AC charging station: An approach for smart charging development," in *IFAC-PapersOnLine*, vol. 48, no. 4, pp. 55–60, 2015.
28. M. Emre et al., "Task 3.3.1 review of existing power transfer solutions EU seventh framework programme," CORDIS, Hialeah, FL, Rep. Deliverable No. D 3.3.1, 2014.
29. D. Wellisch, S. Kunze, and R. Pöschl, "A modular software implementation for smart charging stations," 2015 International Symposium on Smart Electric Distribution Systems and Technologies *(EDST)*, Vienna, Austria, 2015, pp. 254–259.
30. "Background Open Charge Alliance," Open Charge Alliance, 2009. Available at: https://www.openchargealliance.org/aboutus/background/
31. "Road vehicles: Vehicle-to-grid communication interface—Part 2: Network and application protocol requirements," International Organization for Standardization, Geneva, Switzerland, Rep. ISO/CD 15118-2, 2014. Available at: https://www.iso.org/obp/ui/#iso:std:iso:15118:-2:ed-1:v1:en
32. "Road vehicles: Vehicle to grid communication interface—Part 1: General information and use-case definition," International Organization for Standardization, Geneva, Switzerland, Rep. ISO/DIS 15118-1:2017(E), 2019. Available at: https://www.iso.org/obp/ui/#iso:std:iso:15118:-1:ed-2:v1:en
33. F. Buve, P. Klapwijk, and R. de Leeuw, "OCPP 2.0.1, Part 0: Introduction," Open Charge Alliance, Arnhem, The Netherlands, Rep. 2020-03-31, 2020. Available at: https://www.openchargealliance.org/protocols/ocpp-201/
34. "Road vehicles: Vehicle-to-grid communication interface—Part 3: Physical and data link layer requirements," International Organization for Standardization, Geneva, Switzerland, Rep. ISO 15118-3:2015, 2015. Available at: https://www.iso.org/obp/ui/#iso:std:iso:15118:-3:en

35. M. Parchomiuk, A. Moradewicz, and H. Gawinski, "An overview of electric vehicles fast charging infrastructure," 2019 Progress in Applied Electrical Engineering *(PAEE)*, Koscielisko, Poland, 2019, pp. 1–26.
36. C. Hodge et al., "Vehicle cybersecurity threats and mitigation approaches," National Renewable Energy Laboratory, Golden, CO, Rep. NREL/TP-5400-74247, 2019. Available at: www.nrel.gov/publications.
37. L. Buschlinger, M. Springer, and M. Zhdanova, "Plug-and-patch: Secure value added services for electric vehicle charging," in *Proceedings of the 14th International Conference on Availability, Reliability, and Security*, Canterbury, UK, ACM, 2019, pp. 1–2.
38. Z. Jakó and Á Knapp, "Business scenarios and data flow in NeMo hyper-network," 2018 International Conference on Smart Systems and Technologies *(SST)*, Osijek, Croatia, IEEE, 2018, pp. 139–144.
39. J. Antoun, M. E. Kabir, B. Moussa, R. Atallah, and C. Assi, "A detailed security assessment of the EV charging ecosystem," in *IEEE Netw*, vol. 34, no. 3, pp. 200–207, 2020.
40. R. Bilolikar, A. Magal, N. Lepre, and J. Korsh, "Scaling up electric vehicle charging infrastructure," Natural Resources Defense Council, New York, NY, 2020. Available at: https://www.nrdc.org/sites/default/files/charging-infrastructure-bestparctices-202007.pdf
41. B. Douglas, J. McDonald, N. DeForest, and C. Gehbauer, "Los Angeles Air Force Base Vehicle-to-Grid Demonstration: Final project report," California Energy Commission, Sacramento, CA, Rep. CEC-500-2018-025, 2017.
42. "ECHONET Lite specification, version 1.13," ECHONET Consortium, Tokyo, Japan, 2018. Available at: https://echonet.jp/spec_v113_lite_en/
43. U. Willrett, "Grid integration e-mobility: Developments and challenges," in *Proceedings of the Internationales Stuttgarter Symposium*, Stuttgart, Germany, Springer, 2019, pp. 320–330.
44. V. C. Pham, Y. Makino, K. Pho, Y. Lim, and Y. Tan, "IoT area network simulator for network dataset generation," in *J. Inf. Process*, vol. 28, pp. 668–678, 2020.
45. U. Willrett, "Standards for implementing smart charging," in *ATZ Worldwide*, vol. 122, no. 12, pp. 64–67, 2020.
46. M. Erol-Kantarci and H. T. Mouftah, "Pervasive energy management for the smart grid: Towards a low carbon economy," in *Pervasive Communications Handbook*, Hoboken, NJ: CRC Press, 2017, pp. 251–269.
47. "Open Intercharge protocol, version 2.2," Hubject GMBH, Berlin, Germany, 2018. Available at: https://www.hubject.com/en/downloads/oicp/
48. "Road vehicles: Vehicle to grid communication interface—Part 4: Network and application protocol conformance test," International Organization for Standardization, Geneva, Switzerland, Rep. ISO/STD:15118:4, 2018.
49. "Road vehicles: Vehicle to grid communication interface—Part 5: Physical layer and data link layer conformance test," International Organization for Standardization, Geneva, Switzerland, Rep. ISO/STD:15118:5, 2018.
50. "Road vehicles: Vehicle to grid communication interface—Part 8: Physical layer and data link layer requirements for wireless communication," International Organization for Standardization, Geneva, Switzerland, Rep. ISO/STD:15118:8, 2020.

Index

Note: **Bold** page numbers refer to **tables** and *Italic* page numbers refer to *figures.*

O

P

R

S

T

U

V

W

Z

For Product Safety Concerns and Information please contact our EU
representative GPSR@taylorandfrancis.com
Taylor & Francis Verlag GmbH, Kaufingerstraße 24, 80331 München, Germany

www.ingramcontent.com/pod-product-compliance
Ingram Content Group UK Ltd.
Pitfield, Milton Keynes, MK11 3LW, UK
UKHW022223190926
475702UK00003B/481

* 9 7 8 1 0 3 2 8 4 9 5 9 1 *